Structural Engineering
For First Year Students

396 A

96 A

D1334805

THE McGRAW-HILL INTERNATIONAL SERIES IN CIVIL ENGINEERING

Consulting Editors

Professor F. K. Kong
University of Newcastle upon Tyne

Emeritus Professor R. H. Evans, CBE
University of Leeds

Structural engineering is the science and art of designing and constructing economic and elegant structures to safely resist the forces to which they may be subjected.

Frontispiece: Construction of an *in situ* concrete bridge over the River Rhine without the use of falsework

Front cover The Iron Bridge at Coalbrookdale in Shropshire, England, constructed in 1779, was the first metal bridge ever to be built. Iron was the structural material that made the Industrial Revolution possible and the bridge has become the symbol of that period.

Structural Engineering For First Year Students

Peter Montague

Professor of Civil Engineering
The University of Manchester

and

Roy Taylor

Formerly Senior Lecturer in
Civil Engineering
The University of Manchester

McGRAW-HILL BOOK COMPANY

London · New York · St Louis · San Francisco · Auckland
Bogotá · Guatemala · Hamburg · Lisbon · Madrid · Mexico
Montreal · New Delhi · Panama · Paris · San Juan · São Paulo
Singapore · Sydney · Tokyo · Toronto

Published by
McGRAW-HILL Book Company (UK) Limited
MAIDENHEAD · BERKSHIRE · ENGLAND

British Library Cataloguing in Publication Data
Montague, P. (Peter)
   Structural engineering for first year students.
   1. Structural engineering
   I. Title II.   Taylor, R. (Roy)
   624.1

Library of Congress Cataloging-in-Publication Data
Montague, P. (Peter)
   Structural engineering for first year students/P. Montague and
   R. Taylor
   p.   cm.—(The McGraw-Hill international series in civil
   engineering)
      Bibliography: p.
      Includes index.
      ISBN 0-07-084195-0
      1. Structural engineering.   I. Taylor, R. (Roy)   II. Title.
   III. Series.
   TA640.M66 1989
   624.1—dc 19   88-27555

Typeset by Eta Services (Typesetters) Ltd, Beccles, Suffolk
and printed and bound by M & A Thomson Litho Ltd, Glasgow

Contents

Preface   xiii

Acknowledgements   xv

Part One: Theory and Practice   1

1. Introduction   **3**
  1.1 The structural engineering project   3
  1.2 Some structual terms and definitions   6
     External forces   6
     Internal forces   9
     Stress   10
     Deformation   12
     Strain   12
     Modulus of elasticity   14
  1.3 Types of structural elements   14
     Tension members   14
     Compression members   17
     Bending members   20
     Torsional members   22
     Shear members   23
     A further note on compression members   23
  1.4 Structural complexes   26
     Trusses   27
     Multi-storey buildings   33
     Cable-supported structures   35
     Structures involving mainly bending members   37
  1.5 Some basic concepts relating to structural analysis   40
     Sign convention and coordinate axes   40
     Components of a force   44
     Moment of a force   45
     Equilibrium   46
     Use of matrices   46
     Principle of superposition   48
     Free-body diagrams   49

1.6 Planar statics   50
    Force polygons   50
    Equilibrium of concurrent forces   51
    Moment of a couple   53
    Equivalent force systems   53
1.7 Structural supports   55
    Types of support   56
    Calculation of reactions using equilibrium only   58
    Calculation of reactions using equilibrium and internal conditions 61
    Some special cases   63

2. Structural analysis: mechanics of structures   **68**
A. Statically determinate structures involving direct forces only   68
2.1 Planar pin-jointed frames—forces in members   70
    Calculation of reactions   70
    Method of sections   70
    Method of joints   73
    Method of joints using a matrix approach   75
    Statical determinacy   79
2.2 Planar pin-jointed frames—deflections   82
    Graphical method   84
    Method of strain energy   88
    Introduction to the unit load method—virtual work   91
    Unit load method   95
    Calculation of joint movements using matrices   98
2.3 Planar pin-jointed frames—computer analysis   106
2.4 Space frames   107
    Types of support   108
    Statical determinacy   108
    Equilibrium of concurrent forces   111
    Use of joint coordinates   111
    Example of analysis   112
2.5 Cables   115
    Basic theory   116
    Shape of cable subjected to uniform load   117
    Force in cables   119
    Design considerations in cable structures   119
B. Statically determinate structures involving bending   121
2.6 Beams—calculation of internal forces   121
    Introduction to shear force and bending moment   121
    Bending moment and shear force diagrams   124
    Some standard cases   131
    Relationships between load, shear force and bending moment   131
    Direct method of establishing the bending moment   132

Example of bending moment diagram for multiple loads   138
Example of beam subjected to applied moments   139
The construction of bending moment diagrams using the principle of
superposition   143
Separation of a structure into its components for the construction
of bending moment diagrams   145
2.7 Beams—calculation of displacements   151
Introduction   151
A note on second moment of area   152
Displacements using the differential equation of bending   154
Displacements using the method of strain energy   159
Displacements using the unit load method   161
2.8 Rigid-jointed frames and arches   173
Portal frames   173
Arches   177
C. Statically indeterminate structures   180
Introduction to the problems of analysis   180
Fundamentals of statically indeterminate structures   188
Advantages   188
Disadvantages   189

3. Structural analysis: mechanics of solids   193
A. Stress distributions due to a single internal force   193
3.1 Longitudinal stresses in a beam due to a bending moment   195
Second moment of area and the parallel axis theorem   200
Examples of the calculation of second moments of area   203
Second moment of area and the perpendicular axis theorem   205
Some important points about bending   206
Section moduli of a cross-section   207
Radii of gyration of a cross-section   208
3.2 Shear stresses in a beam due to a shear force   210
Shear stress distribution in a beam of rectangular cross-section   215
Shear stress distribution in an I-section beam   216
3.3 Stress distributions in members of circular cross-section due to a
torque   219
Torsion of a solid circular shaft   219
Torsion of a hollow (thick-walled) circular tube   223
Torsion of a thin-walled, circular tube   223
Transmission of power through circular shafts   224
B. Complex stresses and strains and stress–strain relationships   226
3.4 The analysis of stress   226
Definition of a stress   226
The general state of uniform stress on a three-dimensional element   229
Complementary shear stress   230

Change of stress axes   232
The analysis of plane stress   232
Principal planes, principal stresses and principal axes   237
The values of the principal stresses   239
Which principal stress acts on which principal plane?   241
Location and magnitude of the maximum and minimum shear
stresses   241
The maximum shear stress in terms of the principal stresses   242
The values of the normal (or direct) stresses on planes of maximum
and minimum shear stress   243
Mohr's circle of stress   243
Principal stresses, their magnitudes and locations from Mohr's
stress circle   248
Maximum and minimum shear stresses and their locations from
Mohr's stress circle   249
Special cases of plane stress   251
Three-dimensional stress   253
3.5 The analysis of strain   255
Signs of direct strains and shear strains   256
Strains in terms of displacements   257
Strain array for an element subjected to plane stress   259
Transformation of the strain axes   259
The effect of the direct strains ε_x and ε_y 261
The effect of the shear strain γ_{xy} 261
The combined effect of the direct strains and the shear strain   262
The direct strain ε at any angle θ from the direction of the x-axis   262
The shear strain γ_{xy} on a rectangular element orientated at any angle
θ from the x-axis   263
Mohr's circle of strain   264
Principal strains and principal planes from Mohr's strain circle   267
Maximum and minimum shear strains from Mohr's strain circle   267
3.6 Stress–strain relations (or constitutive relations)   272
Stress–strain relations for elastic, isotropic materials   273
Poisson's ratio   273
A note on plane stress and plane strain   274
Stress–strain relations for direct stress and strain   275
Stress–strain relations for shear   279
The general statement of Hooke's Law   280
Stress–strain relations for plane stress   281
Stress–strain relations for plane strain   282
Volumetric stress–strain relations and the bulk modulus K 282
The relationships between the elastic constants E, G, K, and v 284
3.7 Stresses and strains due to combined forces   287
Beam web subjected to bending and shear   287

Member subjected to bending moment and axial force   291
The non-axially-loaded column   296

4. Structural materials: steel and concrete   **304**
4.1 Introduction to steelwork   304
Properties of steel   304
Steelwork sections   309
Steelwork connections   312
4.2 Problems with steel structures   317
Corrosion   317
Fire protection   318
Local buckling   321
Brittle fracture   323
4.3 Materials for concrete   324
Cement   325
Aggregates   325
Water   326
4.4 Properties of concrete   326
Setting of concrete   326
Strength of concrete   327
Deformation of concrete   329
4.5 Reinforced concrete   331
Development of flexural cracking   331
Bond between the concrete and the reinforcing bars   333
Problem of shear   335
4.6 Elastic analysis of a reinforced concrete section   336
Basis of the theory   337
Position of the neutral axis   338
Transformed section   339
Second moment of area   340
Calculation of stress   340
Modular ratio   341
4.7 Prestressed concrete   343
Principles   344
Prestressing techniques   348
Pretensioning   349
Post-tensioning   350
4.8 Elastic analysis of a prestressed concrete section   351
Introduction to the design problems   351
Example of calculation   354
4.9 Problems in concrete structures   356
Corrosion of reinforcement   356
Cracking due to shrinkage and temperature effects   357
Deterioration of concrete   359

5. Structural design and construction **362**

   5.1 Structural arrangements for resisting vertical loading 362
      Floors with one-way beam system 362
      Floors with two-way beam system 368
      Floors with a triple beam system 370
      Arrangements with cantilever beams 371
   5.2 Structural arrangements for resisting wind loading 374
      Frames with rigid joints 374
      Frames with diagonal bracing 376
      Shear walls 382
      Core structures 384
      Tube system 384
   5.3 Aspects of some specialist structures 386
      Single-storey buildings 386
      Bridges 392
   5.4 Construction 401
      Construction of concrete structures 401
      Construction in steelwork 408
      The choice between steelwork and concrete 411
   5.5 Philosophy of design 412

Appendix A: Answers to problems **416**

Appendix B: Guidance on the procedure for solving selected problems **423**

Part Two: Solutions to Problems **435**

Index **583**

Preface

This book aims to introduce structural engineering to students commencing a degree course in civil engineering, architecture or building. It includes only that part of structural engineering which a student would be expected to cover during his first year at university or college. Indeed, Chapter 1 is essentially a precursor to such a course and it is hoped that students will undertake a thorough study of it *prior* to commencing the course. This would ensure a good understanding of those fundamentals which form the basis of the theory of structures. Such a prior study is considered essential for those students who have not included applied mathematics in their sixth form studies.

In confining attention to structural engineering for first year students, the book will be more elementary than most in this field since they usually aim to cover a two- or three-year period of study. It is, however, much wider in scope than such books, which invariably restrict their attention to one of the various parts of structural engineering such as analysis, design, materials or construction. This book will include aspects of all these topics.

It behoves the authors to explain their point of view regarding the introduction of such wide-ranging aspects of structural engineering to first year students when many degree courses initially confine attention to those sections on structural analysis covered in Chapters 2 and 3. Whilst those sections are certainly an important part of the book, it is essential that students understand from the outset that analysis forms only a part of the design phase, which in turn forms only a part of the structural engineering project. Indeed, it can be said that the real excitement and the glamour of structural engineering comes largely from the planning and design of structures and from their construction. It is therefore an aim of this book to present to students an insight into the wide-ranging nature of structural engineering and of the creative and innovative role of the structural engineer in design and construction.

Moreover, the successes and failures in structural engineering are continually reported and discussed in the weekly and monthly publications of the profession. It is from such reports and discussions that the structural engineer increases his knowledge and develops his understanding of structures. In order that the student can commence this important learning stage as soon as possible, it is desirable that he is introduced to the problems of structural design and of the important structural materials at an early stage.

In the sections on structural analysis, numerous worked examples are included. In addition, many problems are given which students must solve themselves. This is a very important part of study. An understanding of structures and of their behaviour under load cannot be achieved simply by reading, however well written the text or however concentrated the study. Problem solving must be carried out. This is where a textbook normally falls short. When a student cannot begin to solve a problem, knowing the final answer is not much use. This book attempts to provide guidance on problems when it is needed. For example, if a student does not know how to tackle a particular problem, he or she should turn to Appendix B for some pointers. The complete solutions to all the problems are given in Part Two. Even if a student has obtained the correct answer to a problem, it is desirable that the solution is referred to in order to ascertain that the calculations have been carried out in the most efficient manner. The International System of units, commonly referred to as SI units, has been adopted throughout.

Acknowledgements

The authors wish to express their indebtedness to the many organizations which provided the photographs that form such a valuable part of this book. The majority of those concerned with steelwork structures were provided by the British Steel Corporation, the British Constructional Steelwork Association, the Steel Construction Institute (publisher of the journal *Steel Construction Today*), and the Centre Belgo-Luxembourgeois de Information de l'Acier (publisher of the journal *Acier, Stahl, Steel*). The majority of the photographs concerned with concrete structures were provided by the British Cement Association, the Building Research Establishment, and the Concrete Society (publisher of the journal *Concrete Forum*). In addition the following supplied photographs: the Ironbridge Gorge Museum Trust (front cover), Dow-Mac Concrete Limited (Figs 4.57 and 5.63), Cape Boards Limited (Fig. 4.18), Bison Limited (Fig. 5.69), Pozzolanic Lytag Limited (Figs 4.36 and 5.57), Trent Concrete Structures Limited (Fig. 5.62), R. Ward of Tarmac Construction Limited (Fig. 5.65), Manchester University Engineering Department (Figs 4.1, 4.2, 4.4, 4.21 (Dr L. J. Morris), 4.23, 4.39, and 5.72). Thanks are also due to Dr H. Shakir-Khalil for providing material for Fig. 2.27.

PART ONE: Theory and Practice

1. Introduction

1.1 The structural engineering project

It is desirable to begin by outlining the relationship of the topics which can be covered in a degree course with the reality of practical structural engineering.

Subsequent to its initiation and the establishment of the financial and sociological constraints, a structural engineering project can be divided into two main phases—design and construction. These phases can be subdivided as shown in Fig. 1.1.

In the first part of the design phase, the structural engineer will determine the possible structural arrangements that are appropriate to the circumstances. As a

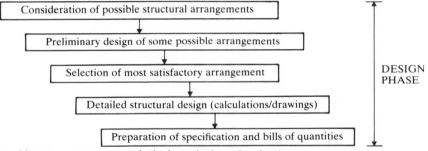

At this stage, contractors are invited to submit tenders for the cost of construction. Following the choice of contractor to undertake the work, the constructional phase can commence.

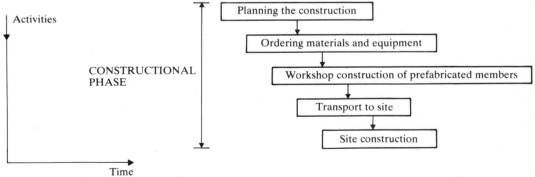

Figure 1.1   Stages in the design and constructional phases of a structural engineering project

3

knowledge of the capacity of the ground to resist the forces from the foundations is often crucial in this regard, an early task will be to initiate an investigation of the soil conditions; this often requires the digging of trial pits or the drilling of boreholes. There is no doubt that the establishment of the possible structural arrangements is the most important part of design and demands a creative and innovative imagination based on many years of experience. It is indeed the highest level of structural engineering practice. Students aspiring to this varied and interesting career must begin to acquire an understanding and knowledge of actual structures as soon as possible. They would be making a first step by undertaking a thorough study of Chapters 4 and 5.

In order to select the particular structural arrangement to be adopted, it may be necessary to carry out a certain amount of preliminary design of several possibilities. In the majority of cases concerned with buildings, the structural engineer will need to collaborate with members of other professions such as the architect and the services engineer. For example, the integration of the heating and ventilation system of a building with its structural system is a major task and the two cannot proceed independently. Needless to say, the circulation and aesthetic requirements of the architect are additional constraints on structural design. Following a joint evaluation of the possibilities, it becomes necessary to select a particular structural arrangement.

Detailed structural design can now commence. This requires calculations to show that the proposed structure will be satisfactory and able to withstand the various loads to which it will be subjected. There are essentially two distinct parts to the calculations:

- application of structural theories to analyse the proposed structure and so determine the forces within the structural members caused by the applied loads;
- application of strength and stress equations and displacement equations to check the properties and details of the proposed members.

Following completion of the design calculations, the detailed drawings and specifications are prepared. From these the quantity surveyor prepares the quantities of materials required. At this stage, contractors are invited to submit tenders for the cost of construction of the structure in accordance with the engineer's drawings and specifications. After the contractor has been chosen, the constructional phase can commence.

Much of the constructional phase (see Fig. 1.1) requires a knowledge and expertise no less demanding than the design phase. In addition to an understanding of structures, management and surveying skills will be particularly relevant. The structural engineer responsible for the design will be represented on site by the resident engineer (colloquially known as the RE) and it is his duty to ensure that the structure is built by the contractor in accordance with the engineer's drawings and specifications.

The analytical part of the detailed design often forms a large proportion of studies in structural engineering at university and college, where it might be presented as in Fig. 1.2. An introduction to the mechanics of structures, defined as the determination

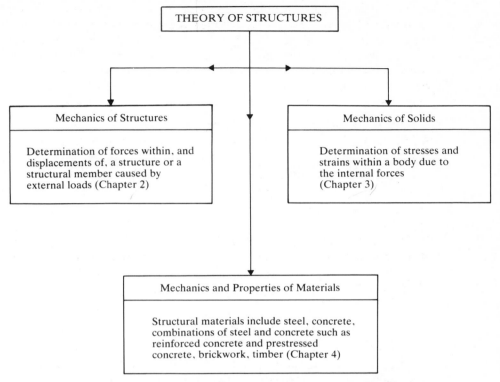

Figure 1.2 Outline of the various sections of the theory of structures

of the internal forces and the displacements of a structure or structural member caused by the applied loads, is given in Chapter 2. An introduction to the mechanics of solids, defined as the determination of stresses and strains within a body, is given in Chapter 3. The mechanics and properties of the two most important constructional materials, steel and concrete, and of their derivatives reinforced and prestressed concrete, are presented in Chapter 4. Needless to say, these studies of the theory of structures are pursued and expanded further in subsequent academic years.

Yet whilst the emphasis of university studies in structural engineering must inevitably be on the theory of structures, students must always be aware that analysis forms only a part of design. Design is essentially creative, requiring abilities of synthesis as well as analysis, and indeed ability in the former is much more important than the latter. Whilst expertise in the creative aspect of structural engineering is developed gradually over a number of years, the student will be introduced to it through the design project. Although such projects are perforce somewhat limited in scope during a first year course, in the subsequent years the design project will be much more open-ended whereby choices of structural form, structural materials, and methods of construction will form important aspects of the exercise. Indeed it is not unusual (and it is certainly the case at the University of Manchester) for students from the Departments of Architecture, Building and Civil Engineering to combine in such design projects, forming an important introduction to the type of collaboration which is necessary in practice in the planning and design of many structures.

1.2 Some structural terms and definitions

A first study of the mechanics of structures requires a knowledge and understanding of that part of applied mathematics known as statics. However, prior to the formal application of statics, it is desirable to start with a qualitative introduction of structural terms and definitions and a pictorial representation of structural concepts.

External forces

It is essential to distinguish between those forces which act externally on a structure or structural member and the internal forces within a member. There are two types of external force:

- forces due to gravity (the mass of the structure itself is acted upon by gravity to cause self-weight or gravity forces which are distributed throughout the structure);
- forces applied at the boundaries (these require physical contact and include the reactions from the supporting ground).

An external force applied to a structural element along the direction of its longitudinal axis is described as *axial*. There are two types of axial force:

- an axial tensile force (a pull, Fig. 1.3a);
- an axial compressive force (a push, Fig. 1.3b).

If a pair of equal and opposite external forces, each of magnitude P, are applied parallel to the longitudinal axis, their lines of action being a distance d apart as in Fig. 1.3c, there is clearly no overall axial effect. It will be apparent, though, that one side of the member is being subjected to compression and the other side to tension. The structural member is said to be subjected to an *applied moment* of value Pd. Such a system of forces frequently occurs at the ends of a structural member and it will generally be more convenient to represent them on subsequent diagrams as shown in Figs 1.3d or 1.3e. Despite being in truth a combination of a pair of forces, it will also be convenient to categorize the applied moment as a *force*. (There is no single comprehensive term which has been universally adopted to cover the various types of force systems. Terms such as *stress-resultant* or *action* are adopted in some texts but the authors have retained the term *force* as their collective noun to encompass moments.)

An external force applied to a structural element perpendicular to the direction of its longitudinal axis (Fig. 1.3f) is described as being *transverse*. For a pair of such equal and opposite forces applied as in Fig. 1.3g, the member is again subject to a *moment*

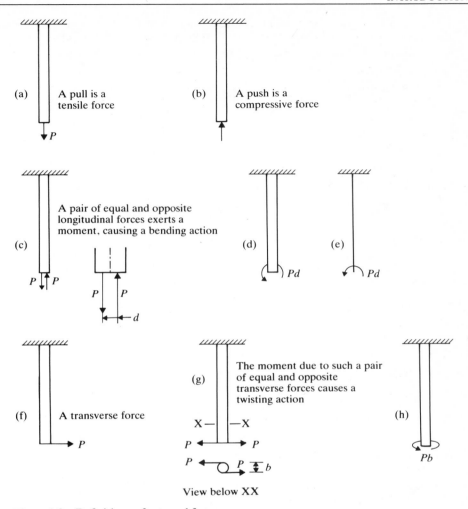

Figure 1.3   Definitions of external forces

(Fig. 1.3h). The deformations resulting from the moments in Figs 1.3d and 1.3h are, however, quite different and are discussed later.

For a structure or structural member in equilibrium (and it will be in equilibrium if it is stationary), then it is obvious that the external forces must balance one another. If the net force in any particular direction were not zero, the body would have an acceleration in that direction. For simple structures this enables us to calculate the external reactive forces applied by the supports to the structure. For example, for the member subjected to an axial force P (Figs 1.3a and 1.3b), the supports will clearly need to apply equal and opposite forces P (Figs 1.4a and 1.4b). Similarly, for members

Forces shown on the member at the position of the supports are those applied by
the support to the member. Forces from the member to the support are equal
and opposite

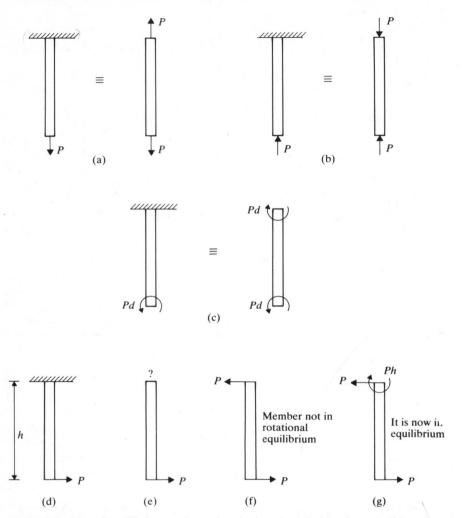

Figure 1.4   Use of equilibrium to determine the forces applied by the supports to the member

subjected to the anticlockwise moment Pd, the supports will need to apply a clockwise
moment Pd (Fig. 1.4c).

The case of the member subjected to the single transverse force P (Fig. 1.4d) is
rather more complicated (Fig. 1.4e). Certainly the supports will need to apply an equal
and opposite transverse force P (Fig. 1.4f), but this alone does not establish

equilibrium as the member would then be subjected to an unbalanced anticlockwise moment of value Ph. It is therefore necessary for the support to apply, in addition to the transverse force P, a clockwise moment of value Ph to the end of the member (Fig. 1.4g).

Internal forces

The externally applied forces are transmitted to the supports through the material of the members, and as a result *internal* forces are caused. For simple structures, these internal forces can be determined by using the principle of equilibrium on part of the member, as it is a fundamental precept that, if a whole body is in equilibrium, then every part is also in equilibrium.

Let us suppose that we require to know the internal force at the section XX in Fig. 1.5a. Consider the equilibrium of the part of the member below the section XX (Fig. 1.5b). To be in equilibrium the internal force at XX must balance the external force on this part of the member. Clearly, then, there must be an upward force P at the section XX (Fig. 1.5c). This is a pull and is therefore a *tensile force*. It is being applied by the top part of the member to the bottom part.

The same answer would be obtained if we considered equilibrium of the part of the member above section XX (Figs 1.5d and 1.5e). For this part to be in equilibrium, there must be a downward force P at the section XX (Fig. 1.5f). This is also a pull and therefore a tensile force. It is being applied by the bottom part to the top part.

It will be apparent that in this example the position of the section XX was not relevant to the determination of the forces, and therefore there exists an internal tensile force of magnitude P at all such cross-sections throughout the length of the member.

As a further example, let us determine the internal forces at section XX in Fig. 1.5g where XX is a distance x from the transverse force P. Equilibrium of the part of the member below XX (Fig. 1.5h) requires that there is a *shear force P* (to balance the externally applied transverse force P) and a *bending moment Px* acting on the section XX. The bending moment is required to counteract the effect of the twin transverse forces P acting at a distance x apart (Fig. 1.5i).

It is of special interest here to show that identical answers are obtained by considering equilibrium of the top part of the member (Figs 1.5j and 1.5k). The answers are given at XX in Fig. 1.5l as will be seen by the fact that the lateral forces cancel out and so do the moments: $P(h - x) + Px - Ph = 0$.

Again note that the two sets of forces shown at XX in Figs 1.5i and 1.5l are equal in magnitude and opposite in sense. Notice also that the clockwise moment at XX in Fig. 1.5i causes compression on the right-hand side of the member and tension on the left-hand side, whilst the anticlockwise moment at XX in Fig. 1.5l also causes compression on the right and tension on the left. This is perhaps seen more clearly by translating the moments back into their equivalent pairs of forces (Figs 1.5m and 1.5n). The effect of the clockwise moment on the one face is identical to the effect of the anticlockwise moment on the other face.

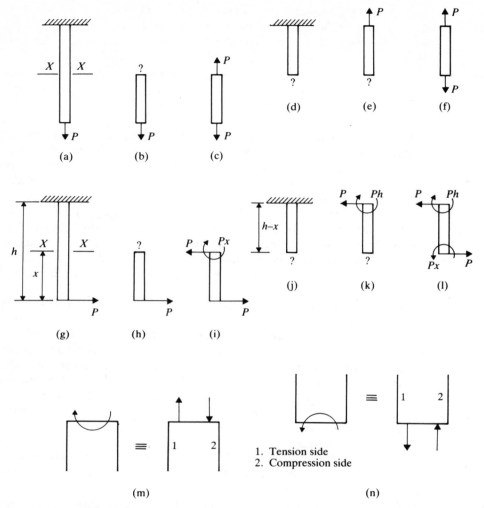

Figure 1.5 Use of equilibrium to determine the internal forces caused within a member by externally applied forces

Stress

The internal forces are distributed through the material of the structural member and it is the knowledge of their distribution which is important in design. The term *stress* is used as a synonym for the intensity of force at a particular point.

For an axially applied tensile force P (Fig. 1.6a), the internal tensile force P at the section XX normal to the direction of P (Fig. 1.6b) will be distributed uniformly across the section (Fig. 1.6c). In other words, there is a uniform stress. The stress due to a

An axially applied tensile force leads to a uniform
distribution of tensile stress on a cross-section

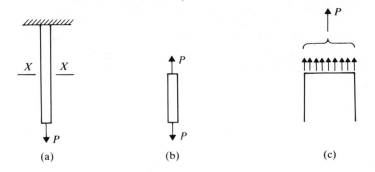

A bending moment on a cross-section causes direct
tensile stresses on one side of the member and
direct compressive stresses on the other. These
will not be uniform as the length of the arrows
in Fig. 1.6g is intended to convey

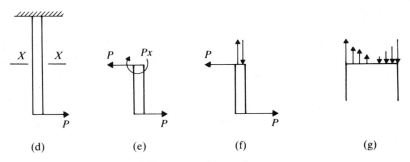

Figure 1.6   Illustration of direct stress in members

force normal to the section under consideration is called a *direct stress*. For this
particular case of uniform stress:

$$\text{direct tensile stress} = \frac{\text{internal force at cross-section}}{\text{cross-sectional area}} = \frac{P}{A}$$

It is most commonly expressed in N/mm^2.

In the case of the member subjected to the transverse load (Fig. 1.6d), there is both a
shear force and a bending moment acting at the section XX (Fig. 1.6e). It will be
recalled that the bending moment can be considered equivalent to a pair of forces

(Fig. 1.6f). Again, these are distributed to cause direct tensile stresses on one side of the cross-section and direct compressive stresses on the other side, but the stresses are not now uniform (Fig. 1.6g). The method of calculating the stress distribution caused by a bending moment will be developed in Chapter 3.

The shear force P acts parallel to the section XX and is distributed over the cross section, giving rise to *shear stresses*. Again, they are not uniform and their calculation is covered in Chapter 3. At this stage it is sufficient to understand that the force P is equal to the summation of these non-uniform stresses.

Deformation

All materials subjected to forces undergo deformation. Whilst in structural materials the deformations are small and usually undetectable to the naked eye, observation of such materials as rubber and plastic gives us an intuitive understanding of behaviour.

It is understood, for example, that the longitudinal fibres of a structural member will extend under the action of an axial tensile force (Fig. 1.7a) and will shorten under an axial compressive force (Fig. 1.7b).

It is also our intuitive knowledge that the member subjected to a transverse force will undergo deflection in that transverse direction (Fig. 1.7c). In this case, however, it is necessary to study the matter further in order to understand the full nature of the deformation. We have already seen that, as a result of the internal bending moments, tensile stresses are caused on one side of the member and compressive stresses on the other. These, in turn, cause an extension of the longitudinal fibres on one side of the member and a shortening of the fibres on the other. Such asymmetrical deformation sets up a curvature of the member, resulting in a transverse deflection.

The type of deformation caused by a shear stress is illustrated in Fig. 1.7d. Thus the total transverse deflection of the member shown in Fig. 1.7c is due partly to the bending effect and partly to the shearing effect, although it is the bending effect which is usually dominant.

A torsional or twisting moment on a member causes the type of deformation shown in Fig. 1.7e, whereby one end of the member rotates relative to the other end so that a line drawn along the member would move as shown.

Strain

Just as the term stress is used to indicate the intensity of force, so the term *strain* indicates the intensity of deformation.

For the case of uniform direct stress throughout the whole of the member, as in Fig. 1.7a, the *direct strain* will also be uniform throughout and is calculated from

$$\text{direct tensile strain } \varepsilon = \frac{\text{extension}}{\text{original length}} = \frac{\delta}{L}$$

Similarly in Fig. 1.7b:

$$\text{direct compressive strain } \varepsilon = \frac{\text{shortening}}{\text{original length}} = \frac{\delta}{L}$$

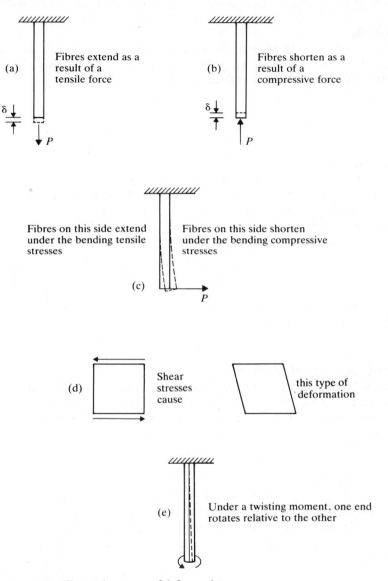

(a) Fibres extend as a result of a tensile force

(b) Fibres shorten as a result of a compressive force

(c) Fibres on this side extend under the bending tensile stresses   Fibres on this side shorten under the bending compressive stresses

(d) Shear stresses cause   this type of deformation

(e) Under a twisting moment, one end rotates relative to the other

Figure 1.7   Illustrating types of deformation

In the case of shear stress, the change in angle in the elements shown in Fig. 1.7d is a measure of *shear strain*. This will be considered in more detail in Chapter 3.

Modulus of elasticity

It has been observed in experiments that, for the majority of materials, up to certain stress levels there is a simple proportionality between stress and strain.

In the case of direct stress and direct strain:

$$\text{stress} \propto \text{strain}$$

thus

$$\frac{\text{stress}}{\text{strain}} = \text{constant } E$$

The constant E is termed the *modulus of elasticity* (and is also known as Young's Modulus). Its value for steel is approximately 205 kN/mm^2 whilst that for aluminium is 70 kN/mm^2. The implication of this is that the deformation of an aluminium structure would be approximately three times that of a similar steel structure at the same level of stress.

In the case of shear stress and shear strain:

$$\frac{\text{shear stress}}{\text{shear strain}} = \text{constant } G$$

The constant G is known as the *shear modulus of elasticity* (and also, rather strangely, as the modulus of rigidity).

1.3 Types of structural elements

Structures are essentially a combination of members which can themselves be categorized depending on their main function. Prior to giving examples of such structural combinations, the characteristics of the various structural elements will be discussed.

Tension members

An important characteristic of a member stressed only in tension (Fig. 1.8a) is that it need have no bending stiffness. In other words, it need not be able to resist transverse forces. This can be illustrated by incorporating a hinge within the member (Fig. 1.8b) which, in the event of a lateral force below the hinge, would offer no resistance to the force and so allow that part of the member to rotate. Clearly, this would not happen under the action of an axially applied tensile force. Indeed, in the event of a slight lateral displacement of the lower part of the member (Fig. 1.8c), the applied force would bring the member back into the line of action of the force. This means that steel cables (which are very flexible in bending) can be used for members stressed only in tension.

Such cables could therefore be used for the situation illustrated in Fig. 1.9a, although it is important to realise that the supports must be capable of resisting the

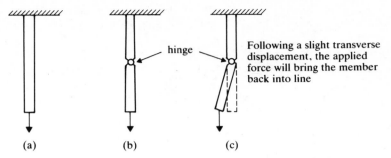

Figure 1.8   Illustrating that tension members need no bending stiffness

inclined pull from the cables (Fig. 1.9b). It is also important to note that the degree of sag in the cable from the supports has an important bearing on the tension force T in the cable. The value of T can be determined by considering vertical equilibrium of the junction of cable and the line of action of the applied load W. Thus

$$2T \sin \theta - W = 0$$

$$T = \frac{W}{2 \sin \theta}$$

From this it is clear that the smaller the angle θ, the larger the value of T. There must therefore be a suitable sag in the cable in order that the force in the cable (and hence

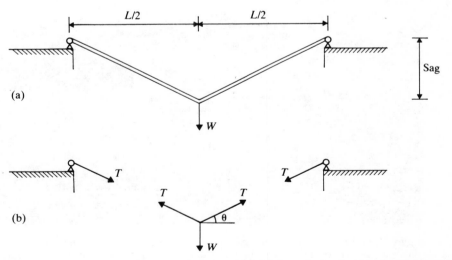

Figure 1.9   An arrangement for supporting a load using members stressed only in tension and therefore requiring only slender members

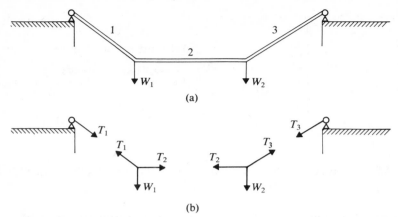

(a)

(b)

Figure 1.10   A cable has sufficient flexibility to adjust its shape to loads applied at different positions so that it is stressed only in tension

Figure 1.11   Maintaining the internal air pressure slightly higher than the external air pressure allows a 'balloon' type roof to be used for temporary structures. For this temporary exhibition hall, the fabric is restrained in position by a series of parallel steel cables (*Source*: Centre Belgo-Luxembourgeois d'Information de l'Acier)

on the supports) is not excessive. If a suspended cable were subjected to loads at different positions, it would simply adjust its shape accordingly (Figs 1.10a and 1.10b).

A sheet of material can act as a three-dimensional flexible tension member, again adjusting its shape to conform to variations in load. An example is the 'balloon' roof (Fig. 1.11) sometimes used for temporary structures—the internal air pressure is increased slightly to create tensile stresses in the sheeting and to maintain its shape. The sheet can also be used in conjunction with cables to form large, tent-type structures.

Compression members

The important difference between tension members and compression members is that the latter (Fig. 1.12a) must have bending stiffness. This can again be illustrated by incorporating a hinge within the member (Fig. 1.12b). Whilst it is conceivable that, with perfect alignment, equilibrium could momentarily be achieved, the arrangement is clearly unstable. A slight angular displacement of the part above the hinge (Fig. 1.12c) would mean that the applied force would now have a moment about the hinge which would increase the rotational displacement to cause failure. The conditions depicted in Figs 1.8b and 1.12b are comparable to the stable and unstable conditions illustrated in Fig. 1.13.

Clearly then in order to resist compression forces the member must have some bending stiffness throughout. An important question in the design of compression members is the degree of stiffness required. Whilst this will be dealt with in some detail during the subsequent years of a degree course, it is essential that an understanding of

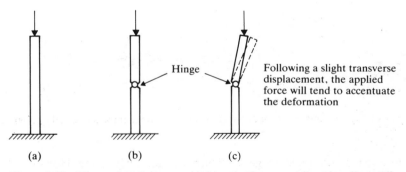

Hinge

Following a slight transverse displacement, the applied force will tend to accentuate the deformation

(a)   (b)   (c)

Figure 1.12   Illustrating that compression members must have bending stiffness

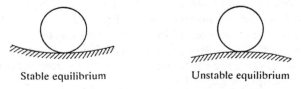

Stable equilibrium   Unstable equilibrium

Figure 1.13   Illustrating the quintessence of stability

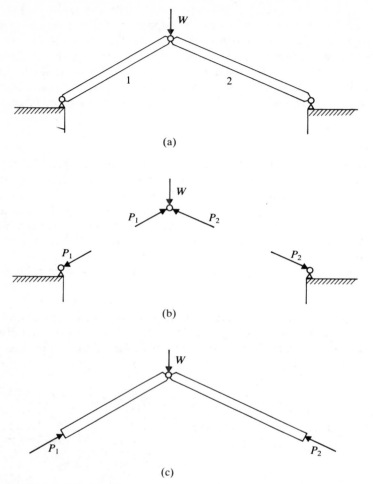

Figure 1.14   An arrangement for supporting a load using members stressed only in compression

the problem is obtained in general terms at an early stage. However, this is best obtained following a knowledge of bending members, and so this particular discussion will be delayed until then.

The arrangement of compression members (or struts as they are often called) with a comparable function to that of the tension members in Fig. 1.9a is shown in Fig. 1.14a. For the asymmetrical case, the forces exerted by the members on the supports and at their junction are shown in Fig. 1.14b—the members, being in compression, are pushing on the connections. There are equal and opposite forces from the connections to the members, as for example in Fig. 1.14c.

The arch (Figs 1.15 and 1.16) belongs to the category of compression members. It

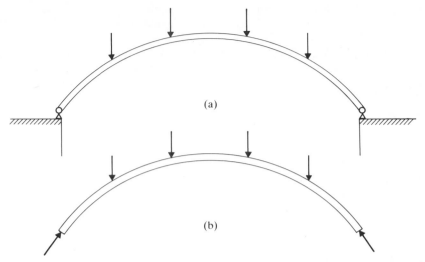

Figure 1.15 Although the arch is primarily a compression member, the direct forces cannot always lie along the axis of the arch and hence bending moments will then also occur

Figure 1.16 An arch bridge (*Source*: British Cement Association)

needs to be much more substantial than the tension-carrying cable in order to have sufficient bending stiffness. Thus, unlike the cable, it does not change its shape under the application of the load. This means that the line of the arch cannot always coincide with the line of the internal forces, and so there will also be a degree of bending involved. The vault and dome are the compressive equivalents of the tensile membrane but again, of course, the dimensions are more substantial.

Bending members

Figures 1.9a and 1.14a illustrated structural arrangements whereby a load positioned between two supports can be carried to the supports. Whilst these are structurally efficient arrangements involving only tension or compression forces within the members, they are too restrictive to deal with numerous practical requirements. For example, the vertical distance between the level of the supports and the level of the load needs to be fairly substantial; moreover, the forces applied to the supports have a horizontal component.

An alternative arrangement which overcomes these disadvantages is the beam (Fig. 1.17a) which supports the load by utilizing its resistance to shear and bending forces. In the way previously indicated in Figs 1.5h and 1.5i, shear forces are set up in the cross-section of the beam, and act in conjunction with appropriate bending moments. Despite the similarities of behaviour between the simply supported beam in Fig. 1.17a and the cantilever in Fig. 1.5g, it is desirable to repeat the arguments regarding behaviour.

Let us consider the case of a symmetrically loaded, simply supported beam having a cross-section which is symmetrical about a horizontal axis. At this type of support there are no moments applied to the beam. (The various types of support are discussed more fully in Section 1.7.) The reactions at the supports will be vertical and each equal to $W/2$ (Fig. 1.17b). What are the internal forces on the cross-section XX at a distance x from the left-hand support (Fig. 1.17c)? In order to balance the upward reaction of $W/2$, there must be a downward shear force equal to $W/2$ on XX. These two transverse forces combine to apply a clockwise moment of $Wx/2$ on this part of the beam, and so there must be a balancing anticlockwise moment of $Wx/2$ on XX (Fig. 1.17d).

This moment implies a longitudinal compressive force acting in the top half of the beam and a longitudinal tensile force acting in the bottom half of the beam (Fig. 1.17e). This, in turn, implies a shortening of the longitudinal fibres in the upper half and a lengthening of the fibres in the lower half. In order that this can occur, the beam must take up a curved form with a consequential downward deflection (Fig. 1.17f). Although such deformation in actual structures is small, it does occur and can be measured using sensitive gauges.

It is necessary at this stage to refer to the nature of the stressing of the fibres due to bending. As the beam considered here is symmetrical about a horizontal axis, it will be clear that the mid-depth of the beam will be at the interface between compression and tension and will itself be unstressed. This is termed the *neutral axis* of the beam. Figure 1.18a shows in exaggerated form the deformation of a short length of the beam, from

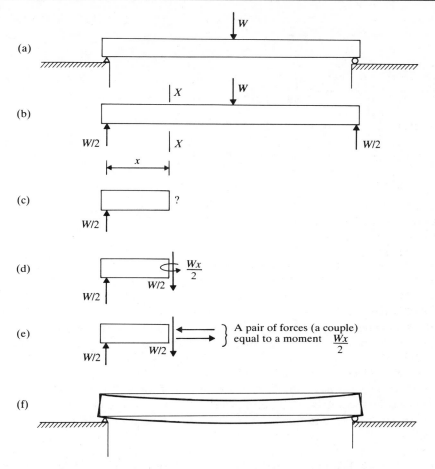

Figure 1.17   The beam utilizes its transverse strength to support the load

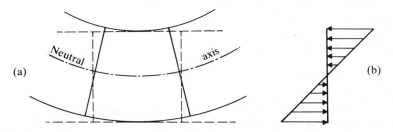

Figure 1.18   Illustrating the development and distribution of longitudinal strains and stresses in a beam caused by bending

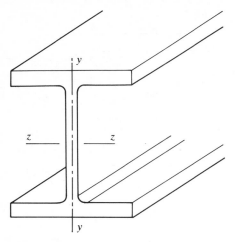

Figure 1.19   The I-shape is a very efficient shape for the cross-section of a beam as it concentrates the material in the region of highest stress

which it will be clear that the outer fibres will extend or shorten to a greater extent than the internal fibres. Indeed the degree of strain of the fibres will depend on their distance from the neutral axis, and the resulting stress distribution for linear elastic materials will be as shown in Fig. 1.18b. (It is implicit in this description that a straight line on the side of the beam would rotate as the beam deflects, and would remain straight. This is one of the basic assumptions of the simple theory of bending.)

Since the top and bottom fibres of a beam will always be the most highly stressed, it follows that in practice one should attempt to concentrate the material of a beam in these regions in order to make the best use of it, and students will recognize that the I-shaped beam (Fig. 1.19) is frequently used in steelwork. Needless to say, such a beam should be used so that it bends about its z–z axis, and it would clearly have a much lower resistance to bending about the y–y axis.

A slab is also a bending member. Indeed in those circumstances where the slab spans one way only, i.e., supported only along two opposite sides as in Fig. 1.20a, it can be likened to a wide beam. However, the slab is rather more versatile and can be used to transmit loads in more than one direction. Figure 1.20b illustrates a slab supported along its four edges with a resulting bending action in two directions. It is a formidable exercise to determine the theoretical bending moments in such cases, but simplified methods are available for use in design.

Torsional members

Torsion occurs in a member whenever the load tends to twist it (Fig. 1.7e). An example of torsion in practice is shown in Fig. 1.21a. When the cantilever beams (supporting, say, a canopy at the entrance to a building) are themselves supported by a beam spanning between widely spaced columns, the spandrel beam is said to be in

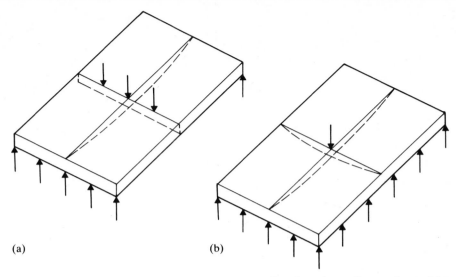

(a)

(b)

Figure 1.20 A slab may undergo bending in two directions depending on the positions of the supports

torsion. The columns are required to exert torsional moments at the ends of the spandrel beam to balance the torsional moments applied by the cantilevers (Fig. 1.21b).

Shear members

In the earlier discussion of the behaviour of members subjected to transverse forces it was stated that of the two internal effects, bending and shearing, the bending effect is usually dominant. Whilst that is true for the great majority of beams, there is a class of structural member in which the shearing effect is particularly important.

A wall-type structural member subject to an applied force in the plane of the wall (Fig. 1.22a) acts as a deep cantilever beam and will transfer the force to the foundations, resulting in a distribution of shear stresses and direct stresses as shown in Fig. 1.22b. The shear stresses in such structures are usually dominant and this is emphasized if the construction includes return walls (Fig. 1.22c). Such an arrangement can now be likened to the I-shaped beam (Fig. 1.19) with the return walls acting as the flanges of the beam to resist the direct stresses caused by the bending effect whilst the wall parallel to the applied force acts as the web of the beam, resisting mainly the shear stresses. Such shear walls play an important part in buildings in resisting the wind loads; they are further discussed in Chapter 5.

A further note on compression members

An understanding of the efficiency of cross-sections in bending (Fig. 1.19) enables us to return to compression members in order to discuss a most important structural phenomenon, namely buckling.

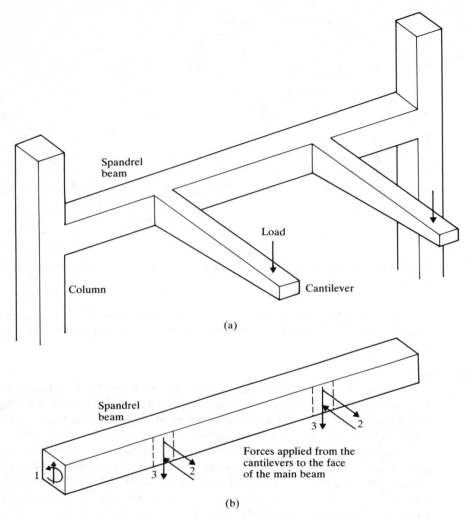

Spandrel beam

Load

Column   Cantilever

(a)

Spandrel beam

3   2

Forces applied from the
cantilevers to the face
of the main beam

1   3   2

(b)

1   Forces applied by column to the end of the spandrel
    beam to balance those applied by the cantilevers
2   The horizontal forces applied by the cantilevers are a
    pull in the top region and a *push* in the bottom
    region. These constitute a torsional moment about the
    longitudinal axis of the spandrel beam
3   The vertical forces cause bending of the beam but, as
    they are eccentric to the longitudinal axis, they also
    cause a twisting moment

Figure 1.21   An example of a torsional member

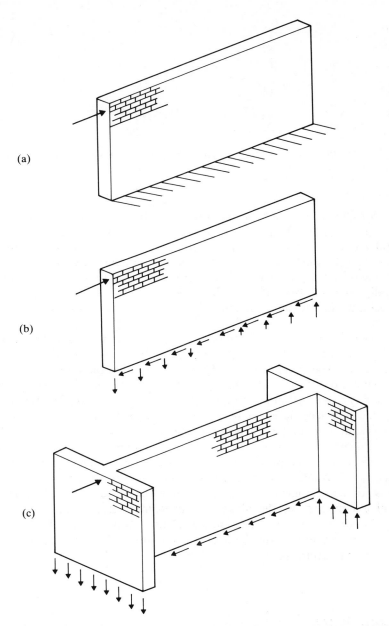

Figure 1.22   The strength and stiffness of walls for forces applied in their plane are frequently used in structures

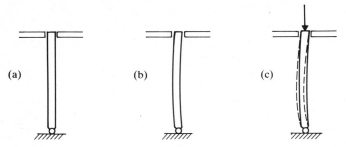

Figure 1.23 Any out-of-straightness of a column will tend to increase under the application of a load

Let us consider a column of a building. We will assume that it is pinned at the bottom (Fig. 1.23a). At its upper end we will assume that the floor will restrain any horizontal movement, but that there is no vertical or rotational restraint to the column. A further assumption is that the column is not perfectly straight. This would be so in practice as a perfectly straight and perfectly axially loaded column is only theoretically possible. Nevertheless, the out-of-straightness of practical columns is quite small and the exaggerated shape shown in Fig. 1.23b is only for purposes of illustration.

It will be clear that the application of a load to the column will, even if it is axial at the ends, be eccentric to the cross-sections throughout most of its length. This will, therefore, cause a degree of bending in the member in addition to the shortening of the column due to the vertical load. Although it might be thought that the eccentricity of the force at any cross-section will be small (as indeed it will), nevertheless if the vertical forces on the column are large the resulting bending moments may be significant, and indeed cause additional lateral deflection (Fig. 1.23c) which in turn cause additional bending moments. The higher the load on a column the greater the additional outward deflection caused by a further increment of load, and a stage may be reached when such outward deflections are the cause of failure of the column.

It follows that, as for members in bending only, the shape of the cross-section has a considerable influence on the ability of the compression member to resist this tendency. Unfortunately the shape so useful in bending (Fig. 1.19) is much less efficient for use in compression as the column would tend to bend about the weaker y–y axis. This important problem is further discussed in Chapter 4.

1.4 Structural complexes

It is the purpose of this section to demonstrate that most structures are essentially a combination of some of the five basic elements previously described, although, as will be shown, some members may be required to perform more than one function.

Trusses

One of the disadvantages of supporting loads using the arrangements in Figs 1.9a and 1.14a is that the forces applied to the supports are inclined in direction, and the supports are therefore required to resist horizontal components of the forces as well as their vertical components. It is, however, possible to eliminate the horizontal forces acting on the supports by incorporating a third member. Thus the arrangement utilising cables to support the load would require a compression member (a strut) as shown in Fig. 1.24a. This resists the horizontal component of the pull on the two top joints by the cable. Only the vertical component of this pull is now required to be resisted by the supports. The types of forces acting on the various joints are indicated in Fig. 1.24b.

For the comparable situation utilizing compression members to support the load, the horizontal components of the forces from the two struts is taken by a tension member (a tie) for which a cable would be suitable (Fig. 1.24c). The types of forces acting on the joints in this case are indicated in Fig. 1.24d.

Although for purposes of illustration it has so far been possible to distinguish in the diagrams between the slender tension members and the more substantial compressive members, as the structures inevitably become more complex this will not continue to be convenient and henceforth both tension and compression members will usually be designated in the diagrams by single lines. In any case, in practice the distinction does often disappear as a particular member which is in tension under a particular load system may well be stressed in compression under an alternative load system.

Attention needs to be drawn here to the assumption that the members are connected through frictionless hinges or pins. These permit relative rotations of the

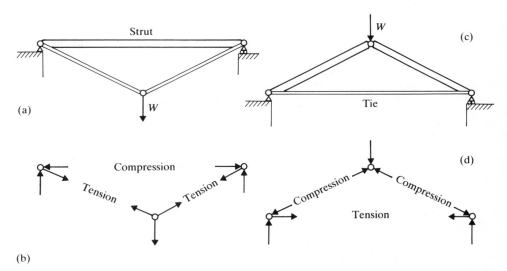

Figure 1.24 Two arrangements of tension and compression members for supporting a load

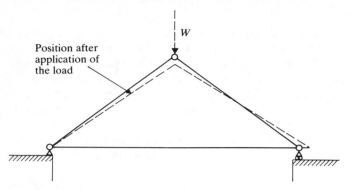

Figure 1.25   The extensions and shortenings of the tensile and compression members of a pin-jointed frame cause a deformation requiring slight relative rotation of the members

members to occur and thereby avoid setting up moments at the ends of the members. That such a rotation is necessary follows from the fact that the members will either shorten or extend slightly as the internal compressive and tensile forces develop under the application of the load, and these small changes in length are accommodated by small changes in the angles between members (Fig. 1.25). Although the joints used in practical trusses are not true frictionless hinges (see Chapter 2), the resulting errors from such an assumption are small.

The importance of the triangulated arrangement of pin-connected members needs to be emphasized. It would not, for example, be possible to use four members connected through pins as shown in Fig. 1.26a. The arrangement is a mechanism which would collapse under the application of the load by the large relative rotation of the members. Incorporation of a diagonal in the frame (Fig. 1.26b) does, however, transform this into a stable triangulated configuration. The basis of the stability of the triangle can be understood from Fig. 1.26c. With one of the joints fixed in position by the supports and another constrained against vertical movement, the position of the third joint is determined simply by the lengths of the members—rotation of member 1 would require the joint to move along the arc a–a whilst rotation of member 2 would require the joint to move along the arc b–b; as these do not coincide, the position of this third joint is fixed and will remain so until variation in the loads causes a change in the member forces and the consequent slight change in the lengths of the members (see Fig. 1.25).

From such a triangular base, one can add two members at a time (Figs 1.26b and 1.26d) to give additional joints constrained in space and forming stable triangulated frames known as trusses. The supports are not required to be at adjacent joints and the arrangement shown in Fig. 1.26e is a structural arrangement frequently found in practice (Fig. 1.27).

It is important to realize that if the truss is loaded only at its joints, then the members of the truss will be stressed only in tension or compression. There is no bending moment in any member.

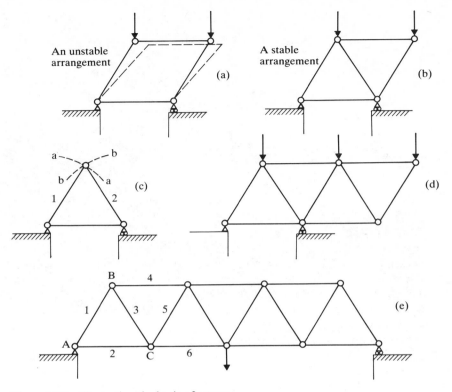

Figure 1.26   Illustrating the basis of a truss

Prior to the detailed analysis of such structures (in Chapter 2), it is of interest to note how easily one can sometimes determine the nature of the forces, whether they are tension or compression, in the members by using simple logic. This is now illustrated.

Each joint of the structure in Fig. 1.26e is in equilibrium under the effect of the forces from the members and from any external forces. (It has already been stated that if a complete structure is in equilibrium under the forces acting on it, so must each part. A joint of a pin-jointed structure can therefore be regarded as a separate body in equilibrium under the forces meeting there.) Thus, starting with a joint at which only two members meet, say joint A in Fig. 1.26e, the situation is as indicated in Fig. 1.28a—there will be an upward force from the supports to the joint and it is required to determine the nature of the forces in the members 1 and 2. In order to be able to balance the upward force from the reaction, the force in member 1 must have a downward component and so it will need to push on the joint (Fig. 1.28b)—member 1 is therefore in compression. It can now be seen that member 2 must pull on the joint (Fig. 1.28c) in order to balance the horizontal component of the push from member 1—member 2 is therefore in tension.

Figure 1.27 The triangulated arrangement of members is a structural form frequently adopted in practice (*Source*: Steel Construction Institute)

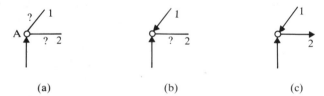

(a)   (b)   (c)

Figure 1.28 The use of simple logic to determine the nature of the forces at joint A of the truss in Fig. 1.26e

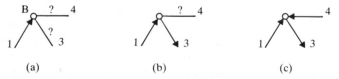

(a)   (b)   (c)

Figure 1.29 The use of simple logic to determine the nature of the forces at joint B of the truss in Fig. 1.26e

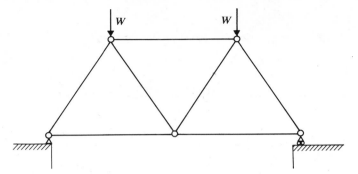

Figure 1.30 Problem 1.2

Coming now to joint B, the situation is as shown in Fig. 1.29a, with the member 1 known to be in compression and therefore pushing on the joint. In order to be able to resist the vertical component of the push, member 3 must pull on the joint (Fig. 1.29b)—member 3 is therefore in tension. As both members 1 and 3 have horizontal components acting on the joint in the same direction, left to right, the force in member 4 needs to balance these by acting right to left and therefore pushing on the joint (Fig. 1.29c)—member 4 is therefore in compression.

Problem 1.1 Use simple logic to determine the nature of the forces in members 5 and 6 of the pin-jointed truss in Fig. 1.26e.

Problem 1.2 Use simple logic to show that the force in each of the two inclined internal members of the symmetrical truss in Fig. 1.30 is zero. Could these members therefore be removed?

It will be understood that the configuration of the members of the truss is a design variable, placing the onus on the structural engineer to suitably shape the structure. Examples of well-established arrangements are shown in Fig. 1.31, but variations of these are possible. For example, the top and bottom members, often referred to as *chords*, need not be parallel.

Problem 1.3 For vertical loads applied only at the joints along the bottom chord, which of the two trusses, Pratt or Howe, is to be preferred? What is the particular advantage of the K truss?

A distinct disadvantage of the planar truss (as indeed with all planar frames) is its tendency to buckle out-of-plane because of the behaviour of the members in compression. As a result, special lateral bracing is often required, a point further discussed in Chapter 5. An alternative is to use a three-dimensional truss (see Figure 1.32) which then has its own in-built stabilizing system. Whilst the analysis of such

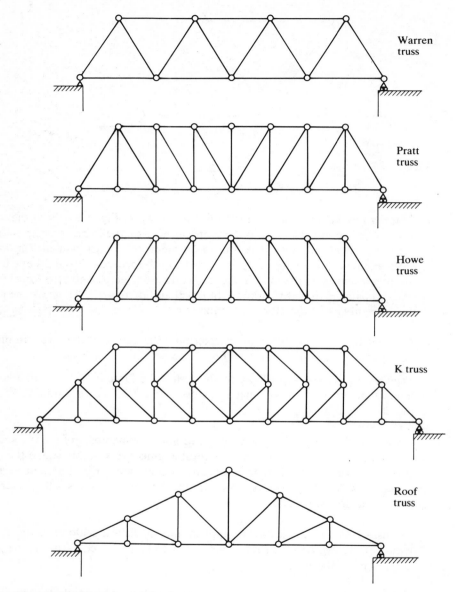

Figure 1.31   Common types of trusses

Figure 1.32 The 3-dimensional triangular arrangement is an inherently stable configuration. Here the space frame is spanning in two directions to give a large span roof structure. It was assembled at ground level and the photograph shows it in the process of being jacked up to its final position (*Source*: Centre Belgo-Luxembourgeois d'Information de l'Acier)

structures becomes a little more involved than that for planar trusses, the same simple principles apply, and in any case the use of computers makes light work of such an analysis. Indeed, as a result there has been a rapid growth in the development of space frames which may span in two (or even three) directions, and these are being increasingly used for covering areas requiring large spans (Fig. 1.32).

Multi-storey buildings

The multi-storey complex (Fig. 1.33) is, from the structural standpoint, frequently a combination of:

- bending members (the floor slabs 1, floor beams 2);
- compression members (the columns 3);
- shear members (end walls 4 or the enclosing walls around staircases or lift shafts).

The floor slabs carry the vertical loads to the floor beams which, in turn, transfer the loads to the columns, and then to the foundations. The horizontal forces from the wind loading acting on the façade are transferred by the floor slabs to the shear walls

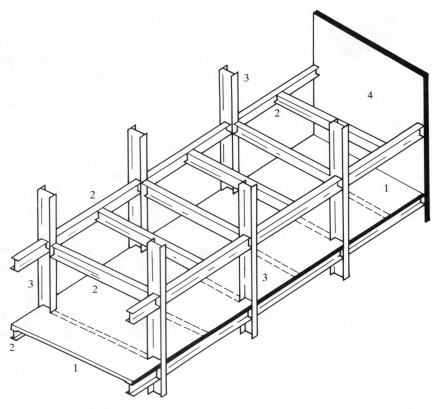

Figure 1.33   Illustrating the structural components of a multi-storey building

(see Chapter 5) which in turn transfer the horizontal forces to the foundations (see Fig. 1.22).

Students should not conclude that such multi-storey buildings are necessarily structurally complicated. It is often possible to consider the various members in isolation and, as a result, *analysis* of the structure is then quite simple and straightforward. This is not to say that *design* is so straightforward, as this involves decisions on the particular structural arrangement to be adopted. These structural arrangements will depend on many aspects of the design, architectural as well as structural. An important structural consideration will be whether the vertical loads (due to gravity) or the horizontal loads (due to the wind) are the dominating forces to be carried, and this will depend largely on the height of the building. A discussion of possible structural arrangements is given in Chapter 5.

The various descriptions of the members accord with their main function and do not preclude them from acting in other ways. For example, although the main function of a column is to act in compression, it is usually required to act also in bending as a result of the arrangement for transferring the load from the beam to the column. Similarly, although the main function of the floor slab is to act as a bending member in carrying the vertical loads to the beams, it also often acts essentially as a shear member in transferring the horizontal forces to the shear walls.

Cable-supported structures

Tension members are exceptionally efficient in transmitting forces as:

– they are not subject to the problem of buckling;
– they are usually subject to a uniform distribution of stress over their cross-section, and often throughout their lengths.

As a result, highly stressed, high-strength steel cables are frequently used in the construction of large span structures. Needless to say, other types of structural element are required to play an essential part. The suspension bridge (Fig. 1.34), for example, is a combination of:

– tension members (the main cables and the hangers);
– compression members (the towers);

Figure 1.34 A suspension bridge (*Source*: Centre Belgo-Luxembourgeois d'Information de l'Acier)

Figure 1.35   A cable-stayed bridge with the cables in a single plane (*Source*: Centre Belgo-Luxembourgeois d'Information de l'Acier)

– bending members (the main girder which spans longitudinally and is supported at the abutments, at the towers, and at the hangers).

The cable-stayed bridge (Fig. 1.35) has some advantages over the suspension bridge in construction and is preferred for the more modest span. Whilst it is also a combination of tension members, compression members and bending members, there are additional structural actions which need to be noted. Although the vertical components of the forces in the cables provide support to the main girders, the

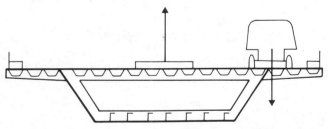

Figure 1.36   For cable-stayed bridges with the supporting cables arranged in a single plane and anchored in the central reservation, the imposed load due to traffic will cause torsion in the main girder

Figure 1.37   Cables are frequently used for supporting the cantilever roof in stadiums (*Source*: British Constructional Steelwork Association)

horizontal component of the forces in the cables must be resisted by the girders acting in compression. Moreover, as the cables are often grouped in a single plane and anchored into the main girder within the central reservation (Fig. 1.36), asymmetrical loading causes large torsional moments which need to be transferred through the girder to the towers. It will be seen, therefore, that although the main girder is required to span only between the positions of the anchorages of the cables in resisting the bending caused by vertical loading, it is required to span the full distance between the two towers in resisting the torsion due to any unsymmetrical loading in this central region. In order to have the necessary torsional strength and torsional stiffness, the hollow-box type of cross-section is essential.

The cantilever roof for use in providing unrestricted viewing in stadiums (Fig. 1.37) is another structure which lends itself to the advantageous combination of tension members, compression members and bending members.

Structures involving mainly bending members

For bridges involving spans smaller than those illustrated in Figs 1.16, 1.34 and 1.35, bending members will frequently suffice for transmitting the loads to the supports. The superstructure of the bridge shown in Fig. 1.38 is essentially a series of such bending members. In this case the beams of the main spans are simply supported at the ends of beams cantilevering a short distance beyond the splayed piers. Such supports are equivalent to pins (see Section 1.7) and do not transmit moment. Thus, where the beams of adjacent spans are simply supported at the position of a pier as in Fig. 1.39a, the load on a particular span has no effect on the beams of adjacent spans

Figure 1.38 Multi-span bridge, but with the beams of the main spans simply supported at the ends of beams cantilevering beyond the splayed piers. The beams vary in depth to give the maximum depth at the positions of maximum bending moment (*Source*: British Cement Association)

(Fig. 1.39b). As a result, slight angular discontinuities occur from one span to the next as the beams deflect under the influence of the loads.

If, however, the beams are constructed so that they are continuous over the supports (as in the superstructure of the bridges in Fig. 1.40), then the load on one span does now affect the other spans (Fig. 1.41a). The continuous nature of the beams means that the deflection of the various spans is such that there is now no angular discontinuity at the supports.

The resulting structural interaction between the various spans of continuous beams is such that bending moments occur at the sections of the beams at the positions of the supports. Although these are internal forces, i.e., no moment is applied by any external

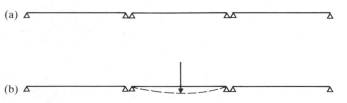

Figure 1.39 Diagrammatic representation of a series of simply supported spans. The bending of one span of the bridge under the effect of load does not affect the adjacent spans

Figure 1.40 Multi-span beam bridges, but with the beams continuous at the piers. In such bridges, the bending moments in the beams are highest at the positions of the piers and so it is structurally advantageous to increase the depths of the beams at these positions as in (a), but nevertheless constructional restraints may make it more economical to adopt a constant depth as in (b). (*Sources*: (a) Steel Construction Institute; (b) Centre Belgo-Luxembourgeois d'Information de l'Acier)

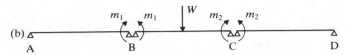

Figure 1.41 (a) Diagrammatic representation of a 3-span continuous beam; the bending of one span of the beam causes bending of the other spans. (b) The continuous beam is equivalent to simply supported beams but with hogging bending moments applied at their ends; the clockwise moment at B is being applied to the end of beam AB by beam BC and similarly the anticlockwise moment at B is being applied to the end of beam BC by beam AB

agency, it may sometimes be preferable for purposes of analysis to separate the spans as shown in Fig. 1.41b whereby the beams of a particular span are now subjected to moments applied at the ends by the adjacent spans. For example, the span BC can be analysed as if it were a simply supported span but subject to external forces W, m_1 and m_2. This procedure of separating a continuous structure into its structural elements with appropriate interaction forces at the ends of the elements is one which is widely used, and it is important that the student becomes fully acquainted with it in the subsequent sections on analysis.

Such continuity of bending action is sometimes used when connecting beams to columns to form a portal frame (Fig. 1.42a). If the beam is so rigidly connected to the column that the angle between the two must remain at 90°, the bending of the beam under the effect of a load would also cause bending of the column (Fig. 1.42b)—the top of the column must rotate by the same amount as the end of the beam.

The importance of such an arrangement is largely to resist horizontal wind forces (Fig. 1.42c), and this is further discussed in Chapter 5. At this stage, it is sufficient to understand that a pinned-based, rectangular frame in which the beam and columns are effectively hinged (Fig. 1.42d) is a mechanism and would not be able to withstand horizontal wind forces, although, as will be shown in Chapter 5, this does not mean that rigid joints between beams and columns are essential.

Needless to say, the columns of portal frames also act as compression members to carry the vertical loads down to the foundations, but for the type of structure for which the rigidly jointed frame is most frequently used, namely single-storey buildings, its bending action is the more important.

1.5 Some basic concepts relating to structural analysis

Sign convention and coordinate axes

The study of the effect of forces on complex structures requires a formalized system of analysis. Although the structures to be analysed in a first year course permit a more

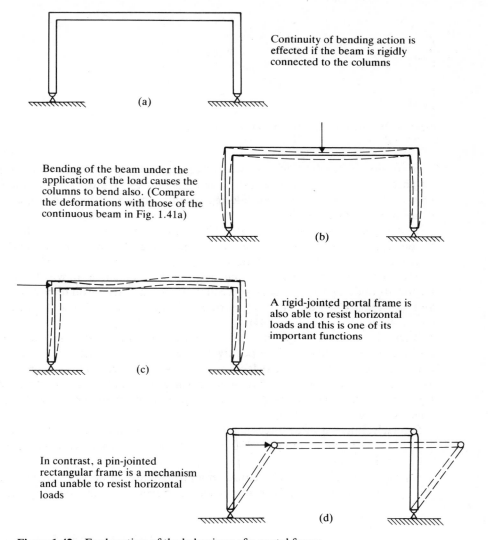

Figure 1.42 Explanation of the behaviour of a portal frame

intuitive approach, it is desirable to use a formal approach from the outset. This necessitates the establishment of a sign convention for forces.

Directions are determined by adopting a rectangular system of perpendicular axes Ox, Oy, Oz (Fig. 1.43). Following the selection of the Ox and Oy directions, the direction for Oz is that in which a right-hand screw along Oz would advance when turned through an angle from Ox to Oy.

Figure 1.43   Sign convention for coordinate axes

In considering *external* forces acting *on* a structure or structural member (or indeed on part of a structural member), due regard is taken of the direction of the forces in relation to these axes when considering equilibrium—and this is discussed in some detail in subsequent sections.

With regard to *internal* forces *within* the members of a structure, in order to be able to designate such forces as positive or negative (for reasons of mathematical convenience only) it becomes desirable to specify the type of face on which the force is acting. For example, a cross-sectional face of part of a member in which the outgoing normal from that part is in the positive direction of the axis is said to be a positive face; and vice versa when the outgoing normal is in the negative direction of the axis the face is a negative face (Fig. 1.44).

Internal forces which act in a positive direction on a positive face are said to be positive. Two negatives make a positive in the sense that an internal force acting in a negative direction on a negative face is also a positive force. It follows from this that an axial tensile force acting within a member will be designated as a positive force (Fig. 1.45a), whilst an axial compressive force acting within a member will be designated as a negative force (Fig. 1.45b).

Similarly, internal shear forces acting on sections of a member will be positive when acting in the directions shown in Fig. 1.46a and negative when acting in the directions shown in Fig. 1.46b.

A positive bending moment acting on a positive face of part of a member tends to rotate that part of the member in a clockwise direction when viewed in the positive

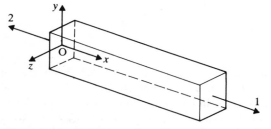

1. Outgoing normal is in the positive *x* direction and hence the face is described as positive

2. Outgoing normal is in the opposite direction to O*x* and hence the face is a negative face

Figure 1.44   Definition of positive and negative faces of a member for use in determining internal forces

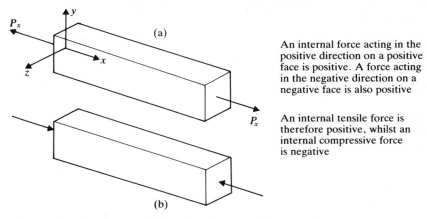

An internal force acting in the positive direction on a positive face is positive. A force acting in the negative direction on a negative face is also positive

An internal tensile force is therefore positive, whilst an internal compressive force is negative

Figure 1.45 Sign convention for direct forces within a member

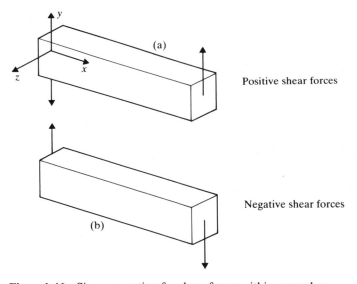

Positive shear forces

Negative shear forces

Figure 1.46 Sign convention for shear forces within a member

direction of the appropriate axis. Similarly, a positive bending moment acting on a negative face tends to rotate that part of the member in an anticlockwise direction. Thus, bending moments acting on sections of a member will be positive when acting in the directions shown in Fig. 1.47a and negative when acting in the directions shown in Fig. 1.47b. To avoid confusion of the directions of the moments on the faces shown in perspective, the moments have been represented by their equivalent twin forces. (Later, double-arrowed lines lying along the axis about which the moment acts will be

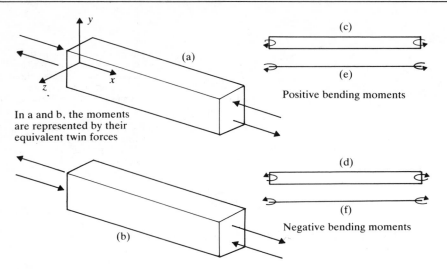

Figure 1.47 Sign convention for bending moments within a member

used to represent the moment vector more concisely, but at this stage the twin forces are retained for clarity.) More usually, the beams will be shown in elevation where the corresponding positive and negative bending moments would be as indicated in Figs 1.47c and 1.47d, or more concisely as in Figs 1.47e and 1.47f.

For the case of horizontal beams it will be observed that when the y-axis is chosen to be upwards, a positive bending moment causes tension in the lower region of the beam and compression in the upper part, and, once the student is more acquainted with the deformational effects of loads, this will usually be the easier method of determining the sign of the bending moment.

Components of a force

To be completely described a force must have:

– magnitude;
– line of action;
– sense (i.e., direction).

These are the properties which define the mathematical concept known as a vector. A vector can be represented by a line drawn to a suitable scale and, in a specific position with regard to the axes, can be used to define a force as in Fig. 1.48—the length of the line represents the magnitude of the force; its position relative to the axes determines the line of action; the arrowhead indicates the sense, or direction, of the vector. This graphical representation of a vector facilitates a visual concept which enables us to understand the idea of components of a force. For example, referring to Fig. 1.48, the components of P in the x, y and z directions are given by P_x, P_y and P_z, these being the projected lengths of P on lines parallel to x, y and z. If α, β and γ are the angles made

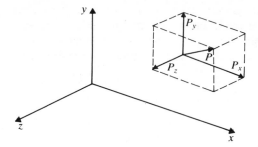

Figure 1.48 Graphical representation of a force and its components

by P with these lines, the components are

$$P_x = P \cos \alpha \qquad P_y = P \cos \beta \qquad P_z = P \cos \gamma$$

That P is equivalent to such a set of components is one of the basic laws of statics and has been established by direct observation and experiment.

Moment of a force

A force on a structure is said to have a moment effect about an axis if it has a tendency to rotate the structure about the axis. The magnitude of the moment is defined as the product of the force and the perpendicular distance from the line of action of the force to the axis. It is usually more easily calculated from the components of the force as the perpendicular distances are then more easily established. For example, if a force P having components P_x, P_y and P_z passes through point $x_1 \ y_1 \ z_1$ (Fig. 1.49), the moments are

- about the x-axis $\quad m_x = P_z y_1 - P_y z_1$
- about the y-axis $\quad m_y = P_x z_1 - P_z x_1$
- about the z-axis $\quad m_z = P_y x_1 - P_x y_1$

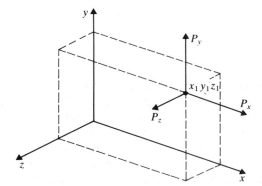

Figure 1.49 If the line of action of a force passes through a point, its components can be considered to act at that point

A positive sign in these expressions indicates that the force tends to cause a clockwise moment about the axis when viewed in the direction of the axis. If the line of the force P also passes through the point x_2 y_2 z_2 then the moments can similarly be calculated using these distances, but the resulting values will be the same.

Equilibrium

When a number of forces $P_1, P_2, P_3 \ldots$ act on a structure, their total effect is

- in the x direction $= P_{x1} + P_{x2} + P_{x3} \ldots = \sum P_x$
- in the y direction $= P_{y1} + P_{y2} + P_{y3} \ldots = \sum P_y$
- in the z direction $= P_{z1} + P_{z2} + P_{z3} \ldots = \sum P_z$

The total moments are

- about the x-axis $= m_{x1} + m_{x2} + m_{x3} \ldots = \sum M_x$
- about the y-axis $= m_{y1} + m_{y2} + m_{y3} \ldots = \sum M_y$
- about the z-axis $= m_{z1} + m_{z2} + m_{z3} \ldots = \sum M_z$

For structures in equilibrium under the action of these forces, it follows from Newton's Second Law of Motion that

- the total effect in each of the three orthogonal directions must equal zero, thus

$$\sum P_x = 0 \qquad \sum P_y = 0 \qquad \sum P_z = 0$$

- the total moments about each of the three axes must equal zero, thus

$$\sum M_x = 0 \qquad \sum M_y = 0 \qquad \sum M_z = 0$$

For a three-dimensional structure there are therefore a total of six equations of equilibrium. For a two-dimensional problem these reduce to three equations of equilibrium—for example, where the structure and the applied forces are all in the x–y plane, the equations of equilibrium are

$$\sum P_x = 0 \qquad \sum P_y = 0 \qquad \sum M_z = 0$$

Use of matrices

For the analysis of large or complex structures, it becomes essential to use computers to manipulate the arithmetic. The best method of arranging the data in a form suitable for processing by the computer is by using matrices. Although such large and complex structures will not be dealt with in a first year course, the method of presenting information in a matrix format will be adopted from the outset in order that students become familiar with the system. Some of the previous arguments are therefore now repeated and the equations presented in matrix form. Their meaning will be clear by a direct comparison with the previous discussion.

If α, β and γ are the angles made by a force with lines parallel to the x, y and z axes, the force is said to have components P_x, P_y and P_z where

$$P = \begin{bmatrix} P_x \\ P_y \\ P_z \end{bmatrix} = \begin{bmatrix} P \cos \alpha \\ P \cos \beta \\ P \cos \gamma \end{bmatrix}$$

When a number of forces P_1, P_2, P_3 ... act on a structure, each has components along the axes:

$$P_1 = \begin{bmatrix} P_{x1} \\ P_{y1} \\ P_{z1} \end{bmatrix} \qquad P_2 = \begin{bmatrix} P_{x2} \\ P_{y2} \\ P_{z2} \end{bmatrix} \qquad P_3 = \begin{bmatrix} P_{x3} \\ P_{y3} \\ P_{z3} \end{bmatrix}$$

The resulting total force vector is

$$P_1 + P_2 + P_3 + \ldots = \begin{bmatrix} P_{x1} \\ P_{y1} \\ P_{z1} \end{bmatrix} + \begin{bmatrix} P_{x2} \\ P_{y2} \\ P_{z2} \end{bmatrix} + \begin{bmatrix} P_{x3} \\ P_{y3} \\ P_{z3} \end{bmatrix} + \ldots$$

$$= \begin{bmatrix} P_{x1} + P_{x2} + P_{x3} + \ldots \\ P_{y1} + P_{y2} + P_{y3} + \ldots \\ P_{z1} + P_{z2} + P_{z3} + \ldots \end{bmatrix} = \begin{bmatrix} \sum P_x \\ \sum P_y \\ \sum P_z \end{bmatrix}$$

The resulting total moment vector is $\begin{bmatrix} m_{x1} \\ m_{y1} \\ m_{z1} \end{bmatrix} + \begin{bmatrix} m_{x2} \\ m_{y2} \\ m_{z2} \end{bmatrix} + \begin{bmatrix} m_{x3} \\ m_{y3} \\ m_{z3} \end{bmatrix} + \ldots$

$$= \begin{bmatrix} m_{x1} + m_{x2} + m_{x3} \ldots \\ m_{y1} + m_{y2} + m_{y3} \ldots \\ m_{z1} + m_{z2} + m_{z3} \ldots \end{bmatrix} = \begin{bmatrix} \sum M_x \\ \sum M_y \\ \sum M_z \end{bmatrix}$$

For a structure in equilibrium, the resultants in each of the three orthogonal directions must be zero, thus

$$\begin{bmatrix} \sum P_x \\ \sum P_y \\ \sum P_z \end{bmatrix} = 0$$

and the moments about each of the three orthogonal axes must also be zero:

$$\begin{bmatrix} \sum M_x \\ \sum M_y \\ \sum M_z \end{bmatrix} = 0$$

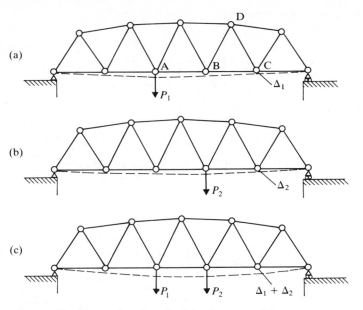

Figure 1.50   Illustrating the principle of superposition

Principle of superposition

For cases where the load–deformation characteristic of a structure is linear (i.e., where the graph relating load and deformation is a straight line), the effects of multiple loads can be determined by considering the effects of the loads separately and super-imposing these effects. An example is shown in Fig. 1.50. If a load P_1 applied at A causes a vertical deflection at C of Δ_1 (Fig. 1.50a), and a load P_2 at B causes a vertical deflection at C of Δ_2 (Fig. 1.50b), then the vertical deflection at C caused by applying P_1 and P_2 together will be $\Delta_1 + \Delta_2$ (Fig. 1.50c). Similarly, the force in any member, say CD, caused by P_1 and P_2 acting together will be the sum of the forces in CD caused when P_1 and P_2 act separately.

The conditions necessary to ensure that a structure is linear are:

- the material of the structure should be on the linear elastic part of its stress–strain relationship;
- there should be no appreciable change of geometry in the structure.

Reference has previously been made to the fact that structural members will undergo a change in length when subjected to an axial force. This implies an inevitable change of geometry. However, the changes of length of structural members are always very small, with the result that the change of geometry of most structures is insignificant. For example, although the extensions of the members in Fig. 1.51a will slightly alter the angle of inclination of the members (Fig. 1.51b), this will not be

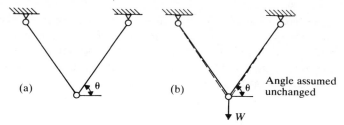

Figure 1.51   Illustrating a situation where changes in length of the members caused by the load have an insignificant effect on geometry

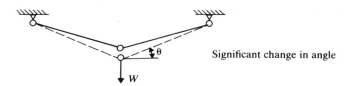

Figure 1.52   Illustrating an arrangement where a significant change of geometry is caused by the load. Such structures are non-linear and the principle of superposition does not apply

sufficient to necessitate allowing for the change in the calculations. The effect of the change in θ upon the member forces will be small enough to ignore.

This is not always the case. The arrangement shown in Fig. 1.52 is an example where the extensions of the members caused by the applied load would cause a sufficient change in the angle of inclination of the members to warrant allowing for it in calculating the forces in the two inclined members. As the angle of inclination of the members now depends on the magnitude of the applied load, the relationship between W and its downward movement (the load–deformation characteristic) is non-linear and the principle of superposition would not apply to such a structure.

Free-body diagrams

A free-body diagram is one which illustrates a part of a structure in which the removed parts have been replaced by idealized forces acting at the contact areas on the isolated part of the structure. Such diagrams are extensively used in structural analysis, and indeed have been used many times already in this book, as for example in Fig. 1.14. It is desirable, however, to draw special attention to their importance and the need for students to develop their skill in drawing free-body diagrams. It is the analyses of free-body diagrams which are the basis for determining the internal forces within structures. Moreover, the use of free-body diagrams facilitates an understanding of the behaviour of structures.

1.6 Planar statics

Statics form the very basis of a study of structures. Whilst the essential principles of statics (namely, the components of a force, the moment of a force, equilibrium) have already been stated, it is desirable to elaborate further in order to ensure a sound foundation for subsequent analysis. Confining our attention to planar problems will also facilitate understanding at this stage.

Force polygons

The reverse of replacing a single force by its components is to combine concurrent forces into their *resultant*. The magnitudes and directions of two forces P_1 and P_2 which act at a single point are shown in Fig. 1.53a, the length of the two arrowed lines representing to some scale the magnitude of the forces. Each of these has components in the x and y directions so that the total effect of P_1 and P_2 is

$$P_1 + P_2 = \begin{bmatrix} P_{x1} + P_{x2} \\ P_{y1} + P_{y2} \end{bmatrix}$$

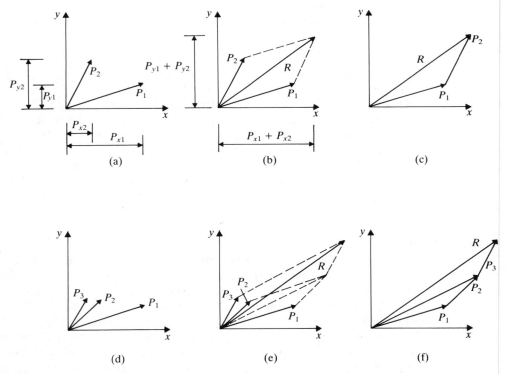

Figure 1.53   The principle of force polygons

It follows that the two forces P_1 and P_2 can be represented by R, the diagonal of the parallelogram in Fig. 1.53b, since this clearly has components:

$$R = \begin{bmatrix} P_{x1} + P_{x2} \\ P_{y1} + P_{y2} \end{bmatrix}$$

In using a graphical method to determine the magnitude and direction of the resultant, it would obviously be easier to draw the triangle in Fig. 1.53c rather than the parallelogram. In doing so, it must be remembered that the forces P_1, P_2 and R are all acting at a particular point.

It will be clear that the resultant of several forces acting at a point (e.g., P_1, P_2 and P_3 in Fig. 1.53d) can be obtained by the successive addition of the vectors—either by drawing the series of parallelograms (Fig. 1.53e) or the series of triangles (Fig. 1.53f). The latter diagram is termed the *polygon of forces*. The total effect of P_1, P_2 and P_3 is

$$P_1 + P_2 + P_3 = \begin{bmatrix} P_{x1} + P_{x2} + P_{x3} \\ P_{y1} + P_{y2} + P_{y3} \end{bmatrix} = \begin{bmatrix} \sum P_x \\ \sum P_y \end{bmatrix}$$

Equilibrium of concurrent forces

The equilibrium conditions for a system of concurrent planar forces reduce to

$$\begin{bmatrix} \sum P_x \\ \sum P_y \end{bmatrix} = 0$$

as it is obvious that the moment of all the forces about their common point is zero.

The graphical implication of these two conditions is that the polygon of forces must close, i.e., the resultant R is zero. This can be used in the graphical analysis of structures. Although the advent of the computer to facilitate calculations has reduced the importance of graphical methods, some knowledge of the procedure is still desirable. A simple example will therefore be given.

It is required to determine the forces in the members in Fig. 1.54a supporting the 10 kN load. The situation depicted in Fig. 1.54b is therefore to be analysed—although the line of action of the forces in the members is known, neither the senses (i.e., the directions) nor the magnitudes of the forces are known.

One commences the triangle of forces by drawing, to a suitable scale, a line representing the 10 kN load (Fig. 1.54c). From the end of this line, a second line representing one of the unknowns (say the force in member 1) is to be added. However, at this stage only its line of action is known—the sense and the magnitude of the line are unknown (Fig. 1.54d). The third line (representing the force in member 2) will follow on from the end of the line representing the force in member 1 when that has been determined. We know that the line which will represent the force in member 2 must terminate at the origin and therefore we are able to indicate its line of action on the diagram (Fig. 1.54e). The point of intersection of the two lines therefore determines the final triangle of forces (Fig. 1.54f). The lengths of the lines 1 and 2 represent the forces in the members 1 and 2, and the senses of the forces are determined by following the lines back to the origin. The resulting forces acting on the

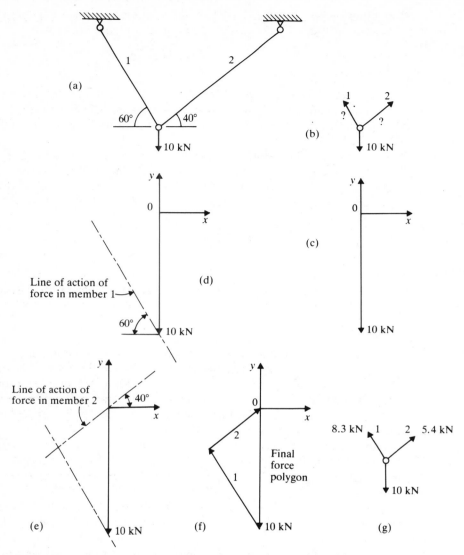

Figure 1.54   A graphical procedure for analysing concurrent forces

joint are then summarized in Fig. 1.54g in which the arrows now represent the sense in which these forces are acting on the joint. This indicates that both members 1 and 2 are 'pulling' on the joint and are therefore in tension.

A simple exercise will demonstrate that, if there were more than two members supporting the 10 kN load in Fig. 1.54, it is impossible to determine the magnitude of the forces in the members by the procedure described here.

There are several corollaries regarding concurrent forces which are worth stating:

1. If there are only three non-parallel external forces acting on a structure, these forces must be concurrent in order to achieve equilibrium.
2. A concurrent three-force system can only be in equilibrium if the three forces lie in the same plane and are such that no two of the forces have the same line of action.
3. A two-force system can only be in equilibrium if both forces are applied along the same line of action, are equal in magnitude, and are opposite in direction.

Moment of a couple

A *couple* consists of a pair of equal and opposite parallel forces—that is, the forces are equal in magnitude but opposite in direction. Such pairs of forces were referred to earlier in Section 1.2 as causing bending and twisting moments. The most important characteristic of a couple is that its rotational effect (the magnitude and direction of the moment) is identical at all points in its plane. In other words, the moment of a couple about a point in the same plane is independent of the distance between the couple and the point.

This is illustrated in Fig. 1.55—a couple acts on the surface of a beam, only a part of which is shown. Each of the pair of forces has a magnitude P and they act at a distance d apart, with one of the forces a distance x from the support. The moment of the two forces about the support is

$$P(d + x) - Px = Pd$$

The value of the moment is seen to be independent of x, and the couple therefore provides a moment effect of Pd at all positions in the plane. The product of one of the forces and the perpendicular distance between them is termed the magnitude of the couple. Such pairs of forces occur frequently in structural analysis—the internal bending moments in beams have been cited as one such example (see Fig. 1.5 m).

Equivalent force systems

A system of forces is said to be statically equivalent to a second system of forces when the vector sums of forces and moments of the two systems are equal. During an analysis, the structural engineer will frequently translate one system of forces to an

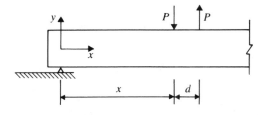

Figure 1.55 A pair of equal and opposite forces constitute a couple and their combined moment about a point is independent of their distance from that point

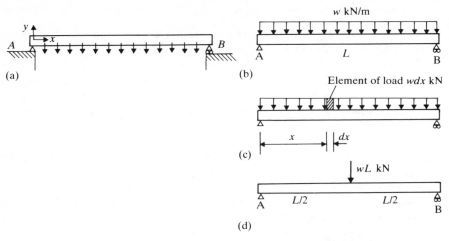

Figure 1.56 The force system equivalent to a uniformly distributed load

equivalent second system of forces. The case of distributed loading is one such example.

Figure 1.56a shows a beam of uniform cross-section. The effect of gravity will cause a uniform distribution of forces along the length of the beam. In subsequent diagrams, such a distribution of forces will be indicated as in Fig. 1.56b. Suppose that, for the purpose of obtaining one of the equilibrium equations, it is required to obtain the moment of this distributed load about A.

Consider initially an element of the load positioned at a distance x from A (Fig. 1.56c). The moment about A of this elemental load is $xw\,dx$. (For many cases there is

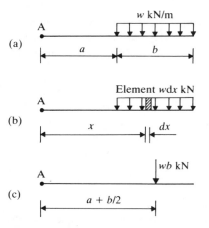

Figure 1.57 Another example of an equivalent force system

no point in following a strict sign convention with regard to direction of moments, and so a negative sign will not be included here despite the anticlockwise rotational effect of the elemental load about A when viewed in the direction of the z-axis.) The moment of the total distributed load about A is now obtained by integrating this expression over the length of the beam:

$$\int_0^L xw \, dx = \left(\frac{wx^2}{2}\right)_{x=0}^{x=L} = \frac{wL^2}{2} = wL \times \frac{L}{2}$$

It follows that the moment of the distributed load about A can be found by replacing the distributed load by a point load equal in magnitude to the total of the distributed load and acting at its centroid (Fig. 1.56d).

That this also applies more generally can be similarly illustrated. For example, in Fig. 1.57a the moment of the uniformly distributed load about A is (Fig. 1.57b)

$$\int_a^{a+b} xw \, dx = \left(\frac{wx^2}{2}\right)_{x=a}^{x=a+b} = \frac{w(a+b)^2}{2} - \frac{wa^2}{2}$$

$$= wb \times \left(a + \frac{b}{2}\right)$$

$$= \text{total load} \times \text{distance of centroid from A}$$

Problem 1.4 Figure 1.58 shows a distributed load which varies linearly from zero at B to w kN/m at C. Show that for moments about A, the statically equivalent force system is a concentrated load of magnitude equal to the total load acting at the centroid of the distributed load. Show that this also applies if A is within BC.

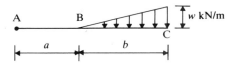

Figure 1.58 Problem 1.4

It is essential to note here that the *effect* of the equivalent load on the beam in Fig. 1.56 is not the same as the *effect* of the actual load. For example, the bending moments which would be caused in the beam by the equivalent load are not identical to those caused by the actual load. The value of using such equivalent loads is restricted in such cases to finding the moment about a point in the plane.

1.7 Structural supports

In this discussion of structural supports, attention is again confined to planar structures.

Types of support

The actual supports to a structure are often complicated and for the purpose of analysis it is necessary to idealize the nature of the supports. One which constrains the structure so that there is no movement of the structure whatsoever at that position is said to be a *fixed support*. For example, Fig. 1.59a shows a beam built-in to a substantial wall which prevents:

– translation of the beam in the x direction;
– translation of the beam in the y direction;
– rotation of the beam about the z-axis.

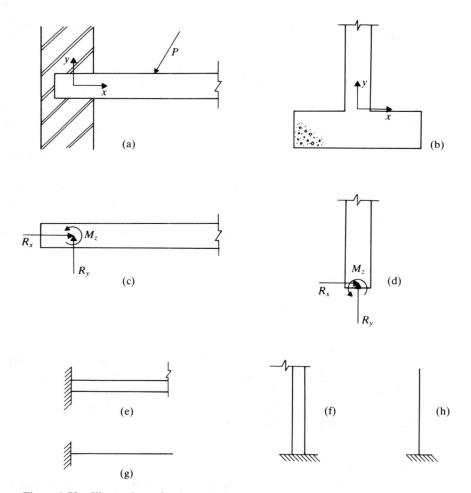

Figure 1.59   Illustrating a fixed support

Similar restraint would apply to the foot of a column if the foundation is sufficiently large (Fig. 1.59b). In order to achieve such restraint, the wall (or foundation) must be able to apply *to the beam* (or column):

- a force along the line of the x-axis (R_x),
- a force along the line of the y-axis (R_y),
- a moment about the z-axis (M_z),

as indicated in Figs 1.59c and 1.59d. There are, therefore, three reactive forces at each fixed support which are unknown at the start of the analysis. In subsequent diagrams, a fixed support will be designated as in Figs 1.59e and 1.59f, or, when the member is represented by a single line, as in Figs 1.59g and 1.59h.

In certain cases it is undesirable to offer rotational restraint to a structure at the points of support—a truss is one example where such restraint is undesirable. Figure 1.60 shows a *pinned* or *hinged support* which prevents:

- translation in the x direction,
- translation in the y direction,

but allows rotation about the z-axis. Although in practice there would be some friction acting as a result of the rotation, this is ignored in analysis and the pinned support (subsequently designated as in Fig. 1.61a or 1.61b) is considered to offer two reactive forces (Fig. 1.61c). It is of interest to note that if the foundation slab to a column is sufficiently small (Fig. 1.61d), it too may be considered to offer little resistance to rotation of the end of the column and would be regarded as a pinned support providing only two reactive forces (Fig. 1.61e).

Figure 1.60 A hinged support to a structure (*Source*: Centre Belgo-Luxembourgeois d'Information de l'Acier)

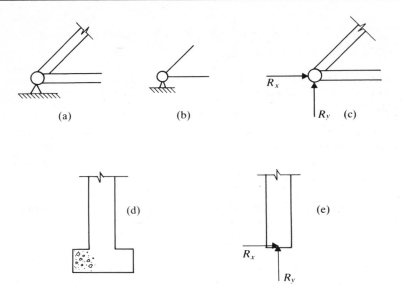

Figure 1.61   Illustrating pinned supports

In some cases it may be desirable for a support to allow translation of the structure in a particular direction as well as rotation. This is particularly important, for example, in permitting long structures to expand under the effect of a rise in temperature, and Fig. 1.62 shows such supports to the superstructure of a bridge. The arrangement shown in Fig. 1.63a allows the structure both to rotate and move horizontally without restraint, but it is totally restrained in the vertical direction. If it is assumed that the roller is frictionless, then the single reactive force must act perpendicularly to the surface on which it rolls (Fig. 1.63b). For larger structures where it is necessary to utilize more than one roller in order to accommodate the vertical force, it becomes necessary to make special provision for the rotation of the structure by incorporating a plate with a curved surface (Fig. 1.63c). A roller support will be indicated as in Fig. 1.63d in subsequent diagrams.

Calculation of reactions using equilibrium only

In the determination of the internal forces of a structure it may be necessary, as a first step, to calculate the reactive forces caused by the known applied loads. Although equilibrium has already been used in an intuitive way to determine the forces at the supports (as in Fig. 1.4), it is now necessary to develop a more formalized approach for use in more complex structures.

For planar structures it has been established that there are three independent equations of equilibrium. It follows that if the nature of the supports of a stable planar structure are such that the total of the reactive forces is three, these can be determined using the equations of equilibrium. If there are more than three reactive forces, it will

Figure 1.62 Roller supports are often used for supporting the superstructure of a long bridge in order that it can accommodate a change in length as the temperature varies (*Source*: Centre Belgo-Luxembourgeois d'Information de l'Acier)

be necessary to acknowledge other conditions in order to determine the reactions. However, initially attention is confined to cases where there are only three reactive forces.

Figure 1.64a shows a bridge type of truss which is supported on a roller at B and a pin at A. The x, y, z axes are initially assumed to pass through A. The three unknown reactive forces caused by the applied loads are shown in Fig. 1.64b—one at the position of the roller and two at the position of the pinned support. (It is most important to understand that the type of support is the indicator of the number and type of reactive forces which can be present at that support.)

Using $\sum P_x = 0$ $\qquad\qquad R_{xa} + 10 = 0$ $\qquad\qquad\qquad$ (1.1)

Thus $\qquad\qquad\qquad R_{xa} = -10 \text{ kN}$

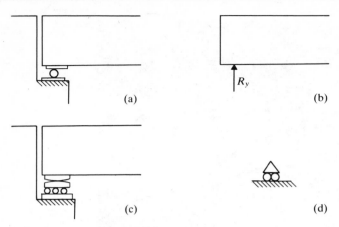

Figure 1.63   Illustrating roller supports

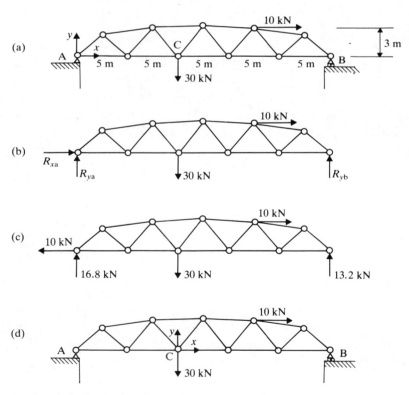

Figure 1.64   Illustrating the determination of the support reactions for a truss

The negative sign indicates that this reaction acts in the opposite direction of that adopted for the x axis.

Using $\sum P_y = 0$ $\qquad\qquad R_{ya} + R_{yb} - 30 = 0$ $\qquad\qquad$ (1.2)

This equation does not enable the value of either R_{ya} or R_{yb} to be determined at this stage, and it is necessary to introduce the third equilibrium equation.

Using $\sum M_z = 0$ $\quad (R_{yb} \times 25) - (30 \times 10) - (10 \times 3) = 0$ $\qquad\qquad$ (1.3)

Thus $\qquad\qquad\qquad\qquad R_{yb} = 13.2 \text{ kN}$

Substituting this value of R_{yb} in Eq. (1.2) yields

$$R_{ya} = 16.8 \text{ kN}$$

A summary of the reactive forces is given in Fig. 1.64c.

There are several important points to note here. Firstly, one can position the three orthogonal axes at any point of the structure, or indeed outside the structure if it is more convenient to do so. To show that this has no influence on the answers, the calculation is now repeated assuming the axes pass through C (Fig. 1.54d). It will be clear that the two equations of equilibrium

$$\begin{bmatrix} \sum P_x \\ \sum P_y \end{bmatrix} = 0$$

will in any case lead to the same equations as (1.1) and (1.2) above. Using $\sum M_z = 0$ now gives

$$(R_{yb} \times 15) - (10 \times 3) - (R_{ya} \times 10) = 0 \qquad\qquad (1.4)$$

Solving the two simultaneous Eqs (1.2) and (1.4) yields the same answers as before.

This, therefore, allows us to take moments about any point in the plane of the truss and so, in order to clarify which point has been chosen, it is desirable to include this in the symbol. Thus in future, instead of using $\sum M_z$ to indicate the rotation equilibrium, the symbol will stipulate the point about which the sum of moments is taken, for example $\sum M_a$ or $\sum M_c$.

It is also important to realize that this facility for adopting any position of the orthogonal axes does not give additional independent equations. The fact that the same answers were obtained indicates this. There remain only three useful equations when considering equilibrium of the whole structure, and this point is particularly important to remember when analysing structures with more than three reactive forces.

Calculation of reactions using equilibrium and known internal conditions

The structure shown in Fig. 1.65a has pinned supports at A and B. Although this implies four reactive forces (Fig. 1.65b), it is possible in this case to combine the three equations of equilibrium for the structure as a whole with a fourth equation obtained from an observation of an internal condition. As a result, the four reactive forces can be calculated.

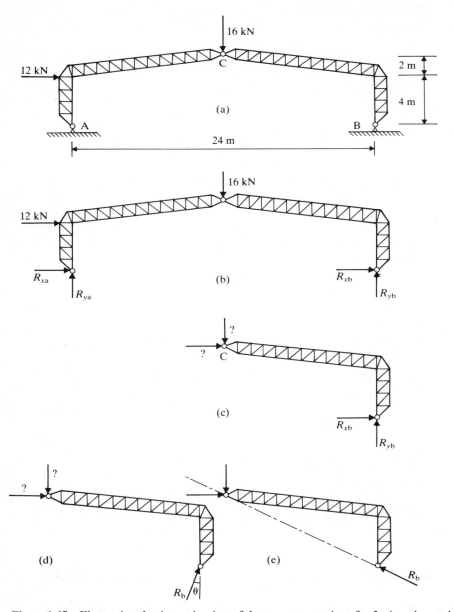

Figure 1.65   Illustrating the determination of the support reactions for 3-pinned portal frames

It will be observed that the structure can be regarded as two separate triangulated frames (AC and BC) connected through a single pin at C. As it is known that a pin-joint cannot resist a moment, the fourth equation is obtained by considering the moments about C of all the external forces acting *on one side* of C, for example, to the right of C as in Fig. 1.65c. Although the forces at C affecting this part of the structure are also unknown at this stage, they do not enter into the equation as they obviously have no moment about C. The four equations are obtained using:

$$\sum P_x = 0 \text{ for the whole structure} \qquad R_{xa} + R_{xb} + 12 = 0 \quad (1.5)$$

$$\sum P_y = 0 \text{ for the whole structure} \qquad R_{ya} + R_{yb} - 16 = 0 \quad (1.6)$$

$$\sum M_a = 0 \text{ for the whole structure} \quad (R_{yb} \times 24) - (16 \times 12) - (12 \times 4) = 0 \quad (1.7)$$

$$\sum M_c = 0 \text{ for forces to the right of C} \qquad (R_{yb} \times 12) + (R_{xb} \times 6) = 0 \quad (1.8)$$

These four equations enable the four reactive forces to be determined. They are

$$R_{xa} = 8 \text{ kN} \qquad R_{ya} = 6 \text{ kN} \qquad R_{xb} = -20 \text{ kN} \qquad R_{yb} = 10 \text{ kN}$$

Again, it is necessary to emphasize that an additional independent equation is not obtained by also considering forces to the left of C. This should be clear from the fact that 'the $\sum M_c = 0$ for the whole structure' could have been used as one of the equations of equilibrium. It follows, then, that having used 'the $\sum M_c = 0$ for forces to the right', the use of 'the $\sum M_c = 0$ for forces to the left' would not constitute an independent condition.

It may be somewhat confusing that one adopts *two* unknown reactive forces at the position of a pinned support when these could be combined into their resultant to give a *single* unknown reactive force. Whilst that is so, it should be realized that the line of action of this single force is not known and hence at the outset of the analysis two unknowns remain, namely R_b and θ (Fig. 1.65d).

In this respect, it may be of interest to note that as R_b is now the only force to the right of C, its line of action must in fact pass through the hinge at C (Fig. 1.65e) in order that its moment about C is zero. Thus, by observation, the line of action of R_b has been established. This observation is, in fact, directly equivalent to Eq. (1.8) above.

The cantilever frame in Fig. 1.66a is another type of structure where observation of the internal conditions enables a fourth equation to be obtained. The four reactive forces at the two pinned supports A and B (Fig. 1.66b) can therefore be determined. In this structure, the fourth equation is obtained by taking moments about C for the part AC. This quickly establishes that in this case R_{ya} is zero (as does also the observation that the line of action of the resultant reaction at A must pass through C). Such observations are often quickly made and there is no need to show formally that R_{ya} is zero. Thus, the starting point of an analysis of such a structure would be as shown in Fig. 1.66c.

Some special cases

If the number of equations available to calculate the reactions exceeds the number of reactive forces, the structure is said to be statically unstable. Figure 1.67a shows a

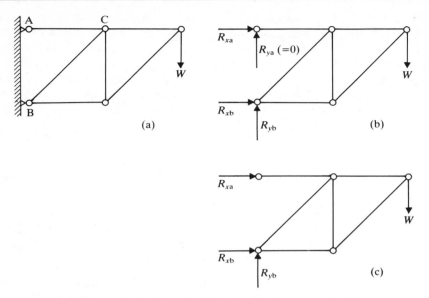

Figure 1.66   In some frames certain of the reactive forces are clearly zero

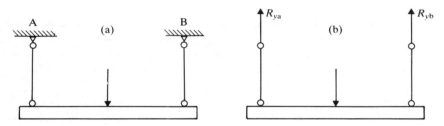

Figure 1.67   If the number of unknown reactive forces is smaller than the number of available equilibrium equations, the structure is unstable

beam supported on hangers. The number of unknown reactive forces is two (after noting that the internal conditions require the horizontal reactions at A and B to be zero, Fig. 1.67b), whilst there are three equations of equilibrium available for the complete structure. Although the structure would be in equilibrium under the action of purely vertical forces, any slight inclination of these applied forces away from the vertical would cause rotation of the hangers and translation of the beam.

A stable structure is not necessarily obtained when the number of unknown reactive forces equals the number of equilibrium equations. This will be clear from Fig. 1.68. The arrangement of the supports and supporting members is therefore important in this regard. The three hangers in Fig. 1.68 can be likened to supporting a beam on three rollers.

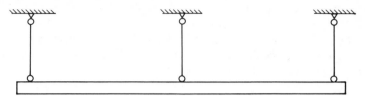

Figure 1.68 A structure is not necessarily stable if the number of unknown reactive forces equal the number of available equilibrium equations

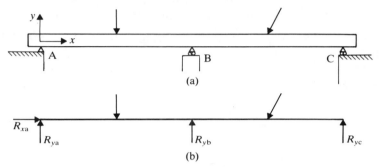

(a)

(b)

Figure 1.69 If for a stable structure the number of unknown reactive forces exceeds the number of available equilibrium equations, the structure is statically indeterminate

A beam supported on one pinned support and two rollers, as in Fig. 1.69a, is stable and is able to resist both vertical and horizontal forces. There are now four reactive forces—two at the pinned support and one at each of the rollers (Fig. 1.69b). There are, however, only the three equations of equilibrium. In this case, no additional equations can be obtained merely by observing the arrangement— there is no part of the structure which can be isolated to enable a fourth equation to be obtained. It is not possible to obtain the four reactive forces from the three equilibrium equations. Such a structure is said to be *statically indeterminate*, i.e., it cannot be solved by the use of static equilibrium alone.

Another example of a statically indeterminate structure is the pin-jointed frame in Fig. 1.70. Although similar to the cantilever frame in Fig. 1.66a, it has an extra

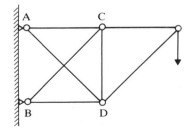

Figure 1.70 Compared with the frame in Fig. 1.66, this frame has an extra diagonal. It is not now possible to obtain the extra equilibrium equation and the frame is statically indeterminate

diagonal. This prevents isolation of part of the structure at the joint C, as the support at A is additionally connected to joint D.

Although the reactive forces of statically indeterminate structures can be calculated, it is now a much more difficult problem requiring calculation of deformations. This will be further explained in Chapters 2 and 3.

Problem 1.5   The line of action of a force of magnitude 14 kN is from A to B, where A and B have coordinates in an orthogonal system of axes x, y, z of 0, 2, 3 and 6, 4, 0, the distance units being metres. Calculate the components of the force in the directions of the axes and also the moment of the force about the x-axis.

Problem 1.6   The roller supporting the bridge truss at B in Fig. 1.71 requires maintenance. For this, the truss is to be rotated slightly about A using the crane as shown. The members of the truss weigh 2 kN/m. Determine whether extra weights need to be added to the crane to ensure a factor of safety against overturning of 1.4.

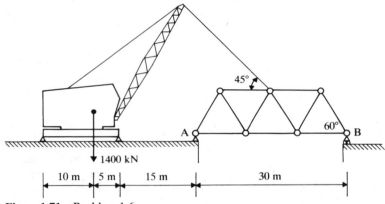

Figure 1.71   Problem 1.6

Problem 1.7   Two balls, each of radius r and weight W, are placed in a tube of radius R as shown in Fig. 1.72. Find the minimum weight of the tube which will ensure stability.

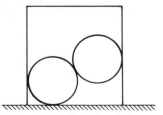

Figure 1.72   Problem 1.7

Problem 1.8   The two bar pin-jointed structure in Fig. 1.73 supports a load W. By considering only external forces, calculate the horizontal and vertical reactive forces at the supports A and B. Explain how it is possible to obtain these four reactive

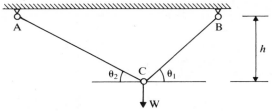

Figure 1.73 Problem 1.8

forces when only three equations of equilibrium apply to the complete system of forces on a planar structure. Show that the resultant of the vertical and horizontal forces at a particular support is at the same angle as the member at that support, and explain why this must be the case. Also obtain expressions for the forces in the members by considering equilibrium at the loaded joint.

Problem 1.9 Explain why, for a body in equilibrium under the action of three forces only, the three forces must be concurrent or parallel. Calculate the horizontal and vertical components of the reactions at the supports of the three-pinned parabolic arch shown in Fig. 1.74. Combine graphically the horizontal and vertical components into their resultants and show that these, with the applied load, satisfy the three-force condition. Show also that the resultant at A acts along a line passing through the pin at C and explain why this must be the case.

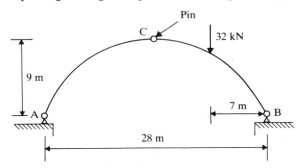

Figure 1.74 Problem 1.9

Problem 1.10 A block of wood of mass M lies on a flat surface as in Fig. 1.75. A horizontal force P acts at the position shown, although this does not cause it to move. Describe all the external forces acting on the block. If P gradually increases, what types of movement are possible?

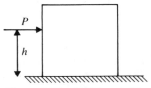

Figure 1.75 Problem 1.10

2. Structural analysis: mechanics of structures

This chapter is devoted to the analysis of simple structures, such analysis being the calculation of the internal forces within the structure caused by the applied loads and also of the displacements of the structure. The term *forces* has been used here in its comprehensive form to include direct forces, shear forces and moments, whilst *displacements* is the general term to include both deflections and rotations.

For the most part we shall be concerned only with those structures which can be solved using statics alone, termed statically determinate structures. However, an introduction will be made to statically indeterminate structures, although their analysis will largely be left until the second year studies.

A. Statically determinate structures involving direct forces only

The reader was introduced to the truss in Chapter 1 where it was explained that bars connected at their ends using a triangulated arrangement form a stable framework, even when the connections are by means of pins which allow the connected members to rotate freely relative to one another. At one time, particularly in the days of cast iron, members were frequently connected via a pin passing through holes at the ends of the members, it being a simple operation to form such holes in cast iron. However, nowadays such pins are never used, the members being connected via bolted or welded joints (Fig. 2.1). Nevertheless, the term 'pin-jointed' is retained and analysis of such structures is frequently carried out assuming frictionless pin connections, since it has been shown both theoretically and experimentally that such an idealization leads to quite accurate results when members are relatively long and slender, as is often the case.

For such structures loaded only at the position of the joint, the forces in the members result only from the forces transmitted via the joints. If the centroidal axes of the members meeting at a joint all intersect at a single point (see Fig. 2.1), the members will then carry an axial force only, and there will be no bending moment or shear force in the members.

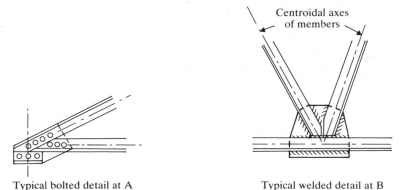

Typical bolted detail at A   Typical welded detail at B

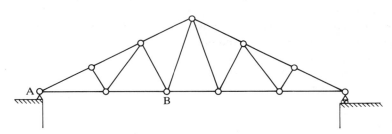

Figure 2.1   Although actual connections in triangulated elements are either bolted or welded, for purposes of analysis pinned connections are assumed

It is true, of course, that the gravitational forces due to the dead load (i.e., the weight) of the members will be distributed along their lengths, and this will cause some bending in the members. However, such loads are normally quite small in comparison with the loads applied at the joints and can be allowed for simply by adopting equivalent point loads at the positions of the joints. An illustration of the manner by which the main loads on a structure are transferred to a truss at the positions of the joints is given in Fig. 5.47a.

Many trusses are simple planar structures but some are three-dimensional in character. However, even with the latter it is often possible to subdivide the structure into planar components for purpose of analysis. The bridge structure in Fig. 5.47 is one such example. The upper and lower chords act as part of the main vertical trusses in resisting the gravitational loads, but also act as part of the horizontal trusses in resisting the wind loads. The analyses are carried out separately assuming planar frames, and the resulting forces are subsequently superimposed.

2.1 Planar pin-jointed frames—forces in members

The determination of the forces in the members of a truss will be illustrated using the frame shown in Fig. 2.2a. Identification of the joints and members will be as in Fig. 2.2b, whereby the joints are indicated by a letter and the members by a number. The force in a member will be designated by the letter F with the appropriate numbered subscripts. For example, F_2 indicates the force in member 2. All such forces are initially assumed to be tensile and therefore pulling on the joints. The subsequent analyses will indicate some of the forces as having a negative value which will mean that they are in fact compressive.

Calculation of reactions

Loads applied to a structure are transmitted via the members of the structure to the ground at the positions of the foundations. The resulting forces exerted from the foundations to the structure are termed *reactions* and from the standpoint of the structure are classified as external forces.

For the calculation of the forces in the members of a truss using simple hand methods, it will usually be necessary to start the analysis by determining the reactive forces. (The exception is the cantilever truss.) Such calculations have been demonstrated in Chapter 1. In this example there are three unknown reactive forces, two at the pin support A (R_{xa} and R_{ya}) and one at the roller support E (R_{ye}) (Fig. 2.2c). It will therefore be possible to calculate their values using the three basic equations of equilibrium for the structure, namely

$$\begin{bmatrix} \sum P_x \\ \sum P_y \\ \sum M_z \end{bmatrix} = 0$$

Using $\sum P_x = 0$ $\qquad\qquad R_{xa} + 15 = 0$

Thus $\qquad\qquad\qquad\qquad R_{xa} = -\underline{15\,\text{kN}}$

Using $\sum P_y = 0$ $\qquad\qquad R_{ya} + R_{ye} - 100 - 100 = 0$

$$R_{ya} + R_{ye} = 200\,\text{kN} \qquad\qquad (2.1)$$

Using $\sum M_a = 0$ $(R_{ye} \times 12) - (100 \times 6) - (15 \times 4) - (100 \times 3) = 0$

$$R_{ye} = \underline{80\,\text{kN}}$$

Substituting this value in Eq. (2.1) $\qquad\qquad R_{ya} = \underline{120\,\text{kN}}$

All the external forces acting on the frame are now known and are summarized in Fig. 2.2d, and this is the starting point for the determination of the internal forces in the members. There are several different procedures which can be followed to achieve this.

Method of sections

It is a fundamental concept of mechanics that, if a body is in equilibrium under a set of external forces, then any part of that body must be in equilibrium under the forces

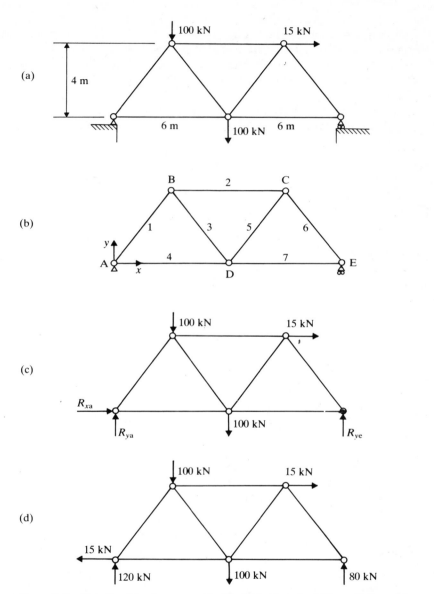

Figure 2.2 Details of the frame used to illustrate the calculation of internal forces

acting upon it, these forces being a mixture of external and internal forces. In the method of sections, part of the frame is isolated by an imaginary cut. For example, consider the part of the frame to the right of the line XX in Fig. 2.3a. This part of the frame is redrawn as the free-body diagram in Fig. 2.3b. The removed part of the frame is replaced by the forces it actually exerts on the free-body part, namely F_2, F_3 and F_4. These are actually the internal forces in the members but, as far as the isolated part CDE is concerned, they are now external forces, just as are the loads at C and D and the reaction at E.

Equilibrium of this free-body is now considered. Again there are three basic equations of equilibrium which will enable the three unknown forces to be determined.

Using $\sum P_x = 0$
$$15 - F_2 - F_3 \cos \theta - F_4 = 0$$
$$F_2 + \tfrac{3}{5}F_3 + F_4 = 15 \qquad (2.2)$$

Using $\sum P_y = 0$
$$80 - 100 + F_3 \sin \theta = 0$$
$$\tfrac{4}{5}F_3 = 20$$
$$F_3 = \underline{25 \text{ kN}}$$

Using $\sum M_d = 0$
$$(15 \times 4) - (80 \times 6) - (F_2 \times 4) = 0$$
$$F_2 = \underline{-105 \text{ kN}}$$

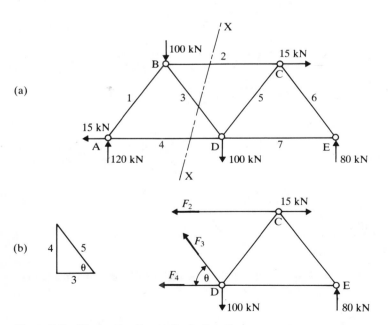

Figure 2.3 Illustrating the method of sections

Moments could have been taken about any point in the plane, but moments about D is most convenient as it eliminates the three forces passing through D (namely F_3, F_4 and 100 kN) from the rotational equilibrium equation.

Substituting the calculated values of F_2 and F_3 into Eq. (2.2) gives

$$F_4 = \underline{105 \text{ kN}}$$

Such a procedure can be used to determine all the member forces, although it is usually limited to the case when the forces in a few specific members are required. When all the member forces are required, the alternative method of joints will be preferable.

Method of joints

Any part of the structure can be isolated to form a free-body. For example, the section XX in Fig. 2.4a has isolated the joint C and this must also be in equilibrium under the forces acting on it, namely the three member forces F_2, F_5 and F_6 and the applied horizontal load of 15 kN (see Fig. 2.4b).

However, it will be recalled (Chapter 1) that for concurrent planar forces there are only two equations of equilibrium, namely

$$\begin{bmatrix} \sum P_x \\ \sum P_y \end{bmatrix} = 0$$

as the equation $\sum M_c = 0$ will clearly not provide an additional useful equation. It is, therefore, not possible at this stage to solve for the three unknown member forces at C. However, there are two joints, A and E, at which only two unknown forces exist. By starting at one of these and solving for the two member forces, it becomes possible to gradually work around the structure and obtain all the member forces.

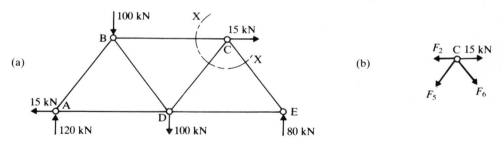

Figure 2.4   Illustrating the method of joints

Figure 2.5   Forces at joint E

For example, consider joint E (Fig. 2.5a):

Using $\sum P_x = 0$ $\quad\quad\quad -F_7 - F_6 \cos\theta = 0$

$$F_7 + \tfrac{3}{5}F_6 = 0 \tag{2.3}$$

Using $\sum P_y = 0$ $\quad\quad\quad 80 + F_6 \sin\theta = 0$

$$\tfrac{4}{5}F_6 = -80$$

$$F_6 = \underline{-100\ \text{kN}}$$

Substituting in Eq. (2.3) $\quad\quad\quad F_7 = \underline{60\ \text{kN}}$

The forces acting on joint E are now known and are summarized in Fig. 2.5b. The force F_7 was found to be positive and therefore in tension and pulling on the joint; the force F_6 was found to be negative and therefore in compression and pushing on the joint.

With the knowledge of the force in member 6, it is now possible to move to joint C where the number of unknown forces has been reduced to two as shown in Fig. 2.6a. In this diagram, it has been acknowledged that member 6 is in compression and therefore pushing on the joint. Thus at joint C:

Using $\sum P_x = 0$ $\quad 15 - 100 \cos\theta - F_2 - F_5 \cos\theta = 0$

$$F_2 + \tfrac{3}{5}F_5 = -45 \tag{2.4}$$

Using $\sum P_y = 0$ $\quad\quad\quad 100 \sin\theta - F_5 \sin\theta = 0$

$$F_5 = \underline{100\ \text{kN}}$$

Substituting in Eq. (2.4) $\quad\quad\quad F_2 = \underline{-105\ \text{kN}}$

The forces acting on joint C are summarized in Fig. 2.6b.

Figure 2.6   Forces at joint C

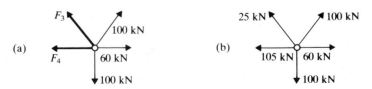

Figure 2.7   Forces at joint D

Moving to joint D, the forces are as shown in Fig. 2.7a.

Using $\sum P_x = 0$ $60 + 100 \cos \theta - F_4 - F_3 \cos \theta = 0$

$$F_4 + \tfrac{3}{5}F_3 = 120 \qquad (2.5)$$

Using $\sum P_y = 0$ $100 \sin \theta - 100 + F_3 \sin \theta = 0$

$$\tfrac{4}{5}F_3 = 20$$

$$F_3 = \underline{25 \text{ kN}}$$

Substituting in Eq. (2.5) $F_4 = \underline{105 \text{ kN}}$

The forces acting on joint D are summarized in Fig. 2.7b.

At this stage only one member force, F_1, remains unknown. To determine this, consider joint B where the forces are as shown in Fig. 2.8a.

Using $\sum P_x = 0$ $25 \cos \theta - 105 - F_1 \cos \theta = 0$

$$\tfrac{3}{5}F_1 = -90$$

$$F_1 = \underline{-150 \text{ kN}}$$

All the member forces have now been determined. However, it is advisable to consider the forces that have been calculated to be acting at the final joint A as a check on the arithmetic. These are shown in Fig. 2.8, where it is easily confirmed that the joint is in equilibrium.

Method of joints using a matrix approach

The method given here has little direct practical application for the analysis of pin-jointed frames, as other matrix methods (discussed later) have been developed which

Figure 2.8   Forces at joint B

Figure 2.9   Forces at joint A

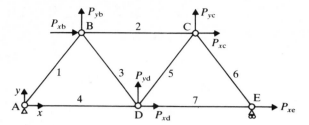

Figure 2.10   General loading case for the matrix technique of the method of joints

are more widely applicable and also generate information on deformation. Nevertheless, it presents a valuable introduction to the usefulness of matrices and computers in structural analysis, and indeed forms the basis of the later matrix methods. At this stage no previous knowledge of matrix arithmetic will be assumed, and this section should not be omitted simply because the subject has not yet been covered in the associated mathematics courses. Indeed, the section is considered a useful precursor to a specialist course in matrices.

The same frame as that used previously will be retained, but instead of the 100 kN and 15 kN loads applied at joints B, C and D, it will be assumed that horizontal and vertical loads act at the joints as in Fig. 2.10. For example, loads P_{xb} and P_{yb} act at joint B. No such loads are applied at joint A as this joint is pinned to the foundations and the loads would not affect the frame; similarly, no vertical load is applied at joint E. The purpose of such loading is to illustrate the generality of the procedure. It will subsequently be easy to return to the original loading by putting the appropriate values of P_x and P_y to -100 kN or 15 kN and the remainder to zero.

The first step is to obtain the seven equilibrium equations involving P_x and P_y—two from each of the joints B, C and D and one from E. From these seven simultaneous equations it will be possible to determine the seven unknown member forces. The equations are similar to those previously determined using the method of joints with the exception that the loads P are included as appropriate.

From joint B	$P_{xb} - \frac{3}{5}F_1 + F_2 + \frac{3}{5}F_3$	$= 0$
	$P_{yb} - \frac{4}{5}F_1 - \frac{4}{5}F_3$	$= 0$
From joint C	$P_{xc} - F_2 - \frac{3}{5}F_5 + \frac{3}{5}F_6$	$= 0$
	$P_{yc} - \frac{4}{5}F_5 - \frac{4}{5}F_6$	$= 0$
From joint D	$P_{xd} - \frac{3}{5}F_3 - F_4 + \frac{3}{5}F_5 + F_7$	$= 0$
	$P_{yd} + \frac{4}{5}F_3 + \frac{4}{5}F_5$	$= 0$
From joint E	$P_{xe} - \frac{3}{5}F_6 - F_7$	$= 0$

These equations are now rearranged to place the known loads P on the right-hand sides of the equations, retaining the unknown member forces F on the left-hand sides. Moreover, to facilitate the subsequent tabular presentation, the equations are suitably spaced.

$$-\tfrac{3}{5}F_1 \quad +F_2 \quad +\tfrac{3}{5}F_3 \qquad\qquad\qquad\qquad = - \ P_{xb}$$

$$-\tfrac{4}{5}F_1 \qquad\qquad -\tfrac{4}{5}F_3 \qquad\qquad\qquad = - \ P_{yb}$$

$$-F_2 \qquad\qquad -\tfrac{3}{5}F_5 \quad +\tfrac{3}{5}F_6 \qquad = - \ P_{xc}$$

$$-\tfrac{4}{5}F_5 \quad -\tfrac{4}{5}F_6 \qquad = - \ P_{yc}$$

$$-\tfrac{3}{5}F_3 \quad -F_4 \quad +\tfrac{3}{5}F_5 \qquad\qquad +F_7 = - \ P_{xd}$$

$$+\tfrac{4}{5}F_3 \qquad +\tfrac{4}{5}F_5 \qquad\qquad = - \ P_{yd}$$

$$-\tfrac{3}{5}F_6 \quad -F_7 \qquad = - \ P_{xe}$$

A tabular presentation of these seven equations would be:

Table 2.1

F_1	F_2	F_3	F_4	F_5	F_6	F_7		
$-\tfrac{3}{5}$	1	$\tfrac{3}{5}$					$= -$	P_{xb}
$-\tfrac{4}{5}$		$-\tfrac{4}{5}$					$= -$	P_{yb}
	-1			$-\tfrac{3}{5}$	$\tfrac{3}{5}$		$= -$	P_{xc}
				$-\tfrac{4}{5}$	$-\tfrac{4}{5}$		$= -$	P_{yc}
		$-\tfrac{3}{5}$	-1	$\tfrac{3}{5}$		1	$= -$	P_{xd}
		$\tfrac{4}{5}$		$\tfrac{4}{5}$			$= -$	P_{yd}
					$-\tfrac{3}{5}$	-1	$= -$	P_{ye}

It must be understood that this is simply a tabular arrangement of the seven equilibrium equations and students should practise reading the equations from the table. This tabular presentation is now rearranged:

$$
\begin{bmatrix}
-\tfrac{3}{5} & 1 & \tfrac{3}{5} & & & & \\
-\tfrac{4}{5} & & -\tfrac{4}{5} & & & & \\
& -1 & & & -\tfrac{3}{5} & \tfrac{3}{5} & \\
& & & & -\tfrac{4}{5} & -\tfrac{4}{5} & \\
& & -\tfrac{3}{5} & -1 & \tfrac{3}{5} & & 1 \\
& & \tfrac{4}{5} & & \tfrac{4}{5} & & \\
& & & & & -\tfrac{3}{5} & -1
\end{bmatrix}
\begin{bmatrix}
F_1 \\ F_2 \\ F_3 \\ F_4 \\ F_5 \\ F_6 \\ F_7
\end{bmatrix}
= -
\begin{bmatrix}
P_{xb} \\ P_{yb} \\ P_{xc} \\ P_{yc} \\ P_{xd} \\ P_{yd} \\ P_{xe}
\end{bmatrix}
$$

This is termed a matrix presentation of the equations. The essential change is that the top horizontal row of the table has become a separate adjacent vertical column—termed a column matrix. Nevertheless, this arrangement represents the same seven previous equations, and again students should practise reading the equations, remembering that the arrangement can be considered equivalent to the previous table.

It is, of course, the intention to calculate the seven unknown member forces F. Whilst it would be a very tedious arithmetical operation to solve the seven simultaneous equations by hand, it will be readily understood that such an operation would be possible. If in fact this were done, the results to three significant figures would be

$$F_1 = -(-0.417P_{xb} - 0.938P_{yb} - 0.417P_{xc} - 0.313P_{yc} - 0.625P_{yd})$$

$$F_2 = -(0.5P_{xb} - 0.375P_{yb} - 0.5P_{xc} - 0.375P_{yc} - 0.75P_{yd})$$

$$F_3 = -(0.417P_{xb} - 0.313P_{yb} + 0.417P_{xc} + 0.313P_{yc} + 0.625P_{yd})$$

$$F_4 = -(-0.75P_{xb} + 0.563P_{yb} - 0.75P_{xc} + 0.188P_{yc} - P_{xd} + 0.375P_{yd} - P_{xe})$$

$$F_5 = -(-0.417P_{xb} + 0.313P_{yb} - 0.417P_{xc} - 0.313P_{yc} + 0.625P_{yd})$$

$$F_6 = -(0.417P_{xb} - 0.313P_{yb} + 0.417P_{xc} - 0.937P_{yc} - 0.625P_{yd})$$

$$F_7 = -(-0.25P_{xb} + 0.188P_{yb} - 0.25P_{xc} + 0.563P_{yc} + 0.375P_{yd} - P_{xe})$$

These seven solutions can be written in matrix form:

$$
\begin{bmatrix} F_1 \\ F_2 \\ F_3 \\ F_4 \\ F_5 \\ F_6 \\ F_7 \end{bmatrix} = -
\begin{bmatrix}
-0.417 & -0.938 & -0.417 & -0.313 & 0 & -0.625 & 0 \\
0.5 & -0.375 & -0.5 & -0.375 & 0 & -0.75 & 0 \\
0.417 & -0.313 & 0.417 & 0.313 & 0 & 0.625 & 0 \\
-0.75 & 0.563 & -0.75 & 0.188 & -1 & 0.375 & -1 \\
-0.417 & 0.313 & -0.417 & -0.313 & 0 & 0.625 & 0 \\
0.417 & -0.313 & 0.417 & -0.937 & 0 & -0.625 & 0 \\
-0.25 & 0.188 & -0.25 & 0.563 & 0 & 0.375 & -1
\end{bmatrix}
\begin{bmatrix} P_{xb} \\ P_{yb} \\ P_{xc} \\ P_{yc} \\ P_{xd} \\ P_{yd} \\ P_{xe} \end{bmatrix}
$$

This square matrix (so called because it has seven rows and seven columns) of coefficients is easily obtained using a computer, the arithmetical operations taking a mere few seconds. It is obtained by operating on the earlier square matrix and is termed the *inverse* of the earlier matrix. It should be noted that the arithmetical operations are carried out without reference to the values of the loads and so, once the inverse has been obtained, the solutions for various combinations of load are quickly determined. For example, the column matrix of loads used in the earlier examples is

$$
\begin{bmatrix} P_{xb} \\ P_{yb} \\ P_{xc} \\ P_{yc} \\ P_{xd} \\ P_{yd} \\ P_{xe} \end{bmatrix} =
\begin{bmatrix} 0 \\ -100 \\ 15 \\ 0 \\ 0 \\ -100 \\ 0 \end{bmatrix}
$$

Substituting this in the above matrix equations and multiplying out gives

$$F_1 = -(0.938 \times 100) + (0.417 \times 15) - (0.625 \times 100) = -150 \text{ kN}$$

$$F_2 = -(0.375 \times 100) + (0.5 \quad \times 15) - (0.75 \quad \times 100) = -105 \text{ kN}$$

$$F_3 = -(0.313 \times 100) - (0.417 \times 15) + (0.625 \times 100) = \quad 25 \text{ kN}$$

$$F_4 = \quad (0.563 \times 100) + (0.75 \quad \times 15) + (0.375 \times 100) = \quad 105 \text{ kN}$$

$$F_5 = \quad (0.313 \times 100) + (0.417 \times 15) + (0.625 \times 100) = \quad 100 \text{ kN}$$

$$F_6 = -(0.313 \times 100) - (0.417 \times 15) - (0.625 \times 100) = -100 \text{ kN}$$

$$F_7 = \quad (0.188 \times 100) + (0.25 \quad \times 15) + (0.375 \times 100) = \quad 60 \text{ kN}$$

Needless to say, these are the same answers as before.

It is of special interest to note that in this matrix approach it was not necessary to pre-determine the reactive forces, as was the case for the previous hand methods. Indeed, it would have been possible to include the three unknown reactive forces with the seven unknown member forces to give a total of ten unknowns. These could have been solved from the ten simultaneous equations obtained from a consideration of $\sum P_x = 0$ and $\sum P_y = 0$ at each of the five joints A, B, C, D and E. The inverse of the resulting 10×10 matrix is still easily obtained using the computer, but would have been too cumbersome to illustrate here.

It is again emphasized that this section has aimed at introducing the first year student to the importance of the computer to the structural engineer and of the importance of matrices within this field. However, the manner in which the equilibrium equations have been used to develop the matrices forms the basis of the more powerful matrix methods introduced later in the section on the calculation of deflections.

Statical determinacy

It is necessary to be able to recognize whether or not a structure is statically determinate. At the start of an analysis, the total number of unknown forces associated with a pin-jointed planar frame depends on the number of members (m) and the number of reactive forces (r). Thus

$$\text{total number of unknown forces} = m + r$$

As there are two equilibrium equations available at each joint, the total number of equations available to solve for the unknowns is $2j$, where j is the number of joints. Thus if the frame is stable and

$$m + r = 2j$$

then the unknown member forces and reactions can be calculated from the equilibrium equations.

(Students are sometimes confused by the earlier statement that, for the planar structure as a whole, there are three equilibrium equations, and they feel that these are additional to the $2j$ equations. This is not so. By counting two equilibrium equations for each joint, including those at the supports, the three equilibrium equations for the entire structure have in effect been included. This can be demonstrated by the student using a simple example.)

In the previous example we saw that

$$m + r = 7 + 3 = 10$$
$$2j = 2 \times 5 = 10$$

which confirms the statical determinacy of the structure.

For the stable frame in Fig. 2.11a:

$$m + r = 13 + 3 = 16$$
$$2j = 2 \times 8 = 16$$

and so it is also statically determinate.

However, $m + r = 2j$ does not necessarily establish a stable structure. For example, in Fig. 2.11b, although

$$m + r = 13 + 3 = 16$$

and

$$2j = 2 \times 8 = 16$$

the frame is clearly unstable as it contains a non-triangulated region.

For the frame in Fig. 2.11c:

$$m + r = 15 + 3 = 18$$
$$2j = 2 \times 8 = 16$$

All the unknown internal forces cannot now be determined using only the equilibrium equations. Such frames are said to be statically indeterminate and are discussed later. (It should be noted, however, that for this frame the unknown *external* forces, namely the three reactions at the two supports, can still be determined using only equilibrium equations.)

For many pin-jointed frames, known as simple frames, their statical determinacy is clear. The *simple frame* is defined as one which, starting from a basic triangle of bars, can be built up by adding two new bars at a time to form a new joint. Such frames can be analysed by the hand methods previously described.

The frame in Fig. 2.11d cannot be so built up by adding two bars at a time. It is known as a *compound frame*, such frames being formed from two or more simple frames joined by a number of bars. Nevertheless the frame is statically determinate as

$$m + r = 27 + 3 = 30$$
$$2j = 2 \times 15 = 30$$

However, it is not possible to solve this frame using only the hand method of joints, although a combination of the method of joints and the method of sections does enable the internal forces to be determined.

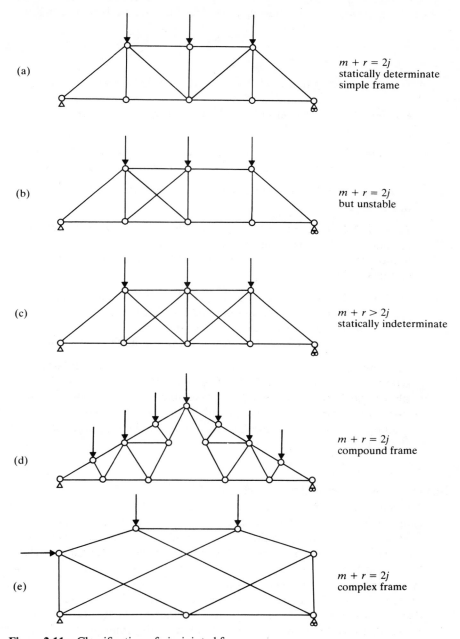

(a) $m + r = 2j$
statically determinate
simple frame

(b) $m + r = 2j$
but unstable

(c) $m + r > 2j$
statically indeterminate

(d) $m + r = 2j$
compound frame

(e) $m + r = 2j$
complex frame

Figure 2.11   Classification of pin-jointed frames

A frame not falling into either of these two categories (i.e., one which is neither simple nor compound) is termed *complex*. An example is shown in Fig. 2.11e. As

$$m + r = 11 + 3 = 14$$

$$2j = 2 \times 7 = 14$$

it is statically determinate but the simple hand methods cannot normally be used.

Problem 2.1   In the pin-jointed frame in Fig. 2.11a, which of the members must have a zero force?

Problem 2.2   For the frame in Fig. 2.12a, use the method of sections to determine the forces in members 1, 2 and 3. Check the values using an alternative method.

Problem 2.3   Obtain the forces in all four members of the cantilever frame shown in Fig. 2.12b.

Problem 2.4   Check that the frame in Fig. 2.12c is statically determinate. Use the method of sections in conjunction with a consideration of equilibrium at joint C to find the forces in members 1, 2, 3 and 4. Why is it not possible to use the method of sections alone?

Problem 2.5   For the frame in Fig. 2.10 used to illustrate the matrix method of joints, use the matrix solution obtained for the general loading case to calculate the force in member 2 for the loading case of vertical loads of 100 kN at each of the joints B, C and D. Check the answer using the method of sections.

Problem 2.6   Show that the frame in Fig. 2.12d is statically determinate. Explain why the simple hand methods of sections and joints cannot be used directly to obtain the forces in the members. Assemble in matrix form the equilibrium equations which would enable the member forces to be calculated for the generalized loading system and assuming the following (x, y) coordinates of the joints: A (0, 0); B (1, 2); C (3, 3); D (5, 3); E (7, 2); F (8, 0).

2.2 Planar pin-jointed frames—deflections

The tensile and compressive forces developed in the members of a pin-jointed frame cause the extension or shortening of the members with a corresponding movement of the joints (see Fig. 1.25). In the previous analyses of a statically determinate frame the member forces were determined without reference to these movements, the reason being that for a well-proportioned structure the movements are too small to have a significant effect on the member forces (see Fig. 1.51). Nevertheless the determination of the deflections of frames is an important part of design. As is explained in Chapter 5, deflections of structures must not exceed values which are considered to be the limits of acceptability. Moreover, as we shall see later, the ability to calculate deflections (or, more generally, displacements, this term including rotations) is an essential part in the analysis of statically indeterminate structures. It is, therefore, necessary that the

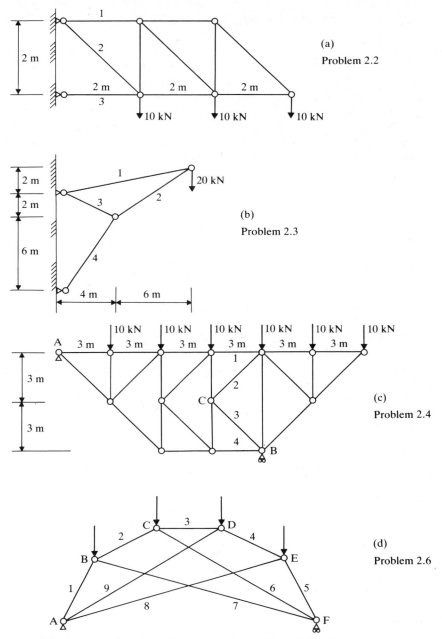

Figure 2.12 Planar pin-jointed frames used in Problems 2.2 to 2.6

structural engineer is able to determine the deflections of frames. There are several methods of determining the deflections.

Graphical method

Although the computer has now completely eliminated this method from practical usage, a brief introduction to it still has some value in presenting to students a pictorial appreciation of the deformation of frames.

Using the definitions in Chapter 1 of stress, strain and the modulus of elasticity, it is possible to calculate the extension δ of a member when subjected to a known tensile force F (Fig. 2.13). Suppose the member has a length L and cross-sectional area A, and the material has a modulus of elasticity E.

$$\text{stress} = F/A$$

$$\text{strain} = \delta/L$$

$$\text{modulus of elasticity } E = \frac{\text{stress}}{\text{strain}} = \frac{FL}{A\delta}$$

Thus
$$\delta = \frac{FL}{AE}$$

The expression for the shortening of a member under the effect of a compression force is, of course, the same.

Consider now the frame in Fig. 2.14a. It is required to find the movement of the loaded joint C.

Equilibrium at joint C enables us to determine the forces in members 1 and 2. These are

$$F_1 = 20\sqrt{2}\,\text{kN}$$

$$F_2 = -20\,\text{kN}$$

Member 1 is in tension and therefore has extended under the effect of the load; member 2 is in compression and has therefore shortened. Thus

$$\delta_1 = \frac{20\sqrt{2} \times 2\sqrt{2} \times 10^3}{200 \times 200} = 2\,\text{mm}$$

$$\delta_2 = -\frac{20 \times 2 \times 10^3}{200 \times 200} = -1\,\text{mm}$$

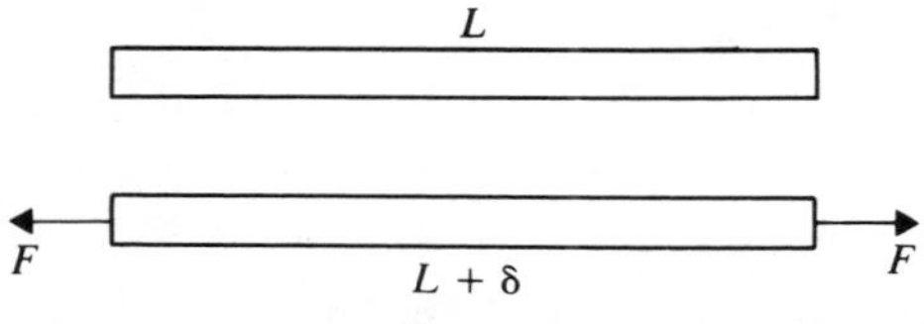

Figure 2.13 Forces developed in the members of a pin-jointed frame cause a change of their length

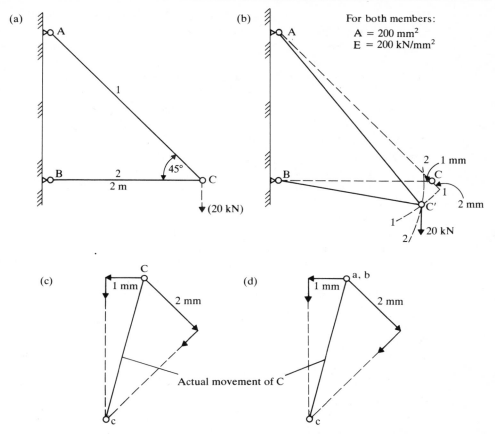

Figure 2.14   Introduction to the graphical method of determining deflections of pin-jointed frames

In order to accommodate these changes in length of the members, the joint C will have to move to the position C′ in Fig. 2.14b, this position being determined by the intersection of the arcs 1–1 and 2–2.

It is important to realize that in Fig. 2.14b the extension and shortening of the two members as given by the position of the arcs are not drawn to scale. The deformations are really too small to be depicted on a diagram drawn to the same scale as the frame and have been exaggerated for clarity. To determine the movement of C graphically, the displacements are therefore drawn to a different scale in a separate diagram as in Fig. 2.14c. This is termed a Williot diagram. In these diagrams it is sufficiently accurate to replace the arcs by straight lines at right angles to the original directions of the members.

It is of interest to note that, although the changes in length of the two members were 2 mm and 1 mm, the actual movement of joint C is approximately 4 mm. For a badly proportioned structure of the type illustrated in Fig. 1.52, the movements become even more exaggerated.

In Fig. 2.14c the line Cc represents the actual movement of joint C. It is also the movement of C relative to the fixed points of the frame, namely A and B. The usual procedure in drawing a Williot diagram is to designate the origin as the fixed points, as in Fig. 2.14d, with the diagram becoming a representation of *relative* movements of the joints. Thus, in Fig. 2.14d, a and b coincide because A and B have zero relative movement. It follows that C moves the same amount relative to A as it does relative to B and, because A and B do not move at all, the line ac (or bc) in Fig. 2.14d is the *actual* movement of C.

For larger frames, the deflections of the various joints are determined one at a time by working from one triangle to the next. The procedure is illustrated by reference to the frame in Fig. 2.15a. It will be assumed that, under the applied load, the changes in length of the members are as indicated on the diagram.

The first triangle of the frame to consider is ABC since this has two fixed points, A and B. The procedure for determining the deflection of joint C relative to A and B is therefore exactly the same as the earlier example and the deflection is given by the line joining a, b and c in Fig. 2.15b. The diagram is, of course, drawn strictly to scale so that the magnitude of the deflection can be measured directly.

The next triangle of the frame for consideration is ACD. The reference points for the movement of D are the fixed position of A and the *new position* of C. Since member AD extends, the joint D will move to a new position somewhere to the right of its original position—in Fig. 2.15b this is represented by the horizontal 8 mm line to the right of the origin a, b and its associated vertical line (representing the arc of D's movement perpendicular to AD). As member CD shortens, the joint D will move downward relative to the new position of C—in Fig. 2.15b this is represented by the vertical 2 mm line below c and its associated horizontal line. The junction of the two lines representing the two arcs determines the position of d. The deflection of D relative to A (and therefore, since this is a fixed point, its actual deflection) is represented by the straight line ad.

The next triangle of the frame for consideration is CDE. The student is left to continue the argument for the completion of the Williot diagram. This indicates that the total movement of E has a vertical component of approximately 37 mm and a horizontal component of 14 mm. (The latter is, in fact, easily determined since it must be the sum of the shortening of the two members BC and CE.) The movements of all the joints determined from this Williot diagram indicate that the resulting deflected form of the frame would be as in Fig. 2.15c.

In both the previous examples it was possible to start the procedure by considering a triangle in which there were two fixed joints. This enabled the position of the third joint to be established. For a truss of the type in Fig. 2.16, this is not possible. Although joint A of triangle ABC is fixed, neither joint B nor joint C is constrained in any way and their new positions cannot now be determined directly. As a result, the procedure for such frames is somewhat more complicated. To start the analysis, an assumption is made regarding the movement of an adjacent joint—for example, joint

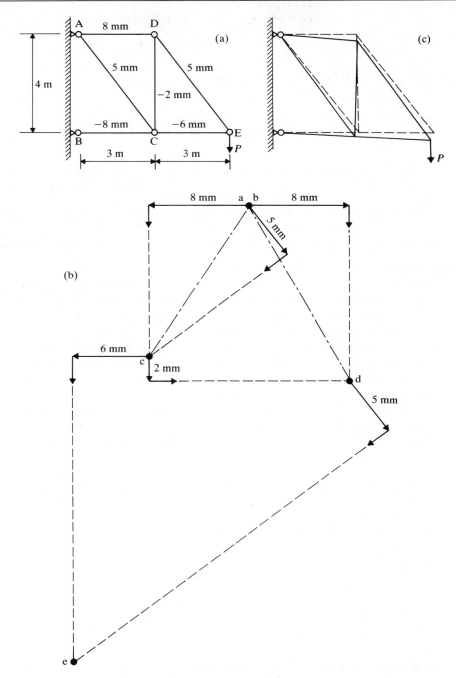

Figure 2.15   Illustrating the graphical method of determining deflections

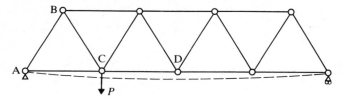

Figure 2.16 The graphical method of determining deflections is more complicated for such frames as there are not two adjacent joints constrained by supports

C could be assumed to move horizontally to the right. Such an assumption will normally prove incorrect, and a correction to the resulting values needs to be applied. This is termed the Mohr-correction and the combined diagram is known as the Williot–Mohr diagram. Although a very neat and technically interesting procedure, it is somewhat complicated and, since it is now only of historical interest and, moreover, does not contribute to a physical understanding of structures, it will not be given here.

Method of strain energy

The joint E in Fig. 2.15c was shown to move downwards and to the left during the application of the vertical load P at E. Since the movement of E has a component in the direction of the load, say Δ_{ye}, the load has performed work.

In determining the amount of work done by the load, it must be understood that the load is considered to be applied gradually and that the deflection of the framework commences as soon as the load starts to come on to the frame. For a linear elastic structure, the load–deflection diagram would be the straight line shown in Fig. 2.17— at the maximum load P the vertical deflection at E is Δ_{ye}.

Suppose at an intermediate stage when the load on the structure has reached a value of p, the vertical deflection at E has reached δ_{ye}. Suppose also that a further small increment of load dp causes the vertical deflection to increase by $d\delta_{ye}$. The work

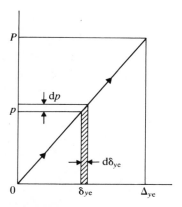

Figure 2.17 The relationship between the load acting on joint E and the vertical deflection at E during the application of the load on the frame illustrated in Fig. 2.15

done by the load during this further increment of load is, to a first approximation, given by $p \times d\delta_{ye}$, which will be seen to equal the shaded area in Fig. 2.17. It follows that as the load increases from zero to P, the work done by the load is equal to the area of the triangle under the complete line; thus

$$\text{work done by the load } P = \frac{P\Delta_{ye}}{2}$$

By the law of the conservation of energy this work is stored in the structure (since for our idealized structure we ignore any dissipation of energy—through heat caused by friction at joints, for example). This storing of work in the members of the framework is termed *strain energy* and can be interpreted as the energy gained by the material as its atomic structure is deformed against the inter-atomic forces. When the strain energy of a structure caused by a single load can be determined, it is possible to use the equality between the work done by the load and the strain energy of the structure to calculate the movement of the load in the direction of the load (which is termed the *corresponding* deflection of the loaded joint).

For the single member shown in Fig. 2.13, if the force F is applied gradually to the member to cause the extension δ, the same argument as above can be used to show that the work W done by the force is

$$W = \frac{F\delta}{2}$$

By the law of the conservation of energy, this must equal the strain energy stored in the member. Since, as was shown earlier

$$\delta = \frac{FL}{AE}$$

the strain energy U_m stored in a member subjected to an axial force F is

$$U_m = \frac{F^2 L}{2AE}$$

and it is clearly irrelevant whether F is a positive or negative value.

It follows that if the six members of a pin-jointed frame similar to that in Fig. 2.15a have forces $F_1, F_2, \ldots, F_6$, cross-sectional areas $A_1, A_2, \ldots, A_6$, lengths $L_1, L_2, \ldots, L_6$ and moduli of elasticity $E_1, E_2, \ldots, E_6$, the total amount of strain energy stored in the structure U_s is

$$U_s = U_1 + U_2 + U_3 + U_4 + U_5 + U_6$$

$$= \frac{F_1^2 L_1}{2A_1 E_1} + \frac{F_2^2 L_2}{2A_2 E_2} + \frac{F_3^2 L_3}{2A_3 E_3} + \frac{F_4^2 L_4}{2A_4 E_4} + \frac{F_5^2 L_5}{2A_5 E_5} + \frac{F_6^2 L_6}{2A_6 E_6}$$

$$= \Sigma \frac{F_i^2 L_i}{2A_i E_i}$$

where the suffix i represents any member i.

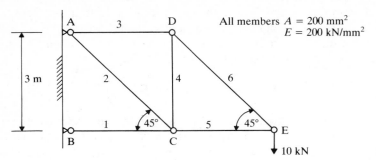

Figure 2.18 Frame used to illustrate the method of strain energy to calculate the deflection of loaded joints

The law of the conservation of energy applied to this structure gives

$$\frac{P\Delta_{ye}}{2} = \Sigma \frac{F_i^2 L_i}{2A_i E_i}$$

Since for a given load P the forces F_i in the members of the framework can be determined, it follows that the vertical deflection Δ_{ye} can be calculated.

An example of the calculation is given for the frame in Fig. 2.18. The calculation is best carried out in the form shown in Table 2.2.

Table 2.2

Member	Force F_i in member kN	Length L_i of member mm	$F_i^2 L_i$
1	-20	3×10^3	1200×10^3
2	$10\sqrt{2}$	$3\sqrt{2} \times 10^3$	$600\sqrt{2} \times 10^3$
3	10	3×10^3	300×10^3
4	-10	3×10^3	300×10^3
5	-10	3×10^3	300×10^3
6	$10\sqrt{2}$	$3\sqrt{2} \times 10^3$	$600\sqrt{2} \times 10^3$

$$\Sigma F_i^2 L_i = (2100 + 1200\sqrt{2}) \times 10^3$$

$$U_s = \Sigma \frac{F_i^2 L_i}{2A_i E_i} = \frac{\Sigma F_i^2 L_i}{2AE} = \frac{(2100 + 1200\sqrt{2})10^3}{2 \times 200 \times 200}$$

$$= \frac{95}{2} \text{ kN mm}$$

Work done by the applied load $= \dfrac{10\Delta_{ye}}{2}$ kN mm

Using the law of the conservation of energy:

$$\frac{10\Delta_{ye}}{2} = \frac{95}{2}$$

$$\Delta_{ye} = \underline{9.5 \text{ mm}}$$

The limitations of this method should be apparent. Firstly, the method can only be used when a single load is applied to the structure—there is only one equation available and therefore only one unknown deflection can be determined from it. Secondly, the deflection that can be obtained is at the position of the single load and then only that component of the deflection in the direction of the load.

The limitations of this method are therefore severe and so it finds little practical use. Nevertheless, it forms a very helpful introduction to a more general field of energy principles which are in fact widely used in structural analysis.

Introduction to the unit load method—virtual work

Figure 2.16 shows a pin-jointed truss loaded at joint C with a load P. The deflected form of the bottom chord caused by the load is also shown. Suppose we need to know the vertical deflection at joint D. It cannot be calculated using strain energy as that method can only determine the vertical deflection at the loaded joint; nor can the graphical procedure previously described be used as we limited that method to frames with two adjacent joints constrained by supports. A more general method for calculating deflections is therefore still required. Such a method is the unit load method and, prior to its illustration, the theory behind it will be demonstrated. It has a touch of magic.

Consider first the single member shown in Fig. 2.19. The member is first subjected to a force F_1 although, as we shall see later in the context of the truss, this force will be imaginary. It will cause a certain extension δ_1 although this will not be of any relevance. The member is now subjected to an additional force F_0 which in turn causes an additional extension δ_0. The work performed *during the application of the force F_0* is

$$\frac{F_0 \delta_0}{2} + F_1 \delta_0$$

Note that there is no $\frac{1}{2}$ in the term giving the work due to the force F_1, as this force maintains its full value throughout the whole of the movement δ_0. The force F_0 is

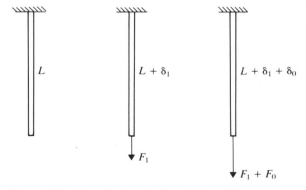

Figure 2.19   Introducing the idea of virtual forces in a member of a pin-jointed frame

gradually applied, increasing slowly from zero to its full value, and hence the $\frac{1}{2}$ appears in its work term.

By the law of the conservation of energy, the *additional* strain energy stored within the member due to the application of the force F_0 is therefore

$$\frac{F_0\delta_0}{2} + F_1\delta_0$$

Since

$$\delta_0 = \frac{F_0 L}{AE}$$

the additional strain energy in the member is

$$\frac{F_0^2 L}{2AE} + \frac{F_0 F_1 L}{AE}$$

These principles are now applied to the pin-jointed frame.

Consider the truss shown in Fig. 2.20a. We need to find the vertical deflection at the joint B caused by the vertical load at A. The load at A has the additional suffix 0 to emphasize that this is the *original* load, the load for which we need the deflection at joint B. The force in typical member i caused by the load P_{a0} is designated F_{i0}, again with the suffix 0 to indicate the case of the original loading. If the vertical deflection at A is Δ_{a0}, then by the law of the conservation of energy

$$\frac{P_{a0}\Delta_{a0}}{2} = \sum \frac{F_{i0}^2 L_i}{2A_i E_i}$$

Consider now the same truss but with only a vertical load X_1 at the joint B as in Fig. 2.20b. This will cause deflections Δ_{a1} and Δ_{b1} at joints A and B but, as in the case of the single member, these values will not come into the reckoning. The force in any member i due to X_1 will be designated F_{i1}.

On to this loaded truss we now superimpose the original loading P_{a0} as in Fig. 2.20c. The *additional deflections* of the truss caused during the application of P_{a0} are identical to those in Fig. 2.20a. Thus the work done by the two loads *during the application of the load P_{a0}* is

$$\frac{P_{a0}\Delta_{a0}}{2} + X_1\Delta_{b0}$$

The *additional* member forces caused by the application of the load P_{a0} are also identical to those caused in the case of the original loading in Fig. 2.20a. Thus by using the earlier arguments, the *additional* strain energy in the members of the framework is

$$\sum \frac{F_{i0}^2 L_i}{2A_i E_i} + \sum \frac{F_{i0}F_{i1}L_i}{A_i E_i}$$

Again using the law of conservation of energy

$$\frac{P_{a0}\Delta_{a0}}{2} + X_1\Delta_{b0} = \sum \frac{F_{i0}^2 L_i}{2A_i E_i} + \sum \frac{F_{i0}F_{i1}L_i}{A_i E_i}$$

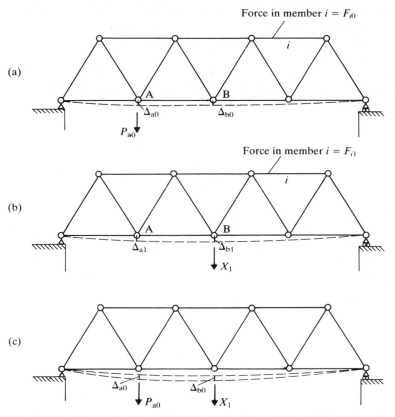

Force in member $i = F_{i0}$

(a)

Δ_{a0} Δ_{b0}

P_{a0}

Force in member $i = F_{i1}$

(b)

Δ_{a1} Δ_{b1}

X_1

(c)

Δ_{a0} Δ_{b0}

P_{a0} X_1

Figure 2.20   Illustrating the principle of virtual work for determining the deflection of pin-jointed frames

but since

$$\frac{P_{a0}\Delta_{a0}}{2} = \sum \frac{F_{i0}^2 L_i}{2A_i E_i}$$

$$X_1\Delta_{b0} = \sum \frac{F_{i0}F_{i1}L_i}{A_i E_i}$$

It might appear at this stage that we still have several unknowns, the arbitrary additional load, the member forces F_{i1}, and the required deflection Δ_{b0}. However, since the member forces F_{i1} are those due to X_1, they are a function of X_1, and indeed are directly proportional to X_1. It is, therefore, possible to assume that the value of X_1 is unity. To emphasize this, the member forces due to $X_1 = 1$ will now be represented by the *lower case* f. Moreover, to simplify the expression, the suffix i will be dropped as the meaning of the symbols will soon be understood. Thus

$$\Delta_{bo} = \Sigma \frac{F_0 f_1 L}{AE}$$

As Δ_{bo} is the deflection at the position of X_1 and in the direction of X_1 (but due to the original load), it is usual to adopt the further simplification of denoting the required deflection by Δ_1. Thus

$$\Delta_1 = \Sigma \frac{F_0 f_1 L}{AE}$$

The next step is to understand the generality of the method. It is not restricted to single loads. For example, Fig. 2.21a shows the truss subjected to several loads and acting in various directions. The member forces F_0 are now those due to this combined loading.

If the horizontal deflection at joint D caused by these loads is required, then, for this second calculation, a horizontal force of $X_2 = 1$ is applied at joint D as in Fig. 2.21b. In other words, the applied unit load *corresponds* to the required deflection. The member forces f_2 are determined for this case of unit loading and the required deflection is

$$\Delta_2 = \Sigma \frac{F_0 f_2 L}{AE}$$

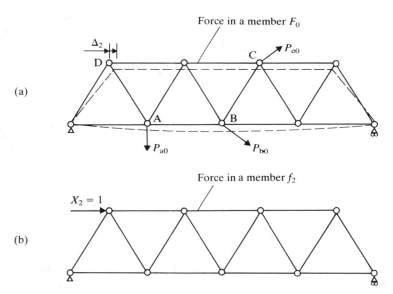

Figure 2.21 Illustrating the procedure of the unit load method for calculating the deflection at a joint of a pin-jointed frame

It is left to the student to repeat the argument on the previous lines to illustrate this generality.

To summarize then, the component of deflection in a particular direction of a joint of a pin-jointed structure is given by

$$\Delta_1 = \sum \frac{F_0 f_1 L}{AE}$$

where Δ_1 is the deflection of the joint in the required direction;
f_1 is the force in any member due to a force $X_1 = 1$ applied at the joint in the required direction;
F_0 is the force in any member due to the original (i.e., actual) loading.

It will be readily understood that the force $X_1 = 1$ is not a force that is actually applied to the structure. It is an imaginary force applied for the purpose of analysis only. The terms *virtual force* and *virtual work* have been used quite aptly to describe the principles involved. The complete subject of virtual work and its attendant theorems is very extensive, but it tends to be somewhat abstract and this unit load version of it will suffice at this stage.

Unit load method

The frame shown in Fig. 2.22a will be used to illustrate the unit load method of calculating deflections. It was used previously to illustrate the calculation of forces. Let us assume that the vertical deflection at joint D is now required. The member forces F_0 are the same as those calculated in the earlier example. The member forces f_1 are those due to $X_1 = 1$ applied as in Fig. 2.22b and can be determined using the method of joints. In this case, symmetry facilitates the calculation. The member forces and properties are summarized in Table 2.3.

Table 2.3

Member	Length L of member, mm	Area A of member, mm^2	Member forces F_0 kN	Member forces f_1	$\dfrac{F_0 f_1 L}{A}$ kN/mm
1	5×10^3	625	-150	-625×10^{-3}	750
2	6×10^3	750	-105	-750×10^{-3}	630
3	5×10^3	625	25	625×10^{-3}	125
4	6×10^3	750	105	625×10^{-3}	525
5	5×10^3	625	100	375×10^{-3}	300
6	5×10^3	625	-100	-625×10^{-3}	500
7	6×10^3	750	60	375×10^{-3}	180

$$\sum \frac{F_0 f_1 L}{A} = 3010 \text{ kN/mm}$$

Thus
$$\Delta_1 = \frac{\sum \dfrac{F_0 f_1 L}{A}}{E} = \frac{3010}{200} = \underline{15 \text{ mm}}$$

Young's Modulus for all members   = 200 kN/mm^2
Cross-sectional area for members 1, 3, 5 and 6 = 625 mm^2
Cross-sectional area for members 2, 4 and 7   = 750 mm^2

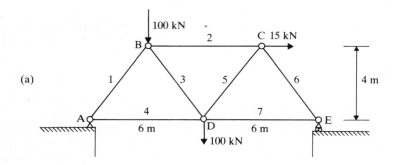

(a)

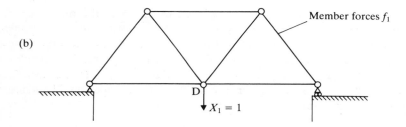

(b)

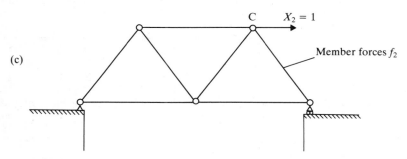

(c)

Figure 2.22   Illustrating the unit load method of calculating the deflection of joints of a pin-jointed frame

The positive value of Δ_1 indicates that the 15 mm deflection is in the same direction as X_1.

It is necessary to observe here that the unit force $X_1 = 1$ has no units. In the earlier manipulation of the equation

$$X_1 \Delta_1 = \sum \frac{F_0 F_1 L}{AE}$$

both sides of the equation were effectively divided by X_1 kN in order to obtain Δ_1 in mm. Thus in using

$$\Delta_1 = \sum \frac{F_0 f_1 L}{AE}$$

both $X_1 = 1$ and the forces f_1 are considered to have no units.

The same frame will now be used to illustrate the calculation of the horizontal deflection at the joint C. The unit load $X_2 = 1$ for this second calculation is as in Fig. 2.22c. The forces F_0 are, of course, the same as before, and the f_2 forces are found again using the method of joints.

Table 2.4

Member	Length L of member, mm	Area A of member, mm^2	Member forces F$_0$ kN	Member forces f$_2$	$\frac{F_0 f_2 L}{A}$ kN/mm
1	5×10^3	625	-150	417×10^{-3}	-500
2	6×10^3	750	-105	500×10^{-3}	-420
3	5×10^3	625	25	-417×10^{-3}	-83
4	6×10^3	750	105	750×10^{-3}	630
5	5×10^3	625	100	417×10^{-3}	333
6	5×10^3	625	-100	-417×10^{-3}	333
7	6×10^3	750	60	250×10^{-3}	120

$$\sum \frac{F_0 f_2 L}{A} = 413 \text{ kN/mm}$$

Thus

$$\Delta_2 = \frac{\sum \dfrac{F_0 f_2 L}{A}}{E} = \frac{413}{200} = \underline{2.07 \text{ mm}}$$

When the extensions of the members of a pin-jointed frame caused by the applied loads have been calculated, the expression for calculating deflections using the unit load method can be suitably amended. For example, since the extension δ_0 of a member is

$$\delta_0 = \frac{F_0 L}{AE}$$

the deflection of a joint of the structure can be determined from

$$\Delta_1 = \sum \delta_0 f_1$$

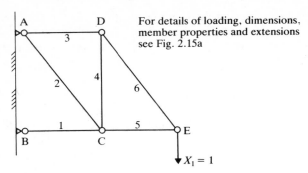

For details of loading, dimensions, member properties and extensions see Fig. 2.15a

Figure 2.23   Example of the unit load method using the frame previously adopted for the graphical method

Suppose, for example, the vertical deflection at E in Fig. 2.15a is required, the unit load $X_1 = 1$ is applied as in Fig. 2.23.

Table 2.5

Member	Extension δ_0 mm	Member force f_1	$\delta_0 f_1$ mm
1	-8	-1.5	12
2	5	1.25	6.25
3	8	0.75	6
4	-2	-1.0	2
5	-6	-0.75	4.5
6	5	1.25	6.25

$$\Delta_1 = \sum \delta_0 f_1 = \underline{37 \text{ mm}}$$

Calculation of joint movements using matrices

When a knowledge of the horizontal and vertical displacements of all the joints of a pin-jointed truss are required, the previous hand methods of calculation are not really suitable and it is essential to adopt a computer-based method using matrices. The procedure demonstrated here forms the basis of methods which are widely used in practice. (They are widely used because of the ease with which the computations can be automated. Moreover, as will be shown, the method is easily extended to include the computation of member forces and, most advantageously, is equally applicable to statically determinate and indeterminate structures.)

The pin-jointed frame previously used to illustrate the calculation of forces will again be adopted. It is shown again in Fig. 2.24a. The horizontal and vertical displacements of the joints caused by the application of the general load system $P_{xb}, P_{yb}, P_{xc} \ldots$ are required. The joint at A, being a pinned support, has zero translational displacement (i.e., $\Delta_{xa} = \Delta_{ya} = 0$); the joint at E, being a roller support, is constrained to move only in the horizontal direction (i.e., $\Delta_{ye} = 0$); each of the joints

Member properties as in Fig. 2.22

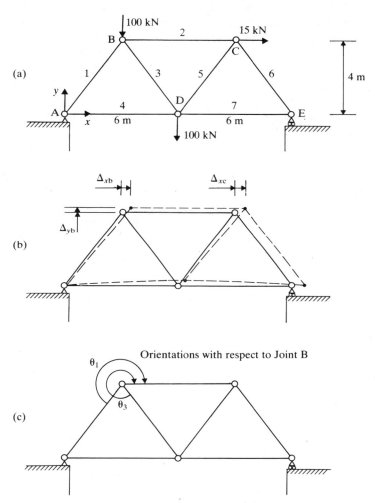

Figure 2.24 Pin-jointed frame used to illustrate the calculation of displacements of joints using a matrix method

B, C and D has both horizontal and vertical displacements. There are, therefore, a total of seven unknown displacements to be determined. They are designated Δ_{xb}, Δ_{yb}, Δ_{xc}, Δ_{yc}, Δ_{xd}, Δ_{yd}, Δ_{xe}, and will be deemed positive when in the positive direction of the appropriate axes (Fig. 2.24b).

The movements are, of course, the result of the small alterations in the lengths of the members caused by the internal forces due to the applied loads. The procedure is to calculate these changes in lengths in terms of the unknown Δ values. These changes in lengths are then translated to member forces by using the known properties of the members (length, cross-sectional area, Young's Modulus). At this stage the seven joint-equilibrium equations are obtained in a manner similar to that demonstrated in Section 2.1, the essential difference now being that the internal forces F are in terms of Δ. Moreover, expressions are developed which allow a more systematic determination of the signs of the forces involved (this being a necessary step towards automatic computation of the forces). By expressing the seven equilibrium equations in matrix form, the computer can be used to determine the values of Δ.

Prior to considering the members of the truss in Fig. 2.24a, it is helpful to illustrate the determination of the force in a general member in terms of the horizontal and vertical displacements of the ends of the member.

The line mn in Fig. 2.25 represents the position of member i of a plane truss in relation to the axes prior to the application of the loads on the truss. The aim here will be to determine expressions for the horizontal and vertical forces exerted by member i on the *joint m* as a result of the applied loads.

The angle θ_i determining the orientation of member i is obtained by rotating about m in a clockwise direction (when viewed in the positive direction of the z-axis) from the positive direction of the x-axis to the line mn, or preferably (since it is easier to visualize) by *rotating clockwise about m from mn to the positive direction of the x-axis when viewed as in Fig. 2.25.*

As a result of applying the loads to the truss, member i moves to the new position m′n′—the displacements of m are Δ_{xm} and Δ_{ym} whilst at n the displacements are Δ_{xn}

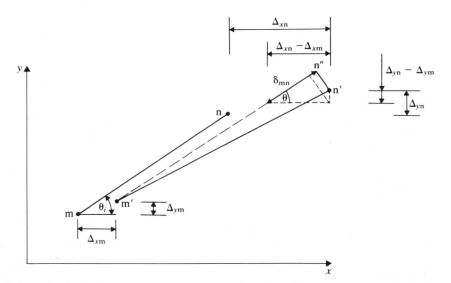

Figure 2.25   Illustrating the calculation of the extension of a member in terms of the end displacements

and Δ_{yn}. (The displacements shown in Fig. 2.25 have been exaggerated for clarity. In particular, it must be remembered that the rotation of the member will, in reality, be very small.)

For ease in determining the change in length of the member, m'n' has been rotated to m'n'', parallel to the original direction mn. Because of the small angle of rotation of the line, n'n'' can be regarded as being normal to m'n''.

It will now be clear from the diagram that the increase in length of the member, δ_i, is given by

$$\delta_i = (\Delta_{xn} - \Delta_{xm}) \cos \theta_i + (\Delta_{yn} - \Delta_{ym}) \sin \theta_i$$

Using the relationship

$$\text{Young's Modulus } E = \frac{\text{stress}}{\text{strain}} = \frac{F/A}{\delta/L}$$

the tensile force in member i is given by

$$F_i = \left(\frac{AE}{L}\right)_i \times \delta_i$$

$$= \left(\frac{AE}{L}\right)_i ((\Delta_{xn} - \Delta_{xm}) \cos \theta_i + (\Delta_{yn} - \Delta_{ym}) \sin \theta_i)$$

This is therefore also the expression for the force acting on the *joint m* from member i and acting in the direction from m to n. The components of this force in the x and y directions are obtained by multiplying by $\cos \theta$ and $\sin \theta$ respectively, and the resulting expressions are

$$F_{xi} = \left(\frac{AE}{L}\right)_i ((\Delta_{xn} - \Delta_{xm}) \cos^2 \theta_i + (\Delta_{yn} - \Delta_{ym}) \sin \theta_i \cos \theta_i)$$

$$F_{yi} = \left(\frac{AE}{L}\right)_i ((\Delta_{xn} - \Delta_{xm}) \cos \theta_i \sin \theta_i + (\Delta_{yn} - \Delta_{ym}) \sin^2 \theta_i)$$

(The forces applied by member i to the joint at the end n are clearly equal and opposite, and so the horizontal and vertical components at n can easily be obtained by applying a negative sign to the corresponding values at m.)

The above expressions apply to each member connected to *joint m*. The joint will be in equilibrium under the horizontal and vertical forces from the various members and the known loads P_{xm} and P_{ym} applied at joint m.

Using $\sum P_x = 0$ $P_{xm} + \sum F_{xi} = 0$

Using $\sum P_y = 0$ $P_{ym} + \sum F_{yi} = 0$

These give two equations in which the unknowns are the horizontal and vertical displacements of the ends of the members meeting at joint m. A similar consideration of the other joints of the truss allow other equations to be developed, and the equations together will permit the computation of the various displacements.

Applying these principles to the frame in Fig. 2.24a, the development of the equilibrium equations will be demonstrated for *joint B*. Thus in the following, joint B corresponds to joint m in the general expressions, whilst the joints at the other ends of the members meeting at B (namely, joints A, C and D) will correspond in turn to joint n in the general expressions.

As in the earlier examples for this frame, the values of AE/L (known as the axial stiffness) for all the members is 25 kN/mm.

The orientations of the members 1, 2 and 3 meeting at joint B (see Fig. 2.24c) are such that

$$\sin \theta_1 = -\tfrac{4}{5} \qquad \cos \theta_1 = -\tfrac{3}{5}$$

$$\sin \theta_2 = 0 \qquad \cos \theta_2 = 1$$

$$\sin \theta_3 = -\tfrac{4}{5} \qquad \cos \theta_3 = \tfrac{3}{5}$$

Hence

$$F_{x1} = 25\left((0 - \Delta_{xb})\frac{9}{25} + (0 - \Delta_{yb})\frac{12}{25}\right)$$

$$= -9\Delta_{xb} - 12\Delta_{yb}$$

$$F_{x2} = 25[(\Delta_{xc} - \Delta_{xb})1 + (\Delta_{yc} - \Delta_{yb})0]$$

$$= -25\Delta_{xb} + 25\Delta_{xc}$$

$$F_{x3} = 25\left((\Delta_{xd} - \Delta_{xb})\frac{9}{25} + (\Delta_{yd} - \Delta_{yb})\left(-\frac{12}{25}\right)\right)$$

$$= -9\Delta_{xb} + 12\Delta_{yb} + 9\Delta_{xd} - 12\Delta_{yd}$$

The equilibrium equation $P_{xm} + \sum F_{xi} = 0$ at joint B therefore results in

$$-43\Delta_{xb} + 25\Delta_{xc} + 9\Delta_{xd} - 12\Delta_{yd} = -P_{xb}$$

Similarly

$$F_{y1} = 25\left((0 - \Delta_{xb})\frac{12}{25} + (0 - \Delta_{yb})\frac{16}{25}\right)$$

$$= -12\Delta_{xb} - 16\Delta_{yb}$$

$$F_{y2} = 25[(\Delta_{xc} - \Delta_{xb})0 + (\Delta_{yc} - \Delta_{yb})0]$$

$$= 0$$

$$F_{y3} = 25\left((\Delta_{xd} - \Delta_{xb})\left(-\frac{12}{25}\right) + (\Delta_{yd} - \Delta_{yb})\frac{16}{25}\right)$$

$$= 12\Delta_{xb} - 16\Delta_{yb} - 12\Delta_{xd} + 16\Delta_{yd}$$

The equilibrium equation $P_{yb} + \sum F_{yi} = 0$ results in

$$-32\Delta_{yb} - 12\Delta_{xd} + 16\Delta_{yd} = -P_{yb}$$

These are the first two of seven simultaneous equations required for solving for the seven unknown displacements. The remaining five are obtained by considering equilibrium at C, D and E. The complete set of the seven equations summarized in matrix form are

$$
\begin{bmatrix}
-43 & 0 & 25 & 0 & 9 & -12 & 0 \\
0 & -32 & 0 & 0 & -12 & 16 & 0 \\
25 & 0 & -43 & 0 & 9 & 12 & 9 \\
0 & 0 & 0 & -32 & 12 & 16 & -12 \\
9 & -12 & 9 & 12 & -68 & 0 & 25 \\
-12 & 16 & 12 & 16 & 0 & -32 & 0 \\
0 & 0 & 9 & -12 & 25 & 0 & -34
\end{bmatrix}
\begin{bmatrix}
\Delta_{xb} \\
\Delta_{yb} \\
\Delta_{xc} \\
\Delta_{yc} \\
\Delta_{xd} \\
\Delta_{yd} \\
\Delta_{xe}
\end{bmatrix}
= -
\begin{bmatrix}
P_{xb} \\
P_{yb} \\
P_{xc} \\
P_{yc} \\
P_{xd} \\
P_{yd} \\
P_{xe}
\end{bmatrix}
$$

As in the earlier matrix example, the computer can operate on the coefficients in the 7×7 matrix without reference to the values of the loads to give the inverse, so leading to the seven equations for the values of Δ. These are

$$
\begin{bmatrix}
\Delta_{xb} \\
\Delta_{yb} \\
\Delta_{xc} \\
\Delta_{yc} \\
\Delta_{xd} \\
\Delta_{yd} \\
\Delta_{xe}
\end{bmatrix}
= -0.01
\begin{bmatrix}
-6.28 & 2.62 & -4.28 & 1.88 & -3.00 & 3.00 & -4.00 \\
2.62 & -6.66 & 1.12 & -2.97 & 2.25 & -5.38 & 3.00 \\
-4.28 & 1.12 & -6.28 & 0.38 & -3.00 & 0 & -4.00 \\
1.88 & -2.97 & 0.38 & -6.66 & 0.75 & -5.38 & 3.00 \\
-3.00 & 2.25 & -3.00 & 0.75 & -4.00 & 1.50 & -4.00 \\
3.00 & -5.38 & 0 & -5.38 & 1.50 & -9.63 & 3.00 \\
-4.00 & 3.00 & -4.00 & 3.00 & -4.00 & 3.00 & -8.00
\end{bmatrix}
\begin{bmatrix}
P_{xb} \\
P_{yb} \\
P_{xc} \\
P_{yc} \\
P_{xd} \\
P_{yd} \\
P_{xe}
\end{bmatrix}
$$

If we now assume the set of real loads applied to this structure in the earlier example (Fig. 2.24a), namely

$$
\begin{bmatrix}
P_{xb} \\
P_{yb} \\
P_{xc} \\
P_{yc} \\
P_{xd} \\
P_{yd} \\
P_{xe}
\end{bmatrix}
=
\begin{bmatrix}
0 \\
-100 \\
15 \\
0 \\
0 \\
-100 \\
0
\end{bmatrix}
$$

the displacements corresponding to these loads can be obtained by multiplying the matrices to give

$$
\begin{bmatrix}
\Delta_{xb} \\
\Delta_{yb} \\
\Delta_{xc} \\
\Delta_{yc} \\
\Delta_{xd} \\
\Delta_{yd} \\
\Delta_{xe}
\end{bmatrix}
=
\begin{bmatrix}
6.26 \\
-12.2 \\
2.06 \\
-8.41 \\
4.20 \\
-15.0 \\
6.60
\end{bmatrix}
$$

The values of the displacements are in mm. The negative signs mean that the particular displacement is in the opposite direction to the positive direction of the corresponding axis. For example, the vertical displacement at D, Δ_{yd}, is in the opposite direction to the positive direction of the y-axis, namely, downward as we would expect.

It should be mentioned that, for such problems, the calculation of the displacement via the computation of the inverse is inefficient and the equations would be solved by the computer directly. However, the computation of the inverse is useful when the effects of multiple sets of loads are to be analysed and it serves our purpose to illustrate that procedure.

As part of the process in determining the values of Δ, expressions for the member forces in terms of Δ were obtained. Now that the values of Δ are known, the member forces can be obtained by substituting the values of Δ into these expressions. For example

$$
F_1 = 25\left[(0 - 6.26)\left(-\frac{3}{5} \right) + (0 + 12.2)\left(-\frac{4}{5} \right) \right] = 150 \text{ kN}
$$

This is the same answer as obtained previously.

Although at first sight it may appear a somewhat cumbersome method, it must be understood that the arithmetical operations illustrated can be carried out entirely by the computer. However, the further refinements which permit the computation to be automated will not be dealt with at this stage. The great advantage of the method is in being applicable to both statically determinate and statically indeterminate structures, and this will be demonstrated in a later section of this chapter.

Problem 2.7   Use the unit load method to calculate the vertical deflection at joint D of the frame in Fig. 2.22a for the loading case of vertical 100 kN loads at each of the joints B, C and D. Check the value using the matrix inverse given in the earlier example.

Problem 2.8   In the pin-jointed cantilever frame in Fig. 2.26a, all the members have a cross-sectional area A and a modulus of elasticity E. Use the unit load method to determine the vertical deflection at D and check this using the graphical procedure.

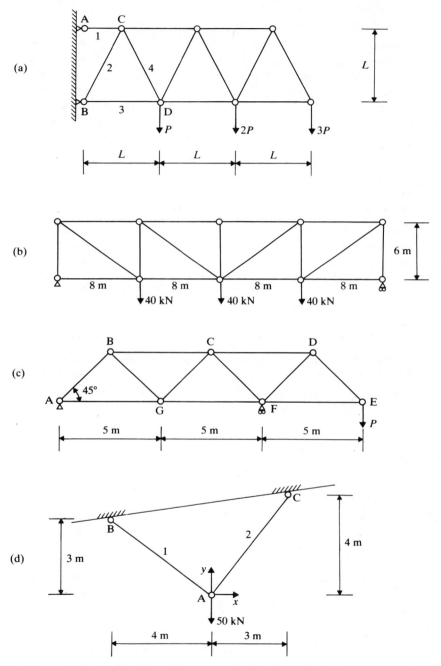

Figure 2.26   Planar pin-jointed frames used in Problems 2.8 to 2.11

Why cannot the strain energy method be used? (Take care not to do unnecessary calculations in the unit load method.)

Problem 2.9   The steel-jointed truss shown in Fig. 2.26b is to be designed to support the loads shown over the span of 32 m. Some of the constraints on design are that (a) all members must be of the same cross-section, (b) the stress in any member shall not exceed 150 N/mm², and (c) the maximum deflection below the supports shall not exceed 40 mm. Determine the minimum cross-sectional area of the members to satisfy these requirements. Assume that E is 200 kN/mm².

Problem 2.10   The designer of the pin-jointed frame in Fig. 2.26c proportioned the members so that, under the design load P, the strain developed in the tension members is 750×10^{-6} whilst in the compression members it is 250×10^{-6}. Explain why he has differentiated in this way between the two types of members. Calculate the vertical deflection of joint E caused by the load P.

Problem 2.11   Both members of the frame in Fig. 2.26d are of steel (Young's modulus 200 kN/mm²), but member 1 has a cross-sectional area of 125 mm² whilst member 2 has a cross-sectional area of 250 mm². The application of the 50 kN load causes joint A to undergo displacements Δ_{xa} and Δ_{ya}. Obtain expressions for the forces in the members in terms of these displacements. Use the expressions to obtain the equilibrium equations at joint A and solve the simultaneous equations for Δ_{xa} and Δ_{ya}. Hence obtain the member forces. Check the forces using the method of joints directly, and check the displacements using the graphical method.

Problem 2.12   Obtain the equilibrium equation corresponding to the x direction for joint C for the frame in Fig. 2.24a in terms of joint displacements. Check the answer with the appropriate equation of the complete set of equilibrium equations given in matrix form on page 103.

2.3 Planar pin-jointed frames—computer analysis

At first sight the method shown earlier for calculating forces in planar pin-jointed frames by initially determining the horizontal and vertical displacements of all the joints will probably seem a very cumbersome and long-winded method. However, it should be understood that all the arithmetical operations involved can be carried out by the computer. Although at this stage students will not be in a position to develop their own computer program which would permit the computation to be automated, if computing facilities and an existing program are available, advantage of this should be taken to obtain experience in their use. The essential steps in a computer program will already be clear. For example, it will be readily understood that, given the coordinates of the joints, the computer can be programmed to:

– calculate the length of the members;
– determine the orientations of the members relative to the axes.

Given the data relating to member properties (Young's modulus, cross-sectional areas), the computer can now develop the joint equilibrium equations in matrix form. Although the actual procedure adopted for computing the coefficients of the matrix which determine these equilibrium equations is outside the scope of this book, this need not deter students from utilizing existing programs when these are available. It only remains here to indicate some of the terminology and notation that may have been adopted in the program. For this, a simple pin-jointed plane frame program has been used to analyse the frame adopted in the earlier examples. Reference should be made to Fig. 2.27a where it will be seen that both joints and members have now been designated by numbers. The type of data required by the computer and the resulting output are given in Fig. 2.27b.

It will have been apparent earlier that it is possible for forces in members of statically determinate structures to be determined without reference to the member properties. However, computer programs utilizing the calculation of displacements as part of the analytical procedure require a knowledge of member properties. In cases of statically determinate structures where these are not known and only member forces are required, any values of the member properties can be adopted as input data for the purpose of the computation.

2.4 Space frames

The *simple* statically determinate planar pin-jointed frames were seen to be structures which could be formed by starting from a basic triangle and adding two members at a time to fix an adjacent joint. The corresponding three-dimensional frames are formed by starting from a basic tetrahedron (say ABCD in Fig. 2.28) and adding three members at a time to fix an adjacent joint. The tetrahedron itself is formed from a basic triangle (ABC in Fig. 2.28) with three additional members connected to the fourth joint at D not in the same plane. The skeletal tetrahedron therefore has six members. As with planar frames, other arrangements of members are possible, again these being termed compound or complex according to the arrangement.

A distinct disadvantage of space frames in comparison with planar frames has, until recently, been the difficulty of connecting the members—the bolting arrangements for such angled members are awkward and site welding is not always satisfactory. However, ingenious connecting devices (Fig. 2.29) have now been developed for the connection of members of modest size. These facilitate the site construction of space frames from prefabricated components and have transformed the economics of such frames for floor and roof construction.

The principles of analysis of statically determinate three-dimensional frames are similar to those discussed for planar frames, and similar methods of analysis are possible. Needless to say, the hand methods of analysis are much more tedious, normally involving many more members and simultaneous equations. The use of matrices and computers is therefore of much greater significance for such analysis. As a result, it is not intended to deal extensively at this stage with the analysis of space frames, although a useful introduction will be made to the use of cartesian coordinates in defining structural joints and member forces.

Young's Modulus for all members = 200 kN/mm²
Cross-sectional area for members 1, 3, 5 and 6 = 625 mm²
Cross-sectional area for members 2, 4 and 7 = 750 mm²

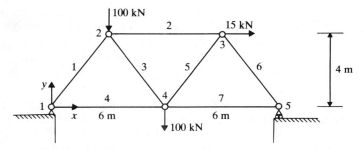

Figure 2.27   (a) Designation of members and joints of the pin-jointed frame analysed using existing computer program

As before, for the purpose of analysis it is assumed that the joints are frictionless hinges (spherical ball and socket type in this case) and transmit only axial forces to the members.

Types of support

The three idealized supports discussed in Chapter 1 for planar structures have their equivalents in space structures. The equivalent of the roller is the *ball*. This permits translation of the structure in two of the three orthogonal directions and offers no rotational restraint whatsoever. There is therefore again only one unknown reactive force, e.g., R_{ya}.

The equivalent of the pin is the *universal joint*. This prevents translation in all three orthogonal directions but offers no rotational restraint. There are therefore three unknown reactive forces, e.g., R_{xa}, R_{ya}, R_{za}.

As with planar structures, a *fixed* support can be provided. This prevents translation in all three orthogonal directions and prevents rotation about the three axes. There are therefore six unknown reactive forces, e.g., R_{xa}, R_{ya}, R_{za}, M_{xa}, M_{ya}, M_{za}.

As can be imagined, other intermediate types of support can be devised. For example, a ball constrained to roll only in the x direction offers two reactive forces, R_{ya} and R_{za}.

Statical determinacy

In the analysis of planar frames by the method of joints, we saw that there were two equilibrium equations available at each joint. These were

$$\begin{bmatrix} \sum P_x \\ \sum P_y \end{bmatrix} = 0$$

Manchester University Plane Pin-Jointed Frame Program

INPUT

Joint Coordinates

No.	x (m)	y (m)	Restraints
1	0.000	0.000	x, y
2	3.000	4.000	—
3	9.000	4.000	—
4	6.000	0.000	—
5	12.000	0.000	y

Members

No.	Joint A	Joint B	Area (mm^2)	E (kN/mm^2)
1	1	2	625	200.000
2	2	3	750	200.000
3	2	4	625	200.000
4	1	4	750	200.000
5	3	4	625	200.000
6	3	5	625	200.000
7	4	5	750	200.000

Loadings

Joint	Direction	Magnitude (kN)
2	y	−100.00
3	x	15.00
4	y	−100.00

OUTPUT

Joint Displacements

Joint	x (mm)	y (mm)
1	0.000	0.000
2	6.267	−12.200
3	2.067	−8.400
4	4.200	−15.000
5	6.600	0.000

Member Forces

No.	Axial force (kN)
1	−150.0
2	−105.0
3	25.0
4	105.0
5	100.0
6	−100.0
7	60.0

Reactions

Joint	x component (kN)	y component (kN)
1	−15.0	120.0
5	0.0	80.0

Figure 2.27   (b) Typical data output from a computer program

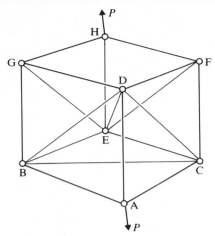

Figure 2.28   Example of a simple space frame

Figure 2.29   A typical device for connecting members of a space frame (*Source*: Centre Belgo-Luxembourgeois d'Information de l'Acier)

In the case of space frames there are three equilibrium equations available at each joint. These are

$$\begin{bmatrix} \sum P_x \\ \sum P_y \\ \sum P_z \end{bmatrix} = 0$$

Thus, if the number of joints in the frame is j, the total number of equilibrium equations available for the analysis is $3j$.

For a stable frame with m members and r reactive forces, if

$$m + r = 3j$$

then the frame is statically determinate.

Equilibrium of concurrent forces

In Chapter 1, the consideration of the equilibrium of a system of concurrent coplanar forces led to a number of useful corollaries. It is worth while stating a similar set for non-coplanar forces.

1. If three non-coplanar members meet at a joint at which there is no external load, then the force in each member must be zero, unless two of the members are collinear.
2. If all members except one meeting at a joint lie in one plane and there is no external load at the joint, then the force in that odd member must be zero.
3. If all but two of the members framing into a joint have zero force and there is no external load at the joint, the force in each of the two members must also be zero unless the two members are collinear.

That these rules follow logically from the equations of equilibrium should be verified by the student.

Use of joint coordinates

In Chapter 1 it was established that a force P in space is equivalent to a set of components acting in the direction of a set of orthogonal axes (see Fig. 1.48). These components are

$$P_x = P \cos \alpha \qquad P_y = P \cos \beta \qquad P_z = P \cos \gamma$$

where α, β, γ are the angles made by P with the lines parallel to the x, y, z axes. The quantitites $\cos \alpha$, $\cos \beta$, $\cos \gamma$ are termed the direction cosines.

If it is known that the force passes through two points which can be defined by coordinates (Fig. 2.30), then the direction cosines can be readily determined:

$$\cos \alpha = \frac{x_2 - x_1}{L} \qquad \cos \beta = \frac{y_2 - y_1}{L} \qquad \cos \gamma = \frac{z_2 - z_1}{L}$$

where

$$L = \sqrt{[(x_2 - x_1)^2 + (y_2 - y_1)^2 + (z_2 - z_1)^2]}$$

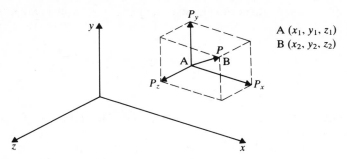

Figure 2.30 Use of joint coordinates facilitates the determination of components of forces

The components are therefore

$$P_x = \frac{x_2 - x_1}{L} P \qquad P_y = \frac{y_2 - y_1}{L} P \qquad P_z = \frac{z_2 - z_1}{L} P$$

Example of analysis

Figure 2.31 shows a tripod-shaped space frame supported through universal joints at A, B, C. The joint D is subject to loads

$$P_x = 10 \text{ kN} \qquad P_y = -5 \text{ kN} \qquad P_z = -6 \text{ kN}$$

It is required to find the forces P_1, P_2, P_3 in members 1, 2, 3.

As the three members meet at a joint, the forces in the members can be obtained from the three equations of equilibrium available for joint D.

It is first desirable to express the location of the joints in terms of coordinates in order to facilitate obtaining the direction cosines.

These are

$$A(0, 0, 0) \qquad B(6, 0, 1)$$
$$C(3, 0, 5) \qquad D(2, 5, 2)$$

For member 1:

$$\text{Length} = \sqrt{[(0 - 2)^2 + (0 - 5)^2 + (0 - 2)^2]} = \sqrt{33}$$

$$P_{x1} = \frac{(0 - 2)}{\sqrt{33}} P_1 = -0.348 P_1$$

$$P_{y1} = \frac{(0 - 5)}{\sqrt{33}} P_1 = -0.87 P_1$$

$$P_{z1} = \frac{(0 - 2)}{\sqrt{33}} P_2 = -0.348 P_1$$

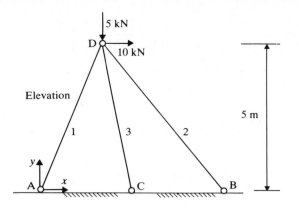

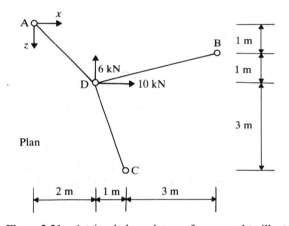

Figure 2.31   A tripod-shaped space frame used to illustrate the method of analysis

For member 2:

$$\text{Length} = \sqrt{[(6-2)^2 + (0-5)^2 + (1-2)^2]} = \sqrt{42}$$

$$P_{x2} = \frac{(6-2)}{\sqrt{42}} P_2 = 0.617 P_2$$

$$P_{y2} = \frac{(0-5)}{\sqrt{42}} P_2 = -0.772 P_2$$

$$P_{z2} = \frac{(1-2)}{\sqrt{42}} P_2 = -0.154 P_2$$

For member 3:

$$\text{Length} = \sqrt{[(3-2)^2 + (0-5)^2 + (5-2)^2]} = \sqrt{35}$$

$$P_{x3} = \frac{(3-2)}{\sqrt{35}} P_3 = 0.169 P_3$$

$$P_{y3} = \frac{(0-5)}{\sqrt{35}} P_3 = -0.845 P_3$$

$$P_{z3} = \frac{(5-2)}{\sqrt{35}} P_3 = 0.507 P_3$$

Considering equilibrium at joint D:

Using $\sum P_x = 0$ $-0.348 P_1 + 0.617 P_2 + 0.169 P_3 + 10 = 0$

Using $\sum P_y = 0$ $-0.870 P_1 - 0.772 P_2 - 0.845 P_3 - \ 5 = 0$

Using $\sum P_z = 0$ $-0.348 P_1 - 0.154 P_2 + 0.507 P_3 - \ 6 = 0$

Solving these three simultaneous equations gives

$$P_1 = 1.92 \text{ kN} \qquad P_2 = -17.28 \text{ kN} \qquad P_3 = 7.89 \text{ kN}$$

The amount of arithmetic involved (mostly solving the simultaneous equations) in order to obtain the member forces for this very simple frame emphasizes the need for computer-based methods for more complex frames. Students with a computer and pin-jointed space frame program available should repeat this example to illustrate the reduction of the analytical tedium available to the present-day structural engineer.

Problem 2.13   The space frame in Fig. 2.28 is in the form of a cube. It is loaded by the twin forces *P* acting along the line of the diagonal. Which two members of the frame must clearly be unstressed under this loading? Prove that the force in member BC also has zero force. Which additional two members will therefore also be unstressed?

Problem 2.14   The pin-jointed cantilever space frame in Fig. 2.32a is supported from a vertical wall at A, B, C, D. Planes ADGE and BCF are horizontal and EFG is vertical. Calculate the forces in the members meeting at F.

Problem 2.15   A pin-jointed space frame is shown in elevation and plan in Fig. 2.32b. Calculate the forces in the members caused by the load of 20 kN applied at E.

Problem 2.16   The loads applied to the joint D of the tripod-shaped frame in Fig. 2.31 cause a displacement of the joint. Use the unit load method to calculate the component of that movement in the *x* direction. (The cross-sectional area of all three members is 250 mm^2 and Young's Modulus is 200 kN/mm^2.)

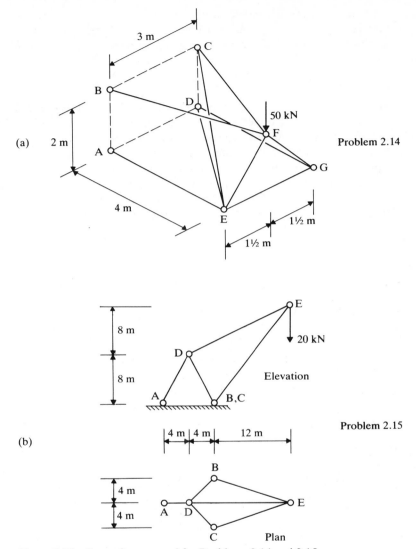

(a)

Problem 2.14

(b)

Elevation

Problem 2.15

Plan

Figure 2.32   Space frames used for Problems 2.14 and 2.15

2.5 Cables

There is no doubt that cables were one of the earliest structural forms as primitive man utilized their efficient load-carrying properties for bridging purposes. Today, cables are widely used in civil engineering structures. They form, for example, the main load-carrying members in suspension and cable-stayed bridges, and are used widely in guyed structures. The majority of these uses involve analyses beyond the scope of this book because of the statically indeterminate nature of the structural forms adopted in practice. Nevertheless, an introduction to the fundamental relationships of the simple cable is desirable in order that the student can appreciate the structural efficiencies of cable systems.

Basic theory

Let us consider the case of a cable supported at two points A and B at the same level and subjected to a number of vertical concentrated loads W_1, W_2, W_3, as shown in Fig. 2.33. The cable is considered to be perfectly flexible in bending so that the bending moment on any section of the cable must be zero, and the cable can only transmit load to the supports by means of tension acting along its length.

At this stage, the phenomenon of bending resistance of a member has not been discussed (see subsequent Sections 2.6, 2.7 and 3.1). Nevertheless, the bending flexibility of a cable will be intuitively understood by a comparison with a cotton

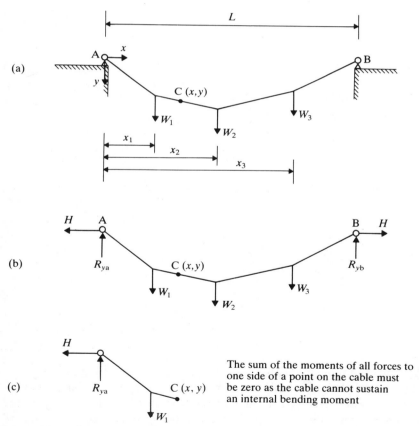

The sum of the moments of all forces to one side of a point on the cable must be zero as the cable cannot sustain an internal bending moment

Figure 2.33   Suspended cable subjected to concentrated loads

thread from which small weights are hung. Clearly, the thread has no resistance to bending forces and, since the weight of the thread will be insignificant in comparison with the loads, it will assume a shape consisting of a series of straight lines between the applied loads. It is such an idealization which has been assumed for the cable in Fig. 2.33a.

As the applied loads are all vertical, the horizontal reactions at A and B must be equal and opposite (since they are the only horizontal forces acting on the structure), and these are denoted by H (Fig. 2.33b). (It is interesting to note the further step along the path of structural understanding that is implied here. Instead of specifying unknown and different horizontal reactive forces acting in the direction of the coordinate axes, we have solved the initial equilibrium equation in our head, i.e., used intuition, and acknowledged the direction the reactive forces must act in order to resist the pull from the cable.) Simple equilibrium will confirm that the *horizontal component* of the tensile force at any section of the cable will also be of value H.

Using $\sum M_b = 0$ for the whole structure:

$$R_{ya}L - W_1(L - x_1) - W_2(L - x_2) - W_3(L - x_3) = 0$$

$$R_{ya} = \frac{\sum(W_n(L - x_n))}{L}$$

Consider any typical point C (coordinates x, y) and in this case positioned between the concentrated loads W_1 and W_2. As the bending moment at C is zero, taking moments of all the forces to the left of C (Fig. 2.33c) gives (see also Section 2.6):

$$R_{ya}x - W_1(x - x_1) - Hy = 0$$

$$Hy = \frac{x\sum(W_n(L - x_n))}{L} - W_1(x - x_1)$$

A study of this expression (in conjunction with the theory of beams in Section 2.6) reveals that, for the cable under consideration, the product of the horizontal component of the cable tension and the vertical distance from a particular point to the line joining the supports equals the bending moment which would occur at the corresponding vertical section of a similarly loaded, simply supported beam of the same span (see Fig. 2.36).

Shape of cable subjected to uniform load

In many practical structures, the suspension bridge for example, the dead load of the suspended superstructure is carried at such numerous positions along the span that it is sufficiently accurate to consider the load on the cable as uniformly distributed across the span. It is desirable to determine the shape that the cable would adopt when subjected to such a load. Figure 2.34 illustrates the typical problem. The symmetry of the arrangement demonstrates that:

– the vertical reactive forces at the supports are $wL/2$;
– the tensile force in the cable at the mid-span section is H.

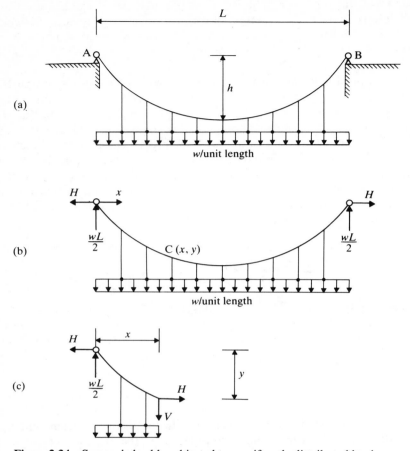

Figure 2.34   Suspended cable subjected to a uniformly distributed load

Application of the earlier proposition at the mid-span section gives

$$Hh = \left(\frac{wL}{2} \times \frac{L}{2}\right) - \left(\frac{wL}{2} \times \frac{L}{4}\right)$$

$$H = \frac{wL^2}{8h}$$

For the general point C, use of the earlier proposition gives

$$Hy = \frac{wLx}{2} - \frac{wx^2}{2}$$

Substituting the previous value of H and rearranging:

$$y = \frac{4hx(L - x)}{L^2}$$

This is a parabolic equation and defines the shape of the cable when subjected to a load distributed uniformly with respect to the horizontal projection of the cable.

Forces in cables

The tensile force T in the cable at any position can be obtained from a knowledge of the horizontal and vertical components at that position. For example, at the point C (Fig. 2.34b) in the cable subjected to a uniformly distributed load:

$$H = \frac{wL^2}{8h}$$

Vertical equilibrium for the forces to the left of C (Fig. 2.34c) gives

$$V = \frac{wL}{2} - wx$$

and acts in the direction shown in Fig. 2.34c.

The tensile force in the cable at C is the resultant of H and V:

$$T_x = \sqrt{(H^2 + V^2)}$$

$$= \sqrt{\left(\left(\frac{wL^2}{8h}\right)^2 + \left(\frac{wL}{2} - wx\right)^2\right)}$$

The maximum tensile force in the cable occurs at $x = 0$, i.e., at the supports, and is given by

$$T_{max} = \sqrt{\left(\left(\frac{wL^2}{8h}\right)^2 + \left(\frac{wL}{2}\right)^2\right)}$$

$$= H\sqrt{(1 - 4\phi^2)}$$

where

$$\phi = \frac{h}{L/2}$$

Design considerations in cable structures

1. As a result of the forces caused by the loads, the cable must undergo an extension. If it can be assumed that this does not significantly alter the geometry, the extension can be calculated by considering an element of the cable and integrating along the length of the cable. In many situations, however, the additional sag of the cables resulting from the extension does have an effect on the internal forces which has to be allowed for in the analysis. For such cases, the analysis is non-linear and although the previous equations would apply, the final value of the sag h would not be immediately known. Non-linear behaviour is particularly significant in the case

of guyed towers or cable-stayed bridges where variations in the shallow sag of the long straight cables have a greater effect.

2. It should be noted that the extensions caused in the cable by the loads are beneficial in that the internal forces are smaller as a result of the additional sag. Moreover, because the cable is stressed only in tension, there is no tendency for buckling to occur. For these reasons, cable systems are structurally very efficient.

3. The dead load (i.e., self-weight) of suspension bridges normally forms the major part of the loading on the cable and would therefore provide an essentially uniformly distributed load, this dictating the shape of the cable. However, the subsequent application of a concentrated imposed load (such as a heavy vehicle) would tend to cause the cable to change its shape. Clearly, significant changes would not be acceptable in bridge structures. In order to minimize such effects, the imposed load is transmitted from the deck to the cable via a stiffening girder. The stiffening girder ensures that a concentrated load on the deck is distributed to a number of hangers for transmission to the cable. The analysis of such statically indeterminate structures requires the use of sophisticated computation.

Problem 2.17   Show that the proposition 'the product of the horizontal component of the cable tensile force at any point and the vertical distance from that point

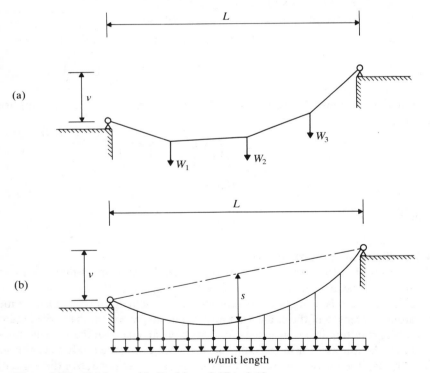

Figure 2.35   Cables used in Problems 2.17 to 2.19

to the line joining the supports equals the bending moment at the corresponding section of a similarly loaded simply supported beam of the same span' applies to the cable in Fig. 2.35a in which one support is at a height v above the other support.

Problem 2.18 Calculate the difference between the vertical reactions at the supports for the cable in Fig. 2.35b.

Problem 2.19 Calculate the maximum tensile force in the cable arrangement in Fig. 2.35b if $v = 10$ m, $s = 12$ m, $L = 48$ m, $w = 100$ kN/m.

Problem 2.20 Show that a suspended uniform cable acting under its own weight only will adopt a catenary shape.

B. Statically determinate structures involving bending

It was shown in Chapter 1 that, when a load is applied transversely to a member, shear forces and bending moments are set up at sections within the member. The calculation of these internal forces and of their effects is now illustrated. Our main attention will be directed to the simple beam but it will also extend to rigid-jointed frames and arches, although only statically determinate examples will be illustrated at this stage.

In reality, most beam and frame structures are three-dimensional. Nevertheless, for purposes of analysis it is often possible to divide such structures into planar sub-structures and it is therefore quite logical to confine our attention to planar problems.

2.6 Beams—calculation of internal forces

The case of a simply supported beam with a concentrated load positioned at mid-span was discussed in Section 1.3. The more general case of loading shown in Fig. 2.36a will now be considered for this further introduction to the calculation of internal forces in beams.

Introduction to shear force and bending moment

As was the case for pin-jointed frames, the first step in many problems is to calculate the reactive forces at the supports. For the simply supported beam in Fig. 2.36a subjected to the known loads W_1, W_2 and W_3, the number of unknown reactive forces at the start of the analysis is three—two at the pin at A (R_{xa} and R_{ya}) and one at the roller at B (R_{yb}). It will, therefore, be possible to calculate these using the three equations of equilibrium for planar structures:

$$\sum P_x = 0 \qquad \sum P_y = 0 \qquad \sum M_z = 0$$

It will immediately follow from the use of $\sum P_x = 0$ that, because we are concerned in this example with loads applied only transversely to the beam (i.e., vertical loads), the value of R_{xa} is zero. The *external* forces to which the beam is subjected are therefore the loads W and the two vertical reactions R_{ya} and R_{yb} (Fig. 2.36b). By using the other two equilibrium equations, the values of R_{ya} and R_{yb} can be determined in terms of the

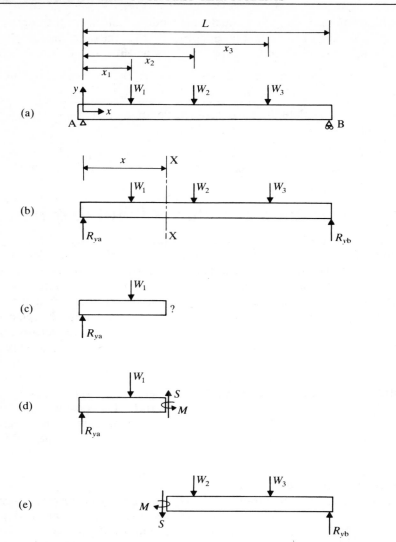

Figure 2.36   Introduction to the determination of internal forces in beams

known loads W_1, W_2 and W_3. However, for this particular example, these values of R_{ya} and R_{yb} will not be determined, but it must be understood that in the subsequent analysis R_{ya} and R_{yb} can be regarded as known.

Let us assume that we require to determine the internal forces acting on the typical cross-section XX in Fig. 2.36b. In this example, section XX is between W_1 and W_2 and at a distance x from support A.

In a similar way to that used in the method of sections for pin-jointed frames, we now consider equilibrium of a part of the beam to one side of the section XX. We will therefore consider the part to the left as in Fig. 2.36c.

The forces acting on this left-hand part of the beam at XX are a shear force S and a bending moment M. These forces are being applied by the right-hand portion of the beam to this left-hand part. They are shown in Fig. 2.36d acting in the directions which establishes these internal forces as positive forces. (A subsequent analysis may determine that the actual internal forces act in the opposite directions to those shown, and they would then be designated as negative.)

The sign convention for internal forces was discussed fully in Chapter 1 and is therefore only summarized here. Firstly, it should be recalled that the cut section shown in Fig. 2.36d is a positive surface (see Fig. 1.44); the positive direction of S on such a surface is in the positive direction of the appropriate axis, in this case the y-axis. As regards the bending moment, for the case of horizontal beams and when the y-axis is chosen to be upwards, a positive moment causes tension in the lower fibres of the beam and compression in the upper fibres of the beam. The arc representing M in Fig. 2.36d can always be viewed as the equivalent of twin forces, a push on the upper region of the beam (i.e., compression) and a pull on the lower region (i.e., tension) (compare Fig. 1.47a).

The magnitudes of S and M are now determined by considering the equilibrium of the free body shown in Fig. 2.36d. For this purpose, the forces previously termed *internal forces of the beam* become, in effect, *external forces of the free body*. In considering equilibrium of these 'external forces', it is now necessary to acknowledge the direction of the forces in relation to the x, y and z axes. In this case, R_{ya} and S act in the positive direction of the y-axis whilst W_1 acts in the negative direction. Thus, using $\sum P_y = 0$ gives

$$R_{ya} + S - W_1 = 0$$

$$S = W_1 - R_{ya}$$

The shear force S is seen to be equal in value, but opposite in sense, to the algebraic sum of all external forces acting to the left of the section XX. That it is also the algebraic sum of the forces to the right can be proved by considering equilibrium of the part of the beam to the right as in Fig. 2.36e. The internal forces S and M have again been drawn in their positive directions—the forces are now acting on a negative face and therefore the positive direction of S is opposite to the positive direction of the y-axis; again, the moment M is applying a push to the upper fibres. Again, using $\sum P_y = 0$ for the 'external forces' acting on this free-body:

$$R_{yb} - W_2 - W_3 - S = 0$$

$$S = R_{yb} - W_2 - W_3$$

That this is identical to the earlier value of S can be seen by using $\sum P_y = 0$ for the whole beam (Fig. 2.36b):

$$R_{ya} + R_{yb} - W_1 - W_2 - W_3 = 0$$

which on rearranging gives

$$R_{yb} - W_2 - W_3 = W_1 - R_{ya}$$

It follows that the shear force at a section of a beam is equal in value, but opposite in sense, to the algebraic sum of all forces acting on one side of that section.

To obtain the value of M, consider again the equilibrium of the part of the beam shown in Fig. 2.36d. Using $\sum M_z = 0$, assuming an axis passing through a point on the face XX (and henceforth referred to as $\sum M_{xx} = 0$):

$$W_1(x - x_{1'}) + M - R_{ya}x = 0$$

$$M = R_{ya}x - W_1(x - x_1)$$

Again, the identical value could be obtained by considering the forces to the right of XX. It follows that the bending moment at a section is equal in value, but opposite in sense, to the sum of the moments of all external forces to one side of the section.

Bending moment and shear force diagrams

At this stage, it is necessary to note an essential difference between the members of a pin-jointed structure and those members of a structure involving bending. Whereas in the former the force in a member is constant throughout its length, in the latter the bending moment and shear force can vary along the length of the member. In the previous case, for example, the expression for M included the variable x; moreover, different expressions for both S and M would have been obtained if the section XX had been between, say, W_2 and W_3.

The variations of force along the length of a member can be represented in graphs in which the values of the shear force and bending moment are plotted as ordinates against the distance x along the beam as abscissa. A number of examples will be given to illustrate these shear force and bending moment diagrams.

EXAMPLE 2.1

The 10 m simply supported beam shown in Fig. 2.37a is subjected to a concentrated load of 80 kN positioned 4 m from the left-hand support. It is required to construct the shear force and bending moment diagrams.

The first step is to determine the two vertical reactions R_{ya} and R_{yb} (Fig. 2.37b).

Using $\sum M_b = 0$ gives $R_{ya} = \dfrac{80 \times 6}{10} = 48$ kN

Using $\sum P_y = 0$ gives $R_{yb} = 80 - 48 = 32$ kN

Using $\sum M_a = 0$ verifies that the arithmetic is correct

Consider a section XX between the support A and the concentrated load and at a distance x from A (Fig. 2.37c). The forces acting on the part of the beam to the left of XX are as in Fig. 2.37d. (After a number of examples in which these free body diagrams are drawn separately, the student should practice dispensing with this aid

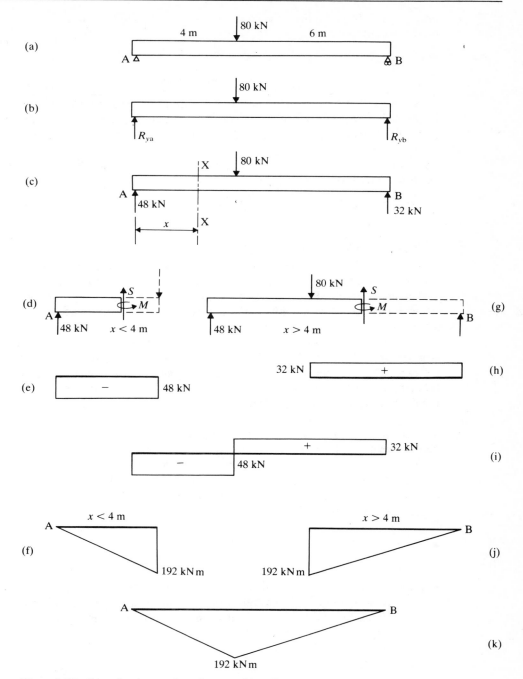

Figure 2.37   Introduction to shear force and bending moment diagrams

and obtain the equilibrium equations by visualizing such free-bodies in the imagination.)

Using $\sum P_y = 0$ $\qquad\qquad$ $48 + S = 0$

$$S = -48 \text{ kN}$$

This value of S is seen to be independent of x and so is constant over this region of the beam, namely for $0 \leqslant x \leqslant 4$. The diagram representing this constant value over this particular region is shown in Fig. 2.37e. Being a negative value, it has been drawn below the base line.

Before completing the shear force diagram, we will pursue the equilibrium of this left-hand part further in order to obtain M.

Using $\sum M_{xx} = 0$ $\qquad\qquad$ $M - 48x = 0$

$$M = 48x$$

The value of M is seen to be dependent on x. Again, this expression only applies within the range $0 \leqslant x \leqslant 4$ m. The value varies linearly from zero at $x = 0$ to 192 kN m at $x = 4$ m. The diagram representing this variation over this region is shown in Fig. 2.37f.

It is necessary to draw attention here to a dilemma—although M has a positive value, it has been drawn below the line. Although not strictly in accordance with our sign convention, this is in accordance with tradition. Most structural engineers prefer to draw the bending moment diagram on the tension side of the member and so this will be adhered to here. Not all text-books adopt this method however, and students must be aware of the dichotomy. To avoid the confusion which can occur when referring to positive and negative bending moments, it becomes preferable to describe the bending of beams as either 'sagging type bending' as in Fig. 2.41b or 'hogging type bending' as in Fig. 2.41a. Nevertheless, for some purposes it is necessary to retain a formal sign convention for internal bending moments, the case of calculating deflections using the differential equation of bending (Section 2.7) being one example.

To complete the shear force and bending moment diagrams, it is now necessary to consider the section XX between the concentrated load and the support B, but again a distance x from support A, i.e., $4 \leqslant x \leqslant 10$ m. Figure 2.37g summarizes the forces to the left of this section.

Using $\sum P_y = 0$ $\qquad\qquad$ $S + 48 - 80 = 0$

$$S = 32 \text{ kN}$$

The shear force diagram for this part of the beam is shown in Fig. 2.37h. Again, the shear force is constant over this region, but this time is positive and is therefore plotted above the base line. Combining the diagrams in Figs 2.34e and 2.34h gives Fig. 2.34i, the shear force diagram for the whole beam.

Using $\sum M_{xx} = 0$ $\quad$ $80(x - 4) + M - 48x = 0$

$$M = 48x - 80(x - 4)$$

As this expression is linear in x, the bending moment diagram for this region of the beam will be a straight line varying from $M = 192\,\text{kN m}$ at $x = 4\,\text{m}$ and $M = $ zero at $x = 10\,\text{m}$ (Fig. 2.34j). Combining the diagrams in Figs 2.37f and 2.37j gives Fig. 2.37k, the bending moment diagram for the whole beam.

EXAMPLE 2.2

For the second example, the 10 m span simply supported beam will be assumed to be subjected to a uniform distributed load of $40\,\text{kN/m}$ (Fig. 2.38a).

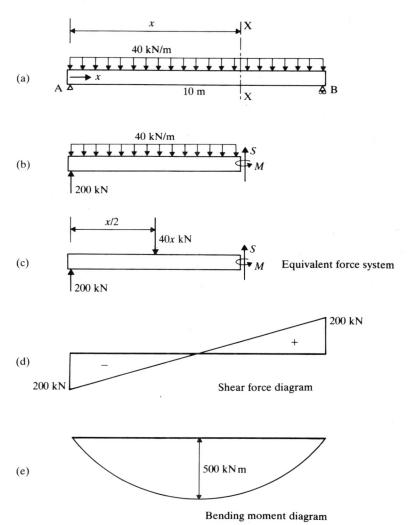

Bending moment diagram

Figure 2.38 Illustrating the determination of shear force and bending moment diagrams for a uniformly loaded simply supported beam

The symmetry of the arrangement allows the two vertical reactions at A and B to be quickly calculated, each being half the total load, namely 200 kN. To obtain the shear force and bending moment diagrams, a typical section XX is again considered at a distance x from A. The forces on the part of the beam to the left of XX are shown in Fig. 2.38b.

It will be recalled from Chapter 1 that, *for the purpose of considering equilibrium*, the distributed load can be considered as its equivalent force system of a concentrated load of $40x$ kN acting at a position $x/2$ from support A as in Fig. 2.38c. (Again, after the initial examples, students should dispense with the need for drawing this additional free-body diagram by visualizing it in the imagination.)

Using $\sum P_y = 0$ $200 + S - 40x = 0$

$$S = -200 + 40x$$

There is no discontinuity in the loading which limits the range of applicability of this expression and it applies over the full length of the beam. The value of S is linear in x, varying from -200 kN at $x = 0$ to 200 kN at $x = 10$ m. The resulting shear force diagram is shown in Fig. 2.38d.

Using $\sum M_{xx} = 0$ $M + 40x\left(\dfrac{x}{2}\right) - 200x = 0$

$$M = 200x - 20x^2$$

It will be noted that $M =$ zero at both $x = 0$ and $x = 10$ m—and, of course, that is to be expected at the ends of simply supported beams. In between, the curve representing the bending moment has a parabolic shape reaching a maximum ordinate of 500 kN m at mid-span (Fig. 2.38e).

EXAMPLE 2.3

The cantilever beam in Fig. 2.39a is rigidly fixed at the support B and subjected to a uniformly distributed load of 20 kN/m throughout its 5 m length.

For cantilevers, it is not necessary to calculate the reactive forces in order to determine the internal forces as it is always possible to consider equilibrium of the part of the beam towards the free end. Thus, for the forces at XX distance x from A, consideration is given to the part of the beam in Fig. 2.39b.

Using $\sum P_y = 0$ $S - 20x = 0$

$$S = 20x$$

The shear force on an internal cross-section of the beam therefore varies linearly from zero at $x = 0$ to 100 kN at $x = 5$ m. The shear force diagram is shown in Fig. 2.39c.

Using $\sum M_{xx} = 0$ $20x\left(\dfrac{x}{2}\right) + M = 0$

$$M = -10x^2$$

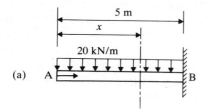

(a)

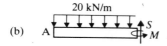

(b)

(c) Shear force diagram

(d) Bending moment diagram

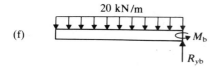

(e)

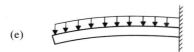

(f)

Figure 2.39   The shear force and bending moment diagrams for a uniformly loaded cantilever

The bending moment varies parabolically from zero at $x = 0$ to $-250\,\text{kN m}$ at $x = 5\,\text{m}$ as shown in Fig. 2.39d. The bending moment diagram, being everywhere negative, is plotted above the base line as the tension side of the beam is now the upper region. Apart from the sign of the bending moment indicating the tension side, the deflected form of the beam shown in Fig. 2.39e is intuitively known and so the tension side, the side of the beam on the outside of the curve, is similarly apparent.

Although the reactive forces at B were not needed to determine the internal forces, it is important to realize that there are two forces at B—the upward force R_{yb} and the moment M_b. These are external forces applied by the supporting wall to the beam as in Fig. 2.39f.

$$\text{Using } \sum P_y = 0 \qquad\qquad R_{yb} - (20 \times 5) = 0$$

$$R_{yb} = 100\,\text{kN}$$

$$\text{Using } \sum M_B = 0 \qquad\qquad M_b + \left(20 \times 5 \times \frac{5}{2}\right) = 0$$

$$M_b = -250\,\text{kN m}$$

These conform to the shear force and bending moment values previously determined for the section of the beam adjacent to B.

Table 2.6   Shear force and bending moment diagrams for some standard cases

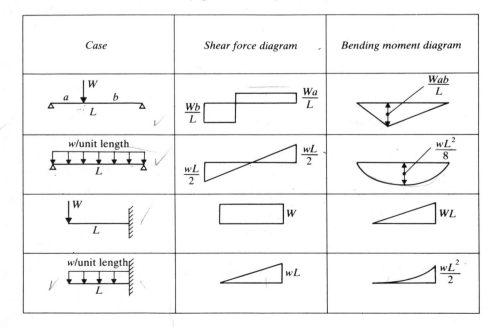

Some standard cases

The cases of concentrated loads and uniformly distributed loads on simply supported beams and cantilevers occur frequently in practice. Needless to say, it is not necessary to carry out an analysis from first principles each time. It is sufficient to refer to the standardized solution as, for example, in Table 2.6. A knowledge of these is essential and must become second nature to all students of structural engineering. Students should begin by verifying the values given in the table.

Relationships between load, shear force and bending moment

The diagram in Fig. 2.40 represents an infinitesimally short length of a beam, the intensity of loading at this position being w/unit length. The value of the shear force and the bending moment acting on the left-hand face of the element, distance x from the origin, are S and M respectively; the values on the other face of the element are $S + dS$ and $M + dM$. These, of course, are internal forces within the beam and have been drawn in the directions to indicate positive values. However, as in earlier examples, these internal forces within the beam become external forces for the purpose of considering equilibrium of the free body, namely the element. For example, the force $S + dS$ acts in the positive direction of the y-axis, whilst the two forces S and $w\,dx$ act in the negative direction of the y-axis. Hence:

Using $\sum P_y = 0$ $\qquad S + dS - w\,dx - S = 0$

$$\frac{dS}{dx} = w$$

Thus the slope of the shear force diagram at a particular position equals the intensity of the loading on the beam at that position. It follows that in the regions where w is zero, the shear force diagram is horizontal.

Using $\sum M_{xx} = 0$ $\qquad (M + dM) + (S + dS)\,dx - M - w\,dx\,\frac{dx}{2} = 0$

Neglecting the products of infinitesimally small quantities and rearranging gives

$$\frac{dM}{dx} = -S$$

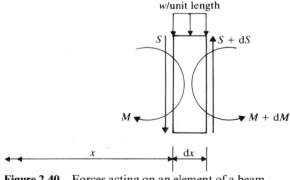

w/unit length

Figure 2.40   Forces acting on an element of a beam

Thus the slope of the bending moment diagram at a particular position equals the value of the shear force at that position. (Remember that with our method of plotting bending moment diagrams on the tension side of the beam rather than strictly in accordance with the sign convention, the negative sign in the above expression should be ignored when comparing the sign of the slope of the bending moment at a point with the value of the shear force). Again, it follows that when S is zero, the bending moment diagram is horizontal.

These relationships are useful in sketching and checking bending moment and shear force diagrams. The latter relationship is particularly useful in obtaining maximum values of bending moment.

Direct method of establishing the bending moment at a section

The previous examples of establishing the bending moment at a section involved:

- drawing the free body and indicating the forces acting on it;
- consideration of equilibrium of the free-body;
- establishment of the equilibrium equation involving M;
- rearrangement of the equation to give M.

It is now desirable to short-circuit this procedure to enable the moment M to be obtained more directly.

The method now presented is based on our intuitive knowledge of the behaviour of the cantilever. For example, we have seen previously that a downward force such as P in Fig. 2.41a causes a hogging type curvature (tension in the upper fibres of the beam) and a bending moment in the beam adjacent to the support equal to $-Ps$.

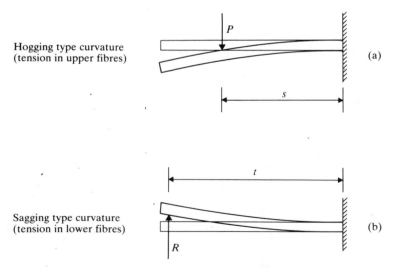

Figure 2.41   The bending of a cantilever is intuitively understood and other situations can be determined from this understanding

In contrast, an upward force such as R in Fig. 2.41b causes a sagging type curvature (tension in the lower fibres) and a bending moment adjacent to the support equal to $+ Rt$.

Once these two simple cases have been understood, other situations can be considered as various combinations of these. For example, for the beam previously considered in the introduction (Fig. 2.42a), suppose we again require the bending

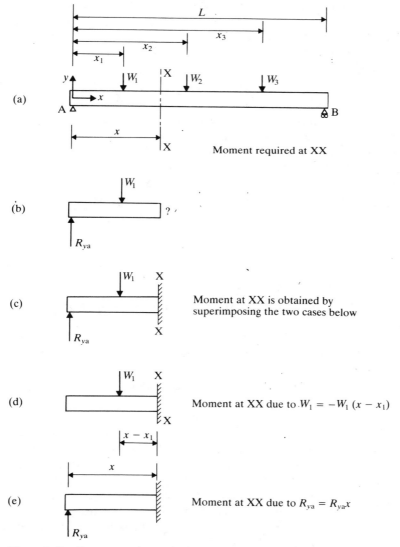

Figure 2.42 Demonstration of the 'imaginary' procedure for determining the bending moment at a section directly

moment at a section between W_1 and W_2, a distance x from the support at A. The free-body to the left of section XX (Fig. 2.42b) can, for purposes of determining the forces at XX, be considered as a cantilever fixed at XX (Fig. 2.42c). This cantilever is subject to the two known forces—the load W_1 and the reaction R_{ya}. The bending moment at XX caused by these two forces can be obtained by considering their effects separately and then superimposing these values. Thus from Fig. 2.42d

bending moment at XX caused by $W_1 = -W_1(x - x_1)$

and from Fig. 2.42e

bending moment at XX caused by $R_{ya} = R_{ya}x$

Superimposing these values:

bending moment at XX $= R_{ya}x - W_1(x - x_1)$

The next step to improve the procedure is to eliminate the need for the separate diagrams of the free-body and the 'cantilevers'. From this stage, it will be assumed that the 'cantilevers' can be visualized in the imagination.

For example, the bending moment at a section between W_2 and W_3 (but again at distance x from support A) would be obtained by visualizing a cantilever subject to the three forces W_1, W_2 and R_{ya}. Thus

$$M_x \atop x_2 \leqslant x \leqslant x_3 = R_{ya}x - W_1(x - x_1) - W_2(x - x_2)$$

Similarly, the bending moment at a section between W_3 and support B (again distance x from support A) would be obtained by visualizing a cantilever subject to the four forces W_1, W_2, W_3 and R_{ya}. Thus

$$M_x \atop x_3 \leqslant x \leqslant L = R_{ya}x - W_1(x - x_1) - W_2(x - x_2) - W_3(x - x_3)$$

It is important to realize that the same value as given by the latter expression would have been obtained by visualizing the cantilever to the right of XX and subject to the single force R_{yb}. Thus

$$M_x \atop x_3 \leqslant x \leqslant L = R_{yb}(L - x)$$

Needless to say, one chooses to consider either forces to the left or forces to the right, depending on which is the more convenient. However, one advantage of consistently considering forces to the left is the facility by which single equations can be adopted to represent the bending moment at any section of the beam. It will be noted that in the expressions for the bending moments within the three successive regions between the various concentrated forces, certain terms are common, but after passing a particular load an additional term appears. As a result, it becomes possible to represent the various equations by a single equation using the mathematical technique of *step functions*. Thus

$$M_x \atop 0 \leqslant x \leqslant L = R_{ya}x - W_1[x - x_1] - W_2[x - x_2] - W_3[x - x_3]$$

where the square bracket term is ignored when it is negative. For example, for a section between W_1 and W_2, both $x - x_2$ and $x - x_3$ would be negative, and the expression therefore reduces to

$$\underset{x_1 \leqslant x \leqslant x_2}{M_x} = R_{ya}x - W_1(x - x_1)$$

At a section between the support A and W_1, $x - x_1$ would also be negative and the expression would become

$$\underset{0 \leqslant x \leqslant x_1}{M_x} = R_{ya}x$$

Such generalized bending moment expressions can also be developed for cases which include distributed loading. For example, for the loading case in Fig. 2.43a:

$$\underset{0 \leqslant x \leqslant L}{M_x} = R_{ya}x - \frac{w}{2}[x - c]^2$$

This should be verified by the student by visualizing the appropriate cantilever beams.

The loading case in Fig. 2.43b is slightly more complicated and it is necessary to translate it into a combination of the two types of loading shown in Fig. 2.43c. Thus the partial distributed load has been extended to the support B, but the additional load had been countered by the same load applied in the upward direction, i.e., the net load over this region is still zero. These two separate loads, whilst clearly equal to the original loading, are now each similar to the loading case in Fig. 2.43a for which a generalized expression could be obtained. Hence for the loading case in Fig. 2.43c, and so also for the case in Fig. 2.43b:

$$\underset{0 \leqslant x \leqslant L}{M_x} = R_{ya}x - \frac{wx^2}{2} + \frac{w}{2}[x - d]^2$$

The more general case in Fig. 2.43d is similarly obtained:

$$\underset{0 \leqslant x \leqslant L}{M_x} = R_{ya}x - \frac{w}{2}[x - c]^2 + \frac{w}{2}[x - d]^2$$

The technique will be illustrated by reference to the beam in Fig. 2.44a. It will be observed that the beam cantilevers beyond the support at B. The first step is to calculate the reactions, and their values are shown in Fig. 2.44b. The bending moment equation for the beam can now be obtained:

$$M_x = 148x - 100[x - 3] - \frac{40}{2}[x - 3]^2 + \frac{40}{2}[x - 7]^2 - 100[x - 7]$$

$$+ 372[x - 10] - \frac{40}{2}[x - 10]^2$$

The student should verify this by starting near the left-hand support and successively consider typical sections XX along the beam at distance x from this support.

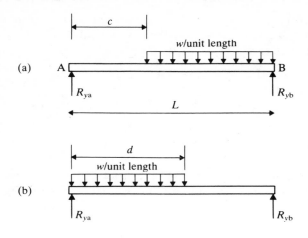

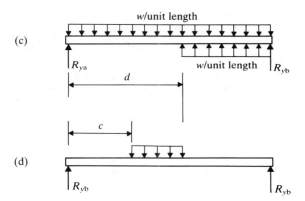

Figure 2.43   Various loading arrangements used to demonstrate the development of generalized bending moment expressions

As will be seen later, the main advantage of writing the equation using the technique of step functions is that it facilitates the determination of deflections when using the differential equation method. It is also advantageous when using the computer to draw bending moment diagrams.

Whilst it could be used for hand methods of obtaining the bending moment diagrams by substituting in suitable values of x, such a procedure would often entail unnecessary arithmetic, and it will generally prove advantageous to revert to separate equations for the purpose of obtaining bending moment values. This is now illustrated.

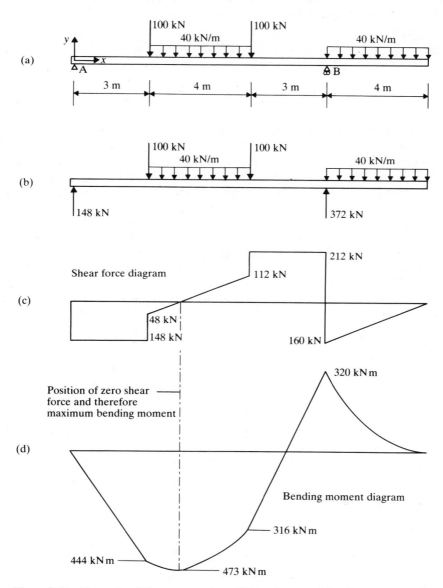

Figure 2.44   Example of the construction of shear force and bending moment diagrams for a non-standard loading case

Example of the bending moment diagram for multiple loads

For the beam in Fig. 2.44b:

$$M_x \atop {0 \leqslant x \leqslant 7} = 148x - 100[x - 3] - \frac{40}{2}[x - 3]^2 \tag{2.7}$$

$$M_x \atop {10 \leqslant x \leqslant 14} = \frac{40}{2}(14 - x)^2 \tag{2.8}$$

The expression in Eq. (2.7) contains square brackets because the region $0 \leqslant x \leqslant 7$ m contains a concentrated load and a discontinuity in the distributed loading. The expression in Eq. (2.8) does not contain square brackets as the region $10 \leqslant x \leqslant 14$ m does not contain such loading discontinuities. Equation (2.8) has only one term (compared with seven of the generalized expression—Eq. (2.6)), and this has been achieved by considering forces to the *right* of the imaginary section within $10 \leqslant x \leqslant 14$ m.

No equation has been presented for the region $7 \leqslant x \leqslant 10$ m. The reason is that since there is no load in this region, the shear force will remain constant and so the bending moment diagram over this region will be a straight line. Hence, knowing the values of the bending moment at the two sections $x = 7$ and $x = 10$ (and these will be known from Eqs (2.7) and (2.8)), the bending moment diagram for this region is determined.

From Eq. (2.7):

$$\text{at} \quad x = 0 \qquad M_x = \quad 0$$
$$x = 3 \text{ m} \qquad M_x = \quad 444 \text{ kN m}$$
$$x = 4 \text{ m} \qquad M_x = \quad 592 - 100 - \quad 20 = 472 \text{ kN m}$$
$$x = 5 \text{ m} \qquad M_x = \quad 740 - 200 - \quad 80 = 460 \text{ kN m}$$
$$x = 6 \text{ m} \qquad M_x = \quad 888 - 300 - 180 = 408 \text{ kN m}$$
$$x = 7 \text{ m} \qquad M_x = 1036 - 400 - 320 = 316 \text{ kN m}$$

From Eq. (2.8):

$$\text{at} \quad x = 10 \qquad M_x = -320 \text{ kN m}$$
$$x = 11 \qquad M_x = -180 \text{ kN m}$$
$$x = 12 \qquad M_x = -80 \text{ kN m}$$
$$x = 13 \qquad M_x = -20 \text{ kN m}$$
$$x = 14 \qquad M_x = \quad 0$$

These values enable the bending moment diagram to be plotted as in Fig. 2.44d. There are several points to note regarding this bending moment diagram:

- in the regions where there is no loading ($0 \leqslant x \leqslant 3$ and $7 \leqslant x \leqslant 10$) the diagram consists of straight lines (compare $\mathrm{d}M/\mathrm{d}x = S$);
- in the regions where there is a uniformly distributed load, the diagram is a parabolic curve;
- the cantilever and the part of the beam between the supports adjacent to the cantilever are subjected to hogging bending (tension in the upper fibres) whilst the remainder is subjected to sagging bending (tension in the lower fibres);
- the precise position and value of the maximum sagging bending moment is not yet known.

In order to determine the maximum sagging bending moment, one utilizes the rule established earlier that, at a position where the shear force is zero, the bending moment diagram is horizontal. The shear force diagram is shown in Fig. 2.44c. It is quickly established (without the use of free-body diagrams) by first considering a section near the left-hand support and gradually working along the beam. It shows that the shear force is zero at a section $3 + 48/40 = 4.2$ m from the left-hand support.

Substituting $x = 4.2$ into Eq. (2.7) gives

$$M_{\max} = 621.6 - 120 - 28.8 = 473 \text{ kN m}$$

It is important to realize that this procedure applies only to the type of maximum shown at $x = 4.2$ m. It does not apply at positions of discontinuity in the shear force diagram as at $x = 10$ m. Such sections may well be the position of the maximum bending moment, but if so this will normally be clear from the diagram.

Example of beams subjected to applied moments

So far, only beams subjected to transverse forces have been considered. A case of loading frequently occurring in practice is that shown in Fig. 2.45a, although it is the form shown in Fig. 2.45b that will be more readily recognized.

To understand the forces that are being applied to the beam by the bracket in Fig. 2.45a, it is advantageous first to study the forces which the beam must apply to the bracket to maintain its equilibrium (Fig. 2.45c).

To prevent translation of the bracket, there must be a counter-balancing force P acting to the right (Fig. 2.45d). This must act on the left face in order that the corresponding equal and opposite force on the beam can be transmitted to the pinned support at A, it being recalled that the roller at B is not capable of resisting horizontal forces.

The twin forces P on the bracket create an anticlockwise moment of value Pd and so to prevent rotation a counter-balancing clockwise moment of Pd must be applied by the beam to the bracket. This moment is applied partly by the left-hand region of the beam to the left-hand face of the bracket and partly by the right-hand region of the beam to the right-hand face, although the relative values of these two partial moments are not known at this stage (Fig. 2.45e).

There must, of course, be equal and opposite forces applied by the bracket to the beam. If the horizontal depth of the bracket is small compared with the length of the

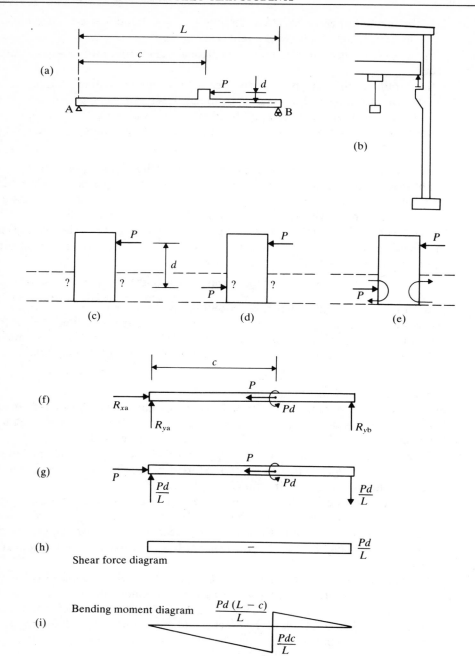

Figure 2.45 Bending moment and shear force diagrams for a beam subjected to an applied moment

beam, we can assume that the moment applied to the beam is concentrated at a single section, and the external forces applied to the beam can be considered to be those shown in Fig. 2.45f. The first step in the analysis is to find the reactions.

Using $\sum P_x = 0$

$$R_{xa} - P = 0$$

$$R_{xa} = P \text{ kN}$$

Using $\sum M_b = 0$

$$Pd - (R_{ya} \times L) = 0$$

$$R_{ya} = \frac{Pd}{L} \text{ kN}$$

Using $\sum P_y = 0$

$$R_{ya} + R_{yb} = 0$$

$$R_{yb} = -\frac{Pd}{L} \text{ kN}$$

The external forces being applied to the beam are therefore as shown in Fig. 2.45g. It will be observed that the two vertical reactive forces form a clockwise (as viewed directly) couple of moment Pd which resists the anticlockwise applied moment Pd. It is important to realize that these reactive forces could have been determined more directly by considering the entire structure shown in Fig. 2.45a (and students should verify this), but the use of the free-body bracket is valuable at this stage when studying the internal forces in the beam.

The shear force diagram for the beam is the rectangle shown in Fig. 2.45h. This is quickly obtained by considering a section adjacent to the support A and then successive sections along the length of the beam.

The bending moment diagram is determined by obtaining the bending moment expressions for the two regions of the beam. Thus, by considering forces to the left of the bracket:

$$\underset{0 \leqslant x \leqslant c}{M_x} = \frac{Pd}{L} x$$

and by considering forces to the right:

$$\underset{c \leqslant x \leqslant L}{M_x} = -\frac{Pd}{L} (L - x)$$

For the left-hand region of the beam:

$$\text{at} \quad x = 0 \quad M_x = 0$$

$$x = c \quad M_x = \frac{Pdc}{L}$$

For the right-hand region of the beam:

$$\text{at} \quad x = c \quad M_x = -\frac{Pd}{L}(L - c)$$

$$x = L \quad M_x = 0$$

The resulting bending moment diagram is therefore as shown in Fig. 2.45i. As one passes the section at which the moment is applied to the beam, there is a jump in the diagram of total value:

$$\frac{Pdc}{L} + \frac{Pd(L-c)}{L} = Pd$$

Such a discontinuity in the diagram equal to the applied moment would have been more apparent if, when considering the region of the beam defined by $c \leqslant x \leqslant L$, we had obtained the bending moment equation from a study of the forces to the left of the section. For example:

$$\underset{c \leqslant x \leqslant L}{M_x} = \frac{Pd}{L} x - Pd$$

This equation is obtained by reference to the 'imaginary' cantilever shown in Fig. 2.46. It is known intuitively that the concentrated force Pd/L causes tension in the bottom fibres and therefore a *positive* bending moment $(Pd/L)x$ adjacent to the 'support'. The applied moment Pd causes a bending moment of Pd adjacent to the 'support', irrespective of the distance between the two. The sign of this bending moment is probably best obtained by comparison with the effect of the concentrated load—the concentrated load is tending to rotate the 'cantilever' clockwise (as viewed directly) whilst causing a positive bending moment; the applied moment tends to rotate the 'cantilever' anticlockwise and therefore the sign of its resulting bending moment adjacent to the 'support' must be *negative*.

It is of interest here to note how the step function is developed to incorporate the effect of an applied moment:

$$M_x = \frac{Pd}{L} x - Pd[x - c]^\circ$$

For values of $x \leqslant c$, the squared bracket becomes negative and is therefore ignored. For values of $x \geqslant c$, the squared bracket to the power zero has a value of unity and so the bending moment value of $-Pd$ will apply over the range $c \leqslant x \leqslant L$ in addition to the term $(Pd/L)x$.

If a moment is applied at the end of the beam, as in Fig. 2.47a for example, the bending moment diagram becomes the triangle shown in Fig. 2.47b. The sign of the

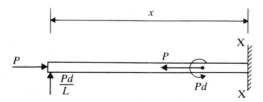

Figure 2.46 'Imaginary' cantilever used to develop the bending moment expression for the beam in Fig. 2.45g

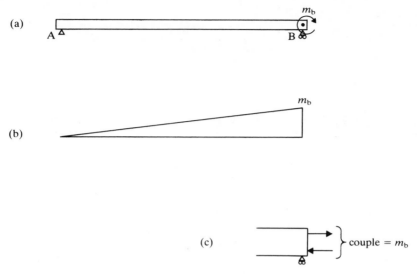

Figure 2.47   Bending moments in a beam subjected to a moment applied at the end of the beam

bending moment for this case is easily determined by remembering that the applied moment is in truth a combination of twin forces (a couple) corresponding in value and direction to the applied moment (Fig. 2.47c). It then follows that for the case shown in Fig. 2.47a, the top fibres are stressed in tension and the bottom fibres in compression. Such moments occur frequently in practice whenever a beam is continuous past a support (compare the beam in Fig. 2.44a).

The construction of bending moment diagrams using the principle of superposition

Reference has been made earlier to the use of tabulated solutions for bending moment diagrams when the loading on a beam falls within a category of standard cases. In practice, beams are frequently subjected to a combination of loads which separately come within the standard cases but together do not. Rather than analyse such a beam from first principles, it may well prove preferable to obtain the tabulated solutions for the separate loads and then to superimpose these to give the overall effect.

A type of combined loading commonly occurring in practice is shown in Fig. 2.48a—a uniformly distributed load over the full length of the beam and a moment applied at the end of the beam. The bending moment diagram for the uniformly distributed load (Fig. 2.48b) is shown in Fig. 2.48c; the bending moment diagram for the end moment (Fig. 2.48d) is shown in Fig. 2.48e. In combining these diagrams, it must be noted that they are of opposite sign—positive and so below the line in the case of the uniformly distributed load, negative and so above the line in the case of the end moment. The diagrams are therefore combined in such a way that where they overlap they cancel each other out.

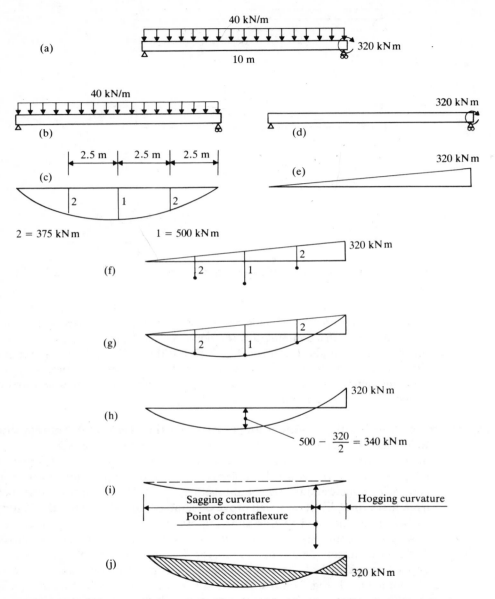

Figure 2.48 Illustrating the principle of superposition for the construction of bending moment diagrams

The construction of the combined diagram is shown in Figs 2.48f and 2.48g. Starting with the triangle for the end moment, the values of the ordinates of the parabola of the distributed load are plotted vertically from the inclined line of the triangle (Fig. 2.48f) to give the combined diagram (Fig. 2.48g). The common areas now cancel one another out, the resulting bending moment diagram for the combined loading being that in Fig. 2.48h. There are several important points to note:

- the maximum sagging bending moment no longer occurs at the mid-span position (the position of zero shear force which determines the position of maximum sagging bending moment is given by $40x = R_{ya}$);
- part of the beam is subjected to sagging bending and part to hogging bending (giving rise to the type of deflected form in Fig. 2.48i—note that the sagging bending changes to hogging bending at the section where the bending moment changes sign, termed the point of contraflexure.)

An alternative method of combining such bending moment diagrams is shown in Fig. 2.48j. The triangle and parabola are drawn using the same base line but on the same side of the line. Again, the common areas cancel out and the the bending moment diagram for the combined loading is that shown shaded. Although its shape seems slightly different from that in Fig. 2.48h, the vertical ordinates at a particular section of the beam are identical, a point which students should verify for themselves. There are clear advantages in such a construction of the combined diagram and many structural engineers prefer it. It has the slight disadvantage for students in that both the sagging and hogging bending moments are on the same side of the original horizontal base line, with the inclined line becoming, in effect, the new base line. In using the procedure, students must remember that the bending moment at a particular section of the beam is given by the *vertical* ordinate of the diagram measured from the inclined line.

Separation of a structure into its components for the construction of bending moment diagrams

The type of structure shown in Fig. 2.49a is frequently used in the design of bridges. The superstructure is supported at the abutments A and F and the piers B and E. The central section CD is simply supported on brackets at the ends of the beams BC and DE which cantilever from the 'holding down' spans of AB and EF. The latter are so called because they must be heavy enough to prevent failure by rotating about the piers, it usually being 'structurally inconvenient' to tie these outer spans down to the abutments.

As a result of the simply supported central section, the structure is statically determinate. This will be clearer from the diagrammatic representation of the bridge in Fig. 2.49b. The total number of reactive forces (including both vertical and horizontal loading effects) is five—two at the pin support at A, and one from each of the rollers at B, E and F. Although there are only three equations of equilibrium from the consideration of the structure as a whole, two additional equations are available from known internal conditions. The nature of the supports to the central section at C

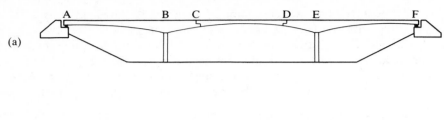

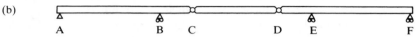

Figure 2.49 Diagrammatic representation of the superstructure of a typical 3-span bridge

and D are such that no bending moment can exist there. Hence, by considering the forces acting to one side at each of these positions, two additional equilibrium equations can be obtained. With the total of five equations, the five reactive forces can be obtained. The structure is therefore statically determinate and it is possible to determine the bending moment diagram using only the principles of equilibrium. The structure and loading in Fig. 2.50a will be used to illustrate the method of drawing bending moment diagrams by splitting the structure up into its components.

The method of splitting up this particular structure is shown in Figs 2.50b, 2.50c and 2.50d. The beam CD can be regarded as simply supported at C and D (Fig. 2.50b), the supporting forces R at C and D coming from the cantilevers BC and DE. There must, therefore, be equal downward forces acting on the cantilevers from the beam CD (Fig. 2.50c).

The cantilever BC is given its support by the holding-down span AB. This will require an upward force and an anticlockwise moment from AB to BC. (Although the representation of the support to the cantilever in Fig. 2.50c is that earlier described as 'fixed', the support actually provided by span AB does allow some rotation of the cantilever at the position B. However, this has no effect on the forces applied from AB to BC.) Action and reaction being equal and opposite, there will be a downward force and a clockwise moment from BC acting on AB at B (Fig. 2.50d).

The loading on CD comes within the standard cases of loading and the bending moment diagram is therefore the parabola in Fig. 2.50e.

Because of symmetry, the supporting forces R are each equal to half the load on CD, namely 50 × 16 = 800 kN. The forces acting on the cantilever BC are therefore those in Fig. 2.50f. Each of the two types of loading is a standard case, and the bending moment diagram for the combined loading case can be obtained by superimposing the triangle for the concentrated load and the parabola for the distributed load to give Fig. 2.50g. In this case, the two separate bending moment diagrams have the same sign and so their ordinates are simply added together.

The reactions at the 'support' to BC (Fig. 2.50h) are:

– an upward force of 800 + (50 × 10) = 1300 kN;
– an anticlockwise moment of 10 500 kN m.

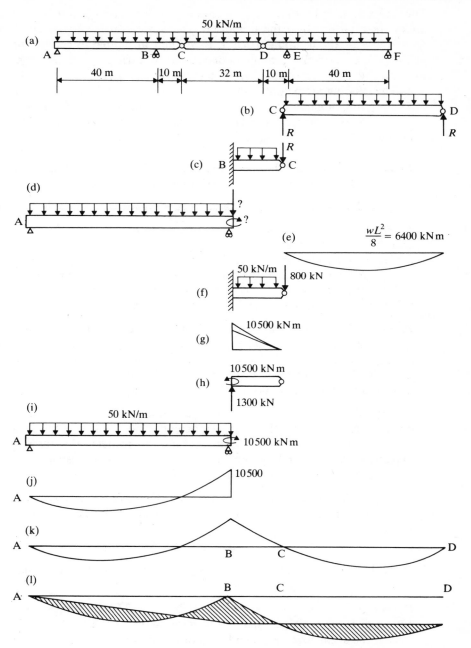

Figure 2.50   Illustrating the separation of a structure into its components for the construction of bending moment diagrams

The forces applied by BC to AB are equal and opposite. The section at which these forces are applied is over the support B. The concentrated force of 1300 kN will not therefore affect the bending of AB as it will be transmitted directly to the pier at B. Thus, as far as the bending moment diagram is concerned, the forces acting on AB are those in Fig. 2.50i. They are similar to those previously discussed in relation to Fig. 2.48. The bending moment diagram is similarly constructed to give that in Fig. 2.50j.

The bending moment diagrams for the separate components are now combined to give that for the complete superstructure in Fig. 2.50k. The alternative method of construction is shown in Fig. 2.50l.

Problem 2.21   Obtain the shear force and bending moment diagram for the beam in Fig. 2.51a and determine the maximum value of the bending moment.

Problem 2.22   Calculate the maximum bending moment in the beam in Fig. 2.48.

Problem 2.23   The simply supported beam in Fig. 2.51b is loaded through the bracket. Calculate the reactive forces directly and use these to obtain the shear force and bending moment diagrams. Indicate on separate diagrams: (a) the forces acting on the bracket, (b) the forces acting on the beam. Write down the single equation for bending moment at a section distance x from the left-hand support and use this to check the bending moment diagram.

Problem 2.24   Obtain the shear force and bending moment diagrams for the beam in Fig. 2.51c. Calculate the bending moment at the position on the beam corresponding to the 40 kN load for the following three cases of loading: (a) the 20 kN load alone, (b) the 40 kN load alone and (c) the 10 kN/m load alone, and then use the principle of superposition to check the value obtained for Fig. 2.51c.

Problem 2.25   Obtain the shear force diagram for the beam in Fig. 2.51d. Write down the expression for the bending moment at any position of the beam distance x from the left-hand end of the beam, and use this to determine the position of the maximum sagging bending moment given that its position is within the length of the distributed load. Verify the answer from the shear force diagram.

Problem 2.26   The designer of the beam in Fig. 2.52a intends to position the supports symmetrically so that the maximum bending moment is as low as possible. Show that this is when

$$a = (\sqrt{(2)} - 1)L$$

Draw the shear force and bending moment diagrams for this case and calculate the positions of zero bending moment.

Problem 2.27   The two-span beam ABC in Fig. 2.52b is statically indeterminate but an analysis indicates that the reaction at B for the loads shown is 180 kN. Obtain the shear force and bending moment diagrams.

Problem 2.28   A simply supported beam of span L is symmetrically loaded with a load that increases linearly from zero at the supports to a maximum of w per unit

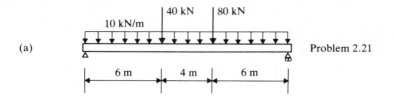

(a)   Problem 2.21

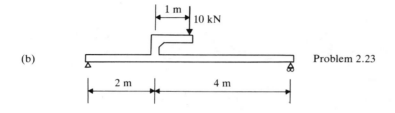

(b)   Problem 2.23

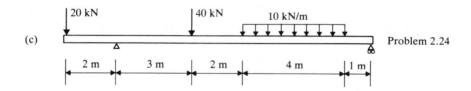

(c)   Problem 2.24

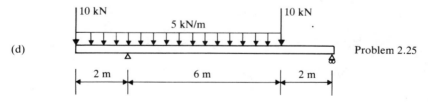

(d)   Problem 2.25

Figure 2.51   Beams used in Problems 2.21 to 2.25

length at mid-span. Obtain expressions for the shear force and bending moment at a section in the left-hand half of the beam, and sketch the shear force and bending moment diagrams. An upward force is now applied at mid-span and increased until the bending moment at that position is zero. Find the new positions and magnitude of the maximum bending moment.

Problem 2.29   The structural form of a flight of reinforced concrete stairs is diagrammatically represented by ABCDE in Fig. 2.52c where DE is a landing

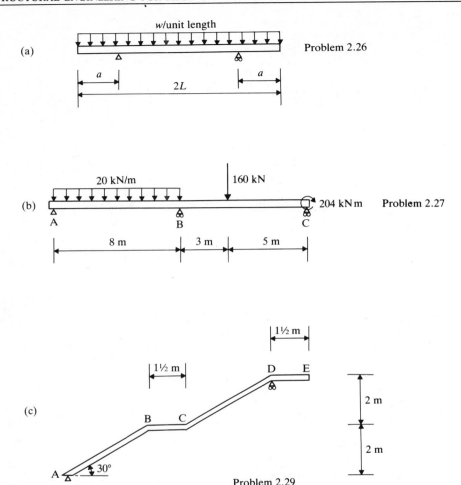

Figure 2.52   Beams used in Problems 2.26 to 2.29

cantilevering beyond the support at D. Assuming the dead load is 6 kN per metre run of stairs, obtain for this load the bending moments at A, B, C, D and E, and also the shear force diagram for the part ABC.

State the effect (a) an imposed load on BC would have on the bending moments in DE and (b) an imposed load on DE would have on the bending moments in BC. Determine whether equal and simultaneous distributed loading on BC and DE would cause the bending moment at B to increase or decrease. Sketch how you would arrange the reinforcement in the concrete.

2.7 Beams—calculation of displacements

The comprehensive term 'displacement' is used to indicate that the movement to be calculated includes *rotation* as well as *deflection*.

Introduction

We know instinctively that a load applied to a beam causes it to deflect. This results largely from the compressive and tensile stresses set up as a result of the bending moments in a beam. For example, the bending moment acting on the element of the beam illustrated in Fig. 2.53 causes compressive stresses in the upper longitudinal

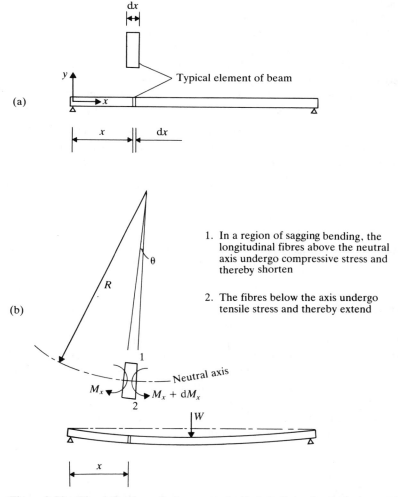

(a)

(b)

1. In a region of sagging bending, the longitudinal fibres above the neutral axis undergo compressive stress and thereby shorten

2. The fibres below the axis undergo tensile stress and thereby extend

Figure 2.53 The deflection of a beam results mainly from the extension and shortening of the longitudinal fibres caused by the bending stresses

fibres of the element and tensile stresses in the lower fibres. This causes the shortening of the upper fibres and extension of the lower fibres, resulting in the deformation of the element to a shape which can be said to be part of a circular curve. The total displacements of the beam result from the combination of deformation of all such elements in the beam.

It will be clear that, as the bending moment varies along the beam, so the deformation of each such element will be different from its neighbours. Indeed, as the stresses will be proportional to the bending moment, so the amount of deformation of an element will be proportional to the bending moment on that element.

Obviously the amount of deformation of an element of a beam will also be dependent on the material of the beam. In this case, the amount of deformation is inversely proportional to the modulus of elasticity of the material.

It will also be apparent that the size and shape of the cross-section of the beam will affect displacements of the beam, as this influences directly the stresses caused in the beam by the bending moments. It will be shown in Chapter 3 that such stresses are inversely proportional to a property of the cross-section termed the second moment of area. An introduction to this property is discussed further below.

To summarize these points, curvature $(1/R)$ of the arc of a deformed element is

- proportional to the bending moment (M) acting on the element;
- inversely proportional to the modulus of elasticity (E) of the material;
- inversely proportional to the second moment of area (I) of the section of the beam.

It will be shown in Chapter 3 that the curvature of a deformed element is given by

$$\frac{1}{R} = \frac{M}{EI}$$

At this stage, we will assume this relationship without proof and use it to develop methods for calculating displacements of beams.

A note on second moment of area

In the introductory discussion on the bending of beams in Chapter 1 (Section 1.3), an intuitive consideration indicated that the longitudinal strain (and therefore longitudinal stress) of a particular fibre was proportional to its distance y from the neutral axis. It is also clear that the moment of the force in the fibre about the neutral axis will be proportional to y. It follows that y^2 will be an important factor in the properties of a beam. This will be proved formally in Chapter 3 where it is shown that stresses at any position in the section of the beam can be related to its second moment of area (I), defined by reference to Fig. 2.54a as

$$I = \int y^2 \, dA$$

where y is the distance from the neutral axis of the element of area dA. The integral is carried out over the whole of the cross-sectional area of the beam.

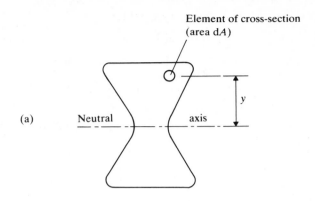

(a)

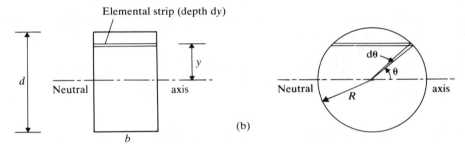

Figure 2.54   An introduction to second moments of area

For standard cross-sectional shapes simple expressions can be developed. For example, for the rectangular section in Fig. 2.54b the neutral axis will be at the mid-depth when the section is subjected to a bending moment in the x–y plane. The second moment of area of the elemental strip parallel to the neutral axis is

$$dI = y^2 \times \text{area of element}$$

$$= y^2 b \, dy$$

Thus
$$I = \int_{-d/2}^{d/2} y^2 b \, dy = \left(\frac{by^3}{3}\right)_{-d/2}^{d/2} = \frac{bd^3}{12}$$

For the circular cross-section in Fig. 2.54b, the position and size of the elemental strip is defined by θ and $d\theta$.

Width of elemental strip	$= 2R \cos \theta$
Thickness of elemental strip	$= R \cos \theta \, d\theta$

$$\text{Area of elemental strip} = 2R^2 \cos^2 \theta \, d\theta$$

$$\text{Distance of strip from neutral axis} = R \sin \theta$$

$$dI = (R \sin \theta)^2 \times \text{area}$$

$$= 2R^4 \sin^2 \theta \cos^2 \theta \, d\theta$$

$$I = R^4 \int_{-\pi/2}^{\pi/2} 2 \sin^2 \theta \cos^2 \theta \, d\theta$$

$$= \frac{R^4}{4} \left(\theta - \frac{\sin 4\theta}{4} \right)_{-\pi/2}^{\pi/2}$$

$$= \frac{\pi R^4}{4}$$

These sections commonly occur in practice and the expressions for their second moments of area must be remembered. The method of calculation for more general cases is dealt with in Chapter 3.

The units of mm⁴ or m⁴ will normally be adopted in calculations, although values for the standard steel sections (Fig. 4.6) are usually quoted in cm⁴.

Displacements using the differential equation of bending

The displacements of a beam caused by an applied load (Fig. 2.55a) can be defined by relation to the direction of the x and y axes (Fig. 2.55b). The symbol v will be used to denote the deflection at a particular point of the beam defined by the distance x from the assumed origin, the positive values of v being in the positive direction of the y-axis. The slope of the beam at a point is determined by the value of dv/dx.

For any given curve related by v and x, the precise relationship for the radius of

(a)

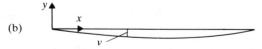

(b)

Figure 2.55 The displacements of a beam caused by loading are defined by relation to its original shape. If it is perfectly straight in the unloaded state, this will coincide with the x-axis

curvature at any point on the curve is given by

$$\frac{1}{R} = \frac{\dfrac{d^2v}{dx^2}}{\left[1 + \left(\dfrac{dv}{dx}\right)^2\right]^{3/2}}$$

However, for the type of structural member we shall be concerned with, v will always be small and so also will dv/dx. As a result, the value of $(dv/dx)^2$ will be very small in comparison with unity and can be neglected. The equation for curvature thereby reduces to

$$\frac{1}{R} = \frac{d^2v}{dx^2}$$

Since we are accepting the relationship $\dfrac{1}{R} = \dfrac{M_x}{EI}$

it follows that $\dfrac{d^2v}{dx^2} = \dfrac{M_x}{EI}$

This is known as the differential equation of bending and applies to all points along the beam. By integrating it once, values of dv/dx (i.e., slopes) can be obtained; by integrating it twice, values of v (i.e., deflection) can be obtained.

For many cases, both E and I will be constant throughout the length of the beam, and for these cases the differential equation is written as

$$EI\,\frac{d^2v}{dx^2} = M_x$$

EXAMPLE 2.4
For the cantilever in Fig. 2.56a:

$$M_x = -Wx$$

E and I constant throughout beam

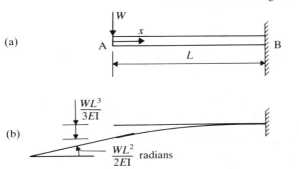

Figure 2.56 Displacements of a cantilever subjected to a concentrated load

Thus
$$EI\frac{d^2v}{dx^2} = -Wx$$

Integrating
$$EI\frac{dv}{dx} = -\frac{Wx^2}{2} + C \qquad (2.9)$$

where C is a constant of integration. In order to find the value of C, it is necessary to substitute a known condition. As this is a cantilever, we know that there can be no rotation at the fixed support B. In other words, when $x = L$, $dv/dx = 0$. Substituting this in Eq. (2.9):

$$0 = -\frac{WL^2}{2} + C$$

$$C = \frac{WL^2}{2}$$

Hence
$$EI\frac{dv}{dx} = -\frac{Wx^2}{2} + \frac{WL^2}{2} \qquad (2.10)$$

From this equation, the slope of the deflected beam at any point along the beam can be determined. At the free-end of the beam (at $x = 0$), for example

$$\left(\frac{dv}{dx}\right)_{x=0} = \frac{WL^2}{2EI}$$

Integrating Eq. (2.10) gives

$$EIv = -\frac{Wx^3}{6} + \frac{WL^2x}{2} + D \qquad (2.11)$$

To find the value of the constant of integration D, the known condition is that $v = 0$ when $x = L$. Substituting this in Eq. (2.11) gives

$$D = \frac{WL^3}{6} - \frac{WL^3}{2} = -\frac{WL^3}{3}$$

Hence
$$EIv = -\frac{W}{6}x^3 + \frac{WL^2}{2}x - \frac{WL^3}{3}$$

From this equation, the deflected form of the beam can be obtained (Fig. 2.56b). The deflection will be a maximum at the free-end of the beam:

$$v_{max} = -\frac{WL^3}{3EI} \qquad \frac{kN\ mm^3}{kN/mm^2\ mm^4} = mm$$

The negative sign simply indicates that it is in the opposite direction to the positive direction of the y-axis—in other words the deflection is downward, as we would expect. (Whilst it is quite permissible to arrange for the y direction to be downward in order to obtain a positive deflection, this would alter the sign of the bending moments from that adopted hitherto and is probably not advisable at this stage.)

EXAMPLE 2.5

For the simply supported beam in Fig. 2.57a subjected to a uniformly distributed load, the vertical reactive forces are each $wL/2$. Thus

$$M_x = \frac{wL}{2} x - \frac{w}{2} x^2$$

and so

$$EI \frac{d^2v}{dx^2} = \frac{wL}{2} x - \frac{w}{2} x^2$$

Integrating

$$EI \frac{dv}{dx} = \frac{wL}{4} x^2 - \frac{w}{6} x^3 + C \tag{2.12}$$

It is intuitively known that the deflected form of the beam will be as shown in Fig. 2.57b. From the symmetry of the problem, it is known that at mid-span ($x = L/2$) the slope of the beam is zero. Substituting this known condition into Eq. (2.12) enables the constant of integration C to be determined:

$$C = -\frac{wL^3}{24}$$

Hence

$$EI \frac{dv}{dx} = \frac{wL}{4} x^2 - \frac{w}{6} x^3 - \frac{wL^3}{24}$$

Integrating

$$EIv = \frac{wL}{12} x^3 - \frac{w}{24} x^4 - \frac{wL^3}{24} x + D \tag{2.13}$$

The deflection is zero when $x = 0$. Substituting this known condition into Eq. (2.13) gives $D = 0$, and the equation for the deflected form of the beam is

$$EIv = \frac{wL}{12} x^3 - \frac{w}{24} x^4 - \frac{wL^3}{24} x$$

(a)

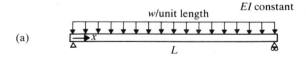

w/unit length

EI constant

L

(b)

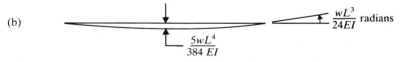

$\frac{5wL^4}{384\,EI}$

$\frac{wL^3}{24EI}$ radians

Figure 2.57   Displacements of a simply supported beam subjected to a uniformly distributed load

It is clear that the maximum deflection occurs at mid-span. Substituting $x = L/2$ gives

$$v_{max} = -\frac{5wL^4}{384EI}$$

EXAMPLE 2.6

For the simply supported beam subjected to a concentrated load (Fig. 2.58a), the single equation representing the bending moment at any point along the beam requires the use of step functions:

$$M_x = \frac{(L-a)}{L} Wx - W[x-a]$$

Thus

$$EI \frac{d^2v}{dx^2} = \frac{(L-a)}{L} Wx - W[x-a] \tag{2.14}$$

It will be recalled that when the squared bracket term is negative, it is ignored.

For purposes of integration, these squared bracket terms are retained throughout and integrated as discrete terms. For example, the integration of $[x-a]$ is with respect to $(x-a)$ and so becomes $[x-a]^2/2$.

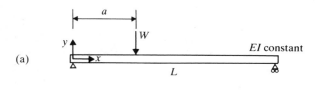

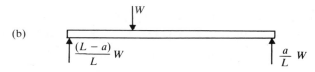

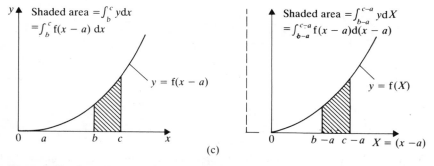

Figure 2.58 Introduction to the integration of equations involving step functions

This procedure is simply equivalent to the transfer of the position of the y-axis for the purpose of the integration (see Fig. 2.58c). Its value lies in the fact that the squared brackets can be retained throughout, thereby clarifying when they need to be ignored.

Integrating Eq. (2.14) gives

$$EI\frac{dv}{dx} = \frac{(L-a)W}{2L}x^2 - \frac{W}{2}[x-a]^2 + C \qquad (2.15)$$

For this asymmetrical case there is no known value of dv/dx and so the constant of integration cannot be determined at this stage. Integrating Eq. (2.15):

$$EIv = \frac{(L-a)W}{6L}x^3 - \frac{W}{6}[x-a]^3 + Cx + D$$

When $x = 0$, $v = 0$, thus $D = 0$

When $x = L$, $v = 0$, thus $C = \frac{W}{6L}(L-a)^3 - \frac{(L-a)WL}{6}$

The complete equation for the deflected form of the beam is therefore

$$EIv = \frac{(L-a)W}{6L}x^3 - \frac{W}{6}[x-a]^3 + \frac{W(L-a)^3}{6L}x - \frac{(L-a)WL}{6}x$$

It is important to realize that the maximum deflection does not necessarily occur at the position of the load, nor at mid-span. The position of the maximum deflection for a particular case can be found by determining the value of x to give $dv/dx = 0$, i.e., the position at which the tangent to the beam is horizontal. This value of x is then substituted into the above equation to obtain the maximum deflection.

If the concentrated load is at mid-span ($a = L/2$), the maximum deflection now occurs at the position of the load and has a value

$$v_{\mathrm{max}} = \frac{WL^3}{48EI}$$

Displacements using the method of strain energy

Although, as with pin-jointed frames, the use of strain energy for the calculation of displacements of beams has severe limitations and has little application in practice, again it is a useful introduction to the powerful unit load (virtual work) method.

If a moment m moves through a small angle $d\phi$, the moment remaining constant during this rotation, the work done by the moment is $m\,d\phi$. This is illustrated in Fig. 2.59a.

If a moment increases gradually from zero to M (Fig. 2.59b) and during this application a total rotation of ϕ occurs, the rotation at any stage being proportional to the applied moment, the work done is $\frac{1}{2}M\phi$. The argument for the $\frac{1}{2}$ is identical to the previous discussion in relation to Fig. 2.17.

It follows that the strain energy in the element of the beam shown in Fig. 2.53b is

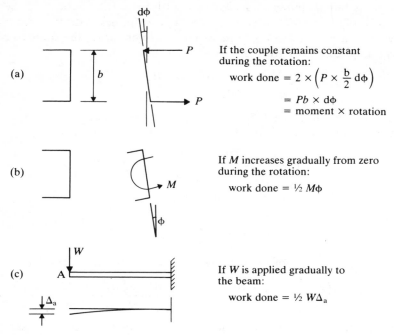

(a)

If the couple remains constant during the rotation:

$$\text{work done} = 2 \times \left(P \times \frac{b}{2} \, \mathrm{d}\phi \right)$$
$$= Pb \times \mathrm{d}\phi$$
$$= \text{moment} \times \text{rotation}$$

(b)

If M increases gradually from zero during the rotation:

$$\text{work done} = \tfrac{1}{2} \, M\phi$$

(c)

If W is applied gradually to the beam:

$$\text{work done} = \tfrac{1}{2} \, W\Delta_a$$

Figure 2.59   Introduction to the method of strain energy

$\frac{1}{2}M_x\theta$. (Although, because of the deflection of the beam, the element undergoes a rotation *in addition* to its deformation, such rotation has no effect on the work done on the element—the work done by the clockwise and anticlockwise moments on the two faces of the element due to this rotation cancel out. The net work done is related to the *difference* in the angles of rotation of the two faces, namely θ.)

As
$$\theta = \frac{\mathrm{d}x}{R}$$

and as previously assumed

$$\frac{1}{R} = \frac{M_x}{EI}$$

the strain energy stored in the element $= \dfrac{M_x\theta}{2} = \dfrac{M_x^2}{2EI} \, \mathrm{d}x$

The moment M_x will normally vary along the length of the beam and so for the total strain energy in a beam a summation procedure is necessary:

$$\text{strain energy in the beam} = \int_0^L \frac{M_x^2}{2EI} \, \mathrm{d}x$$

(In contrast, the strain energy stored in a single member subjected to an axial load is simply $F^2L/2AE$.)

Often, both E and I are constant along the length of the beam and the expression would become:

$$\text{strain energy in a beam} = \frac{1}{2EI} \int_0^L M_x^2 \, dx$$

EXAMPLE 2.7
Referring to the cantilever beam in Fig. 2.56a, at an element distance x from the end of the beam:

$$M_x = -Wx$$

$$\text{Strain energy in the beam} = \frac{1}{2EI} \int_0^L W^2 x^2 \, dx$$

$$= \frac{1}{2EI} \left(\frac{W^2 L^3}{3} \right)$$

$$\text{Work done by the load} = \frac{W\Delta_a}{2}$$

As the work done by the load equals the strain energy stored in the beam:

$$\frac{W\Delta_a}{2} = \frac{1}{2EI} \left(\frac{W^2 L^3}{3} \right)$$

$$\Delta_a = \frac{WL^3}{3EI}$$

this deflection being in the direction of W.

Although the work involved in obtaining this answer was less than that involved in obtaining the same answer using the differential equation, the limitations of the method should again be obvious. It can only be used when the beam is subjected to a single concentrated load and when the deflection required is at the position of the load and is in the same direction of the load. Further examples are not therefore warranted.

Displacements using the unit load method

The basis of the unit load method for calculating displacements of beams is similar to that for calculating deflections of pin-jointed frames. Nevertheless, it is desirable to repeat the fundamentals of the method prior to giving examples showing the procedure.

The deflected form of a beam caused by the load P_0 is shown in Fig. 2.60a. The deflection at A, the position of the load, is Δ_{a0}. Suppose, however, we need to

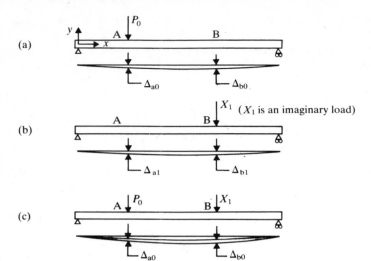

Figure 2.60   Introduction to the basis of virtual work

determine the the deflection Δ_{b0} at the position B. By the law of the conservation of energy we know that

$$\frac{P_0 \Delta_{a0}}{2} = \int_0^L \frac{M_{x0}^2}{2EI}\,dx$$

where M_{x0} is the bending moment on an element at a distance x along the beam caused by the load P_0.

Suppose that, prior to applying the P_0 load, a load X_1 is applied at B causing the deflections shown in Fig. 2.60b—although these deflections will not be relevant to the analysis. Let the bending moment on the typical element distance x along the beam be M_{x1}.

The original load P_0 is now gradually applied to the beam (Fig. 2.60c) to cause additional deflections which are, of course, identical to those shown in Fig. 2.60a. The *additional* rotational deformation of the typical element *during the application of load P_0* is $(M_{x0}/EI)\,dx$.

During the application of P_0, the bending moment M_{x1} on the element due to X_1 remains constant, whilst the additional bending moment due to P_0 increases from zero to M_{x0}. Thus

$$dU = \frac{M_{x0}^2}{2EI}\,dx + \frac{M_{x0}M_{x1}}{EI}\,dx$$

where dU is the additional strain energy stored in the element during the application of the load P_0. Hence

$$U = \int_0^L \frac{M_{x0}^2}{2EI}\,dx + \int_0^L \frac{M_{x0}M_{x1}}{EI}\,dx$$

where U is the additional strain energy stored in the beam during the application of the load P_0.

As the work done by the loads P_0 and X_1 during the application of the load P_0 is $P_0\Delta_{a0}/2 + X_1\Delta_{b0}$, then:

$$\frac{P_0\Delta_{a0}}{2} + X_1\Delta_{b0} = \int_0^L \frac{M_{x0}^2}{2EI}\,dx + \int_0^L \frac{M_{x0}M_{x1}}{EI}\,dx$$

As

$$\frac{P_0\Delta_{a0}}{2} = \int_0^L \frac{M_{x0}^2}{2EI}\,dx$$

$$X_1\Delta_{b0} = \int_0^L \frac{M_{x0}M_{x1}}{EI}\,dx$$

Since M_{x1} is the result of X_1 and directly proportional to it, it is possible to assume that the value of X_1 is unity. The expression can then be simplified to

$$\Delta_1 = \int_0^L \frac{M_0 m_1}{EI}\,dx$$

where Δ_1 is the displacement at the required position and corresponding to the imaginary force X_1,

   M_0 is the bending moment on an element of the beam due to the original load,

   m_1 is the bending moment on the same element due to $X_1 = 1$ acting at the position at which the displacement is required and in the required direction.

EXAMPLE 2.8

The deflection is required for the free-end of the cantilever shown in Fig. 2.61a. For the original loading, the bending moment M_0 on an element distance x from the free-end (Fig. 2.61b) is

$$M_0 = -\frac{wx^2}{2}$$

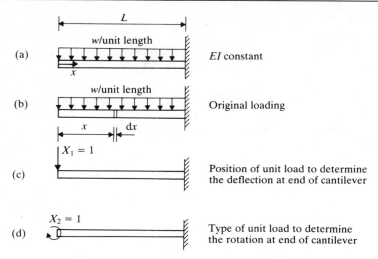

Figure 2.61   Examples of the use of the unit load method for determining the displacements of a cantilever

For the imaginary load $X_1 = 1$ placed at the position where the deflection is required (Fig. 2.61c), the bending moment m_1 on the same element is

$$m_1 = -x$$

Thus

$$\Delta_1 = \frac{1}{EI} \int_0^L M_0 m_1 \, dx$$

$$= \frac{1}{EI} \int_0^L \frac{wx^3}{2} \, dx$$

$$= \frac{wL^4}{8EI}$$

The positive value indicates that it is in the same direction as that assumed for X_1.

EXAMPLE 2.9

If the required displacement is a rotation, the unit load is then a moment. For example, if the rotation at the end of the cantilever beam in Fig. 2.61a is required, the unit load $X_2 = 1$ is applied as shown in Fig. 2.61d.

As before

$$M_0 = -\frac{wx^2}{2}$$

From Fig. 2.61d

$$m_2 = 1$$

Thus
$$\theta = \frac{1}{EI} \int_0^L -\frac{wx^2}{2}\, \mathrm{d}x$$

$$= -\frac{wL^3}{6EI}$$

The negative value indicates that the rotation is in the opposite direction to X_2—whereas X_2 was assumed to be clockwise, the rotation of the end of the beam under the uniformly distributed loading will clearly be anticlockwise.

EXAMPLE 2.10
If the mid-span deflection of the simply supported beam in Fig. 2.62a is required, then:

from Fig. 2.62b
$$M_0 = \frac{wL}{2} x - \frac{w}{2} x^2$$

from Fig. 2.62c
$$m_1 \atop {0 \leqslant x \leqslant L/2} = \frac{x}{2}$$

Acknowledging symmetry:
$$\Delta_1 = \frac{2}{EI} \int_0^{L/2} \left(\frac{wL}{4} x^2 - \frac{w}{4} x^3 \right) \mathrm{d}x$$

$$= \frac{1}{2EI} \left(\frac{wL}{3} x^3 - \frac{w}{4} x^4 \right)_{x=0}^{x=L/2}$$

$$= \frac{5wL^4}{384EI}$$

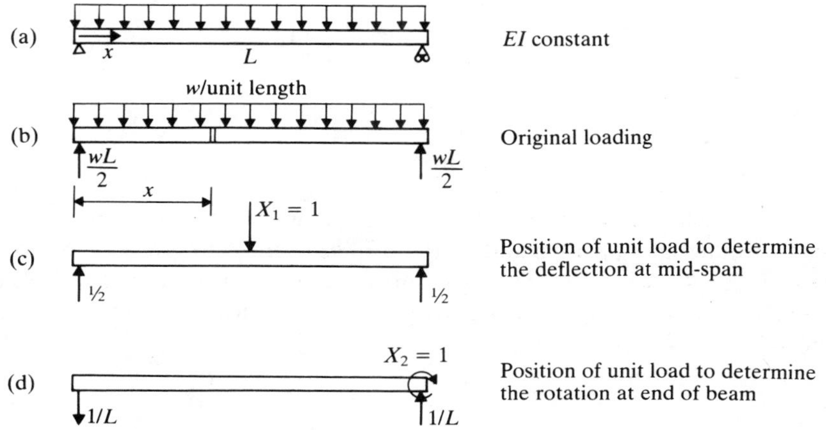

(a) EI constant

w/unit length

(b)   Original loading

$\frac{wL}{2}$ $\frac{wL}{2}$

x $X_1 = 1$

(c)   Position of unit load to determine the deflection at mid-span

½   ½

$X_2 = 1$

(d)   Position of unit load to determine the rotation at end of beam

$1/L$ $1/L$

Figure 2.62   Examples of the use of the unit load for determining the displacements of a simply supported beam

EXAMPLE 2.11
For the calculation of the rotation at the end of the simply supported beam in Fig.
2.62a, the unit load $X_2 = 1$ is applied as in Fig. 2.62d.

As before
$$M_0 = \frac{wL}{2} x - \frac{w}{2} x^2$$

From Fig. 2.62d
$$m_2 = -\frac{x}{L}$$

Thus
$$\phi = \frac{1}{EI} \int_0^L \left(\frac{w}{2L} x^3 - \frac{w}{2} x^2 \right) dx$$

$$= -\frac{wL^3}{24EI}$$

The rotation at B is therefore anticlockwise and has a value of $wL^3/24EI$ (Fig. 2.57b).

It will be understood that, just as with bending moment diagrams for standard cases of loading, the product integration of such standard cases can also be tabulated. The most common are shown in Table 2.7. The previous examples are now repeated using these tabulated values.

EXAMPLE 2.12
The M_0 and m_1 diagrams for the cantilever are shown in Fig. 2.63.

Coefficient from Table 2.7 $= \frac{1}{4}$

$$\Delta_1 = \frac{1}{EI} \times \frac{1}{4} \times L \times \frac{wL^2}{2} \times L$$

$$= \frac{wL^4}{8EI}$$

When the two bending moment diagrams are on the same side of the base line, the product integral is positive and so the displacement is in the same direction as that assumed for X_1; when the bending moment diagrams are on opposite sides of the base line, the product integral is negative and so the displacement is in the opposite direction to that assumed for X_1.

EXAMPLE 2.13
The M_0 diagram is as previously shown in Fig. 2.63a; the m_2 diagram is shown in Fig. 2.63c.

Table 2.7   Coefficients for product integrals $\int_0^L M_0 m_1 \, dx$

M_0 \\ m_1	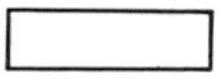	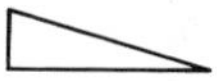		
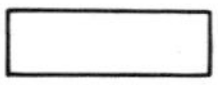	1	½	½	½
	½	⅓	⅙	¼
	½	⅙	⅓	¼
	½	¼	¼	⅓
	⅔	⅓	⅓	5/12
	⅓	¼	1/12	7/48
	⅓	1/12	¼	7/48

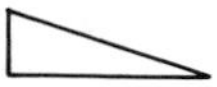

heights b

Heights a

All base lengths $= L$

Product integral $=$ Coefficient $\times L \times a \times b$

Coefficient from Table 2.7 $= -\dfrac{1}{3}$

$$\phi = -\frac{1}{EI} \times \frac{1}{3} \times L \times \frac{wL^2}{2} \times 1$$

$$= -\frac{wL^3}{6EI}$$

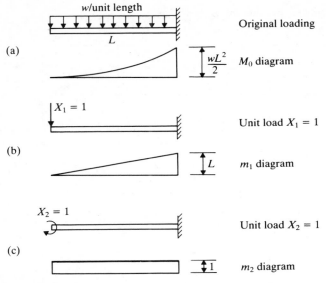

Figure 2.63 Examples of the product integration of bending moment diagrams for a cantilever

EXAMPLE 2.14

The M_0 and m_1 diagrams for the simply supported beam are shown in Fig. 2.64.

Coefficient from Table 2.2 $= \dfrac{5}{12}$

$$\Delta_1 = \frac{1}{EI} \times \frac{5}{12} \times L \times \frac{wL^2}{8} \times \frac{L}{4}$$

$$= \frac{5wL^4}{384EI}$$

EXAMPLE 2.15

The M_0 diagram is as in Fig. 2.64a; the m_2 is shown in Fig. 2.64c.

Coefficient from Table 2.2 $= -\dfrac{1}{3}$

$$\phi = -\frac{1}{EI} \times \frac{1}{3} \times L \times \frac{wL^2}{8} \times 1$$

$$= -\frac{wL^3}{24EI}$$

It will be clear that a mastery of the method facilitates the calculation of

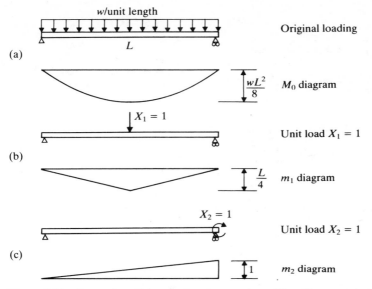

Figure 2.64 Examples of the product integration of bending moment diagrams for a simply supported beam

displacements for standard cases of loading on simple beams. However, the same simple procedure can also be adopted for larger structures where the bending moment diagrams can be split up into a combination of standard cases.

Reference to the bending moment diagram in Fig. 2.50k for the bridge structure in Fig. 2.50a would suggest that the coefficients in Table 2.7 could not be applied in this case. However, by splitting up the loading into a combination of simple cases, use of the coefficients is still possible. This is illustrated by reference to the beam in Fig. 2.65a. Suppose the deflection at C, the end of the cantilever, is required.

The original loading is split into the three cases shown in Fig. 2.65b, c and d giving the three separate M_0 diagrams, termed M'_0, M''_0 and M'''_0. The m_1 diagram due to $X_1 = 1$ at C is shown in Fig. 2.65e.

The principle of superposition gives:

$$EI\Delta = \int_{ac} M'_0 m_1 \, dx + \int_{ac} M''_0 m_1 \, dx + \int_{ac} M'''_0 m_1 \, dx$$

$$= \int_{ab} M'_0 m_1 \, dx + \int_{bc} M'_0 m_1 \, dx + \int_{ab} M''_0 m_1 \, dx$$

$$+ \int_{bc} M''_0 m_1 \, dx + \int_{ab} M'''_0 m_1 \, dx + \int_{bc} M'''_0 m_1 \, dx$$

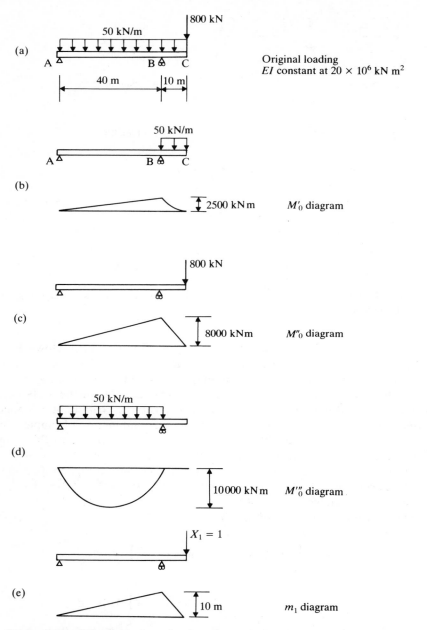

Figure 2.65   Illustrating the reduction of complex bending moment diagrams to a combination of simple diagrams for the purpose of calculating displacements

Using the coefficients from Table 2.7 for the product integrations:

$$\int_{ab} M'_0 m_1 \, dx = \tfrac{1}{3} \times 40 \times 2\,500 \times 10 = 333 \times 10^3 \text{ kN m}^3$$

$$\int_{bc} M'_0 m_1 \, dx = \tfrac{1}{4} \times 10 \times 2\,500 \times 10 = 62 \times 10^3 \text{ kN m}^3$$

$$\int_{ab} M''_0 m_1 \, dx = \tfrac{1}{3} \times 40 \times 8\,000 \times 10 = 1067 \times 10^3 \text{ kN m}^3$$

$$\int_{bc} M''_0 m_1 \, dx = \tfrac{1}{3} \times 10 \times 8\,000 \times 10 = 267 \times 10^3 \text{ kN m}^3$$

$$\int_{ab} M'''_0 m_1 \, dx = -\tfrac{1}{3} \times 40 \times 10\,000 \times 10 = -1333 \times 10^3 \text{ kN m}^3$$

$$\int_{bc} M'''_0 m_1 \, dx = 0$$

Thus
$$\Delta_1 = \frac{(333 + 62 + 1067 + 267 - 1333)10^3}{20 \times 10^6} = 0.02 \text{ m} = 20 \text{ mm}$$

It is desirable to note that the somewhat low value of the calculated deflection results from the counter-balancing effects of the loads—that on the cantilever tends to cause a downward movement of the point C, whilst the load on the span AB tends to cause an upward movement.

Problem 2.30   In order to illustrate the use of three different methods of calculating displacements of beams—the differential equation method, the strain energy method, and the unit load (virtual work) method—a lecturer used the three beams shown in Figs 2.66a, 2.66b and 2.66c, the points marked X being the position at which the deflection was calculated. Identify which method was used for each beam, noting that in the unit load method the coefficients in Table 2.7 were used. Determine the values of the deflections at X in the three beams using the same methods as adopted by the lecturer.

Problem 2.31   A cantilever beam of length L and uniform cross-section has a concentrated load W applied at a distance d from the support. Determine the value of d which would ensure that the deflection at the end of the beam is twice the deflection at the position of the load.

Problem 2.32   A beam AB of length L and flexural rigidity EI is simply supported at A and B. It carries a distributed load which at distance x from A has an intensity of $w_0 \sin \pi x / L$. Show that each reaction is $w_0 L / \pi$.

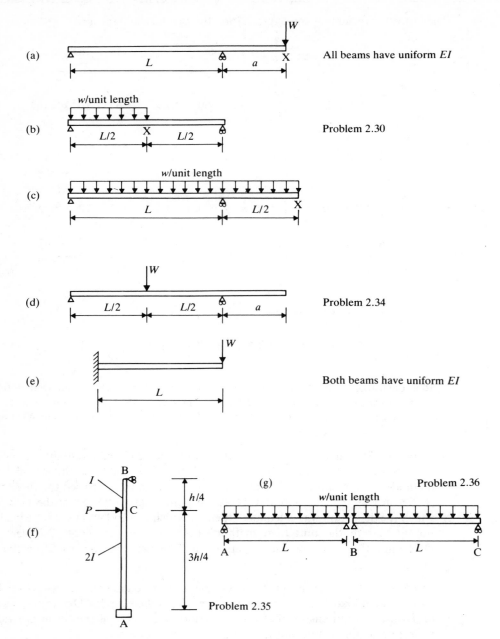

Figure 2.66   Beams used in Problems 2.30 to 2.36

Using the relationships between load intensity and shear force, shear force and bending moment, bending moment and curvature, show that the mid-span deflection is $WL^3/2\pi^3 EI$, where W is the total load on the beam.

Problem 2.33   A cantilever beam of span L is to be designed to support a uniformly distributed loading of w per unit length. However, it is found that the deflection caused by this load at the free-end is excessive when a beam of uniform rectangular section of breadth b and depth d is used. Investigate by what factor this deflection is reduced if the same amount of material is used as a cantilever of the same breadth b but with the depth varying linearly from zero at the free end to $2d$ at the fixed end.

Problem 2.34   Use the unit load method (together with the coefficients in Table 2.7) to determine (a) the deflection at the end of the cantilever of the beam in Fig. 2.66d, (b) the rotation at the same position, and (c) the deflection of the cantilever in Fig. 2.66e at the position midway between support and load.

Problem 2.35   The column AB in Fig. 2.66f can be regarded as simply supported at A and B for horizontal loads applied at C from the crane. There is a change of section at C—the second moment of area above C is I and below C is $2I$. Calculate the horizontal deflection at mid-height D caused by the horizontal load P.

Problem 2.36   The two adjacent beams in Fig. 2.66g are each simply supported and have the same uniform EI. Determine the discontinuity in line as a result of the rotations at B caused by the loads.

2.8 Rigid-jointed frames and arches

Portal frames

The analysis of beams supported at or near both ends has so far been confined to ones simply supported at their soffits. In such cases, no bending moments are transmitted from the supports to the beams. However, if the beams are not merely supported by the columns (or piers) but are also integral with them (the term *rigid-joint* is used to describe this type of connection), then the beam is not free to rotate at the support and bending moments will be transmitted from the support to the beam.

The student must not confuse 'rigid-jointed' support with the 'fixed' support described in connection with Fig. 1.59. With the latter, the supporting member is considered so rigid that no rotation of the beam occurs at the position of the support—see Fig. 2.67a. In the case of the rigid-jointed structure, it is only the connection which is rigid. As a result, *no relative rotation* of the connected members can occur, but overall rotation of the connection can, depending on the stiffness of the supporting member—see Fig. 2.67b.

The fixed-ended beam and the rigid-jointed portal frame shown in Fig. 2.67 are statically indeterminate and so will not be considered further at this stage. However, the type of structure shown in Fig. 2.68 in which the inclined members (termed rafters)

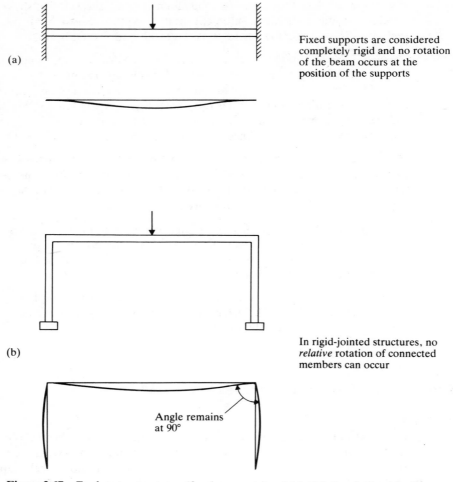

(a)

Fixed supports are considered completely rigid and no rotation of the beam occurs at the position of the supports

(b)

In rigid-jointed structures, no *relative* rotation of connected members can occur

Angle remains at 90°

Figure 2.67 Explanatory note on 'fixed supports' and 'rigid-jointed structures'

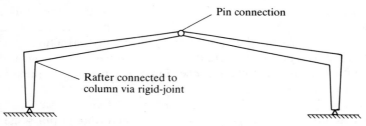

Pin connection

Rafter connected to column via rigid-joint

Figure 2.68 Example of a statically determinate structure incorporating rigid joints

are rigidly connected to the columns is statically determinate and is frequently used for single-storey industrial buildings.

It should be recalled that the four reactive forces at the pinned supports of such three-pinned frames can be obtained by using the three equations of equilibrium for the whole structure and the equation obtained from the known condition of zero moment at the internal pin connection. An example of such a calculation was given in Section 1.7 by referring to the structure in Fig. 1.65. Although the members of that structure were triangulated forms, that did not influence the calculation of the reactive forces.

The structure in Fig. 2.69a will be used to illustrate the determination of bending moment diagrams for statically determinate rigid-jointed frames. The dimensions and loading are identical to those in Fig. 1.65 and therefore the reactive forces will be those determined in Chapter 1. These are summarized in Fig. 2.69b.

It should be noted that, in this example, the external forces of the structure occur only at the ends of members. It therefore follows that the bending moment diagram along a particular member will be a straight line, and it will be necessary to find only

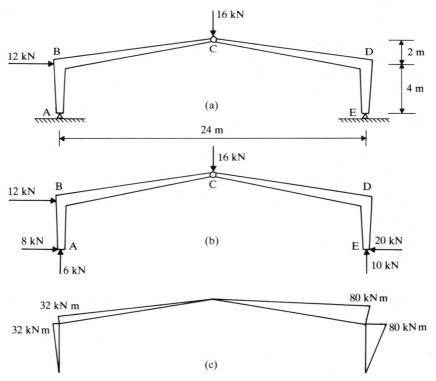

Figure 2.69 Example of the determination of the bending moment diagram for a 3-pinned frame

the bending moment values at the ends of the members. As we already know that at A, C and E, the positions of the pins, the bending moment is zero, it remains to find only the values of the bending moment at B and D in order to draw the bending moment diagram.

At B, by considering the external forces on that part of the structure below B:

bending moment = 8 × 4 = 32 kN m with tension on the outside of the frame

Similarly at D:

bending moment = 20 × 4 = 80 kN m and again tension on the outside of the frame

The bending moment diagram is therefore that shown in Fig. 2.69c.

Problem 2.37   Obtain the bending moment diagram for the structure in Fig. 2.69a when (a) only the vertical 16 kN load is acting and (b) when only the horizontal 12 kN load is acting. Use the principle of superposition to verify the bending moment diagram for the combined loading shown in Fig. 2.69c.

A physical understanding of the development of the horizontal reactive forces in simple portal frames can be obtained from Fig. 2.70. The bending of the beam (the horizontal member) under load tends to cause the columns to splay outward, and if the structure were supported on rollers (Fig. 2.70a) such outward movement would occur. In practice, the supports to the structure are fixed in position, with the result that horizontal forces must develop at the base of the columns. For the case of vertical loads on the beams, these horizontal reactive forces will act inwards (Fig. 2.70b).

It is of importance to note the effect of the inward horizontal reactive forces in reducing bending moments in portal frames. The point is demonstrated in Fig. 2.71. For the case of a beam simply supported on columns and subjected to a uniformly distributed load (Fig. 2.71a), the resulting sagging bending moment distribution will be the parabola having a maximum value of $wL^2/8$ (Fig. 2.71b). The effect of the horizontal reactive forces in the portal frame (Fig. 2.71c) is to cause hogging moments in the beam which, in combination with the parabola, result in the bending moment distribution in Fig. 2.71d. It will be seen that the maximum bending moment is now somewhat lower than the maximum occurring in the simply supported beam, and it is this characteristic which makes the portal frame an important structural form.

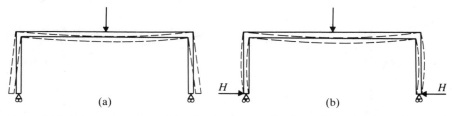

(a)   (b)

Figure 2.70   Demonstration of the development of horizontal reactive forces in simple portal frames

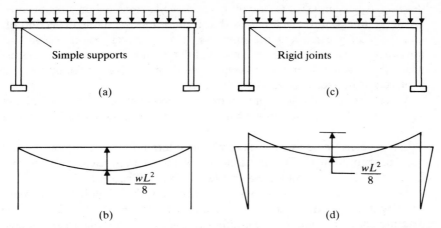

Figure 2.71 Illustrating the effect of the horizontal reactive forces in portal frames in reducing maximum bending moment

Arches

The arch (Fig. 2.72a) has some of the characteristics of the portal frame—its supports are also fixed in position, with the result that vertical loads cause inward acting horizontal reactive forces (Fig. 2.72b). Indeed, the term *arching action* is used more generally to describe structural behaviour in which such horizontal reactive forces

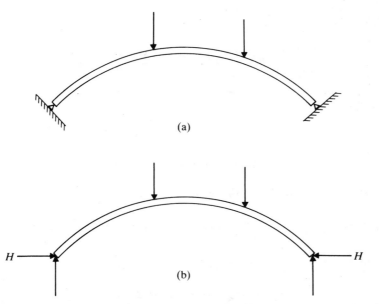

Figure 2.72 The arch has similarities to the portal frame in that horizontal reactive forces occur

develop. An example is the case of the brick wall supported on a beam (Fig. 2.73). The greater stiffness of the wall means that it does not follow the deflection of the beam, and the load is carried to the supports by an 'arching action' within the wall with horizontal forces developing as a result of the friction at the interface between wall and beam. It is this development of the horizontal forces which differentiates the arch from a curved beam (see for example Fig. 5.53).

As with portal frames, three pins are necessary to effect statical determinacy. Calculations for such structures will be demonstrated by reference to the arch in Fig. 2.74a.

Let us determine by way of example the internal forces acting on the section at the quarter span point D. The first step is to determine the reactive forces caused by the load and these are summarized in Fig. 2.74b (see Problem 1.9).

At the section of the arch under consideration, there will be three forces acting—a bending moment M, a shear force S, a direct force F (see Fig. 2.74c). These can be determined by considering the equilibrium of the part of the arch to one side of that section.

It is necessary to determine initially the position and orientation of the section.

For a parabola
$$y = \frac{4hx(L-x)}{L^2}$$

Thus at $x = 7$ m
$$y = \frac{4 \times 9 \times 7(28-7)}{28^2} = \frac{27}{4} \text{ m}$$

For orientation
$$\frac{dy}{dx} = \frac{4h(L-2x)}{L^2}$$

and at $x = 7$ m
$$\beta = \frac{4 \times 9 \times (28-14)}{28^2} = \frac{9}{14} \text{ radians}$$
$$= 36.8°$$

Using $\sum M_d = 0$
$$M + \left(\frac{112}{9} \times \frac{27}{4}\right) - (8 \times 7) = 0$$
$$M = -28 \text{ kN m}$$

i.e., tension on the outside of the arch

Using $\sum P_x = 0$
$$\frac{112}{9} + F\cos\beta - S\sin\beta = 0$$

Using $\sum P_y = 0$
$$8 + F\sin\beta + S\cos\beta = 0$$

Solution of these two simultaneous equations gives

$$F = -14.76 \text{ kN} \quad \text{and} \quad S = 1.05 \text{ kN}$$

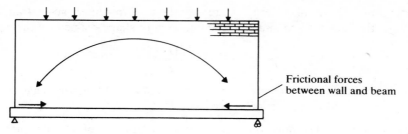

Figure 2.73   'Arching action' can occur within a brick wall supported on a beam

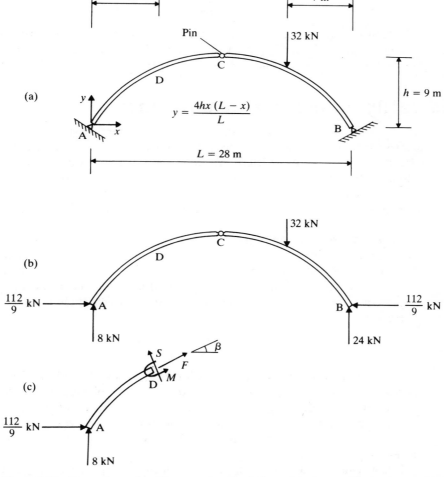

Figure 2.74   Example used to illustrate the calculation of the internal forces in a 3-pinned arch

Problem 2.38   Explain why the arch can be used for longer spans than the beam. Show that, for a three-pinned parabolic arch of shape defined by the equation

$$y = \frac{4hx(L - x)}{L^2}$$

and subjected to a uniformly distributed load of w/unit horizontal length, the bending moment at every section is zero.

Problem 2.39   The horizontal forces at the abutments of a two-pinned parabolic arch of height h and span L caused by a concentrated vertical load W at a horizontal distance x from an abutment is given by

$$H = \frac{5W}{8h}\left(x - \frac{2x^3}{L^2} + \frac{x^4}{L^3}\right)$$

Use this to show that when a uniformly distributed load covers the whole span of such an arch, the bending moment is everywhere zero. The equation determining the shape of the parabolic arch is as in Problem 2.38.

C. Statically indeterminate structures

The analysis has so far been confined to structures termed statically determinate—namely, structures which can be analysed by the use of statics alone. In practice, many structures are statically indeterminate and their analysis is not so straightforward. Although the study of the various methods of analysis of such structures will form an important part of the subsequent years of the undergraduate course, it is desirable to have some basic understanding of statically indeterminate structures at this stage.

Introduction to the problems of analysis

The problems will be introduced by reference to the two-span continuous beam in Fig. 2.75a. It is subjected to vertical loads only and so the use of the equilibrium equation $\sum P_x = 0$ quickly establishes that the horizontal reactive force R_{xa} at A is zero. There remain, therefore, two further equilibrium equations which can be used in the analysis. These are

$$\begin{bmatrix} \sum P_y \\ \sum M_z \end{bmatrix} = 0$$

As there are still three unknown reactive forces R_{ya}, R_{yb} and R_{yc} (Fig. 2.75b), it will not be possible to calculate these using only the equations of equilibrium.

It will be apparent though that, if we can determine any one of these reactive forces, say R_{yb}, from some other condition, it will then be possible to calculate the two remaining reactions and so also the internal forces at sections along the length of the beam.

This will also be clear from Fig. 2.75c in which the original statically indeterminate two-span beam has been transposed to a statically determinate single-span beam of

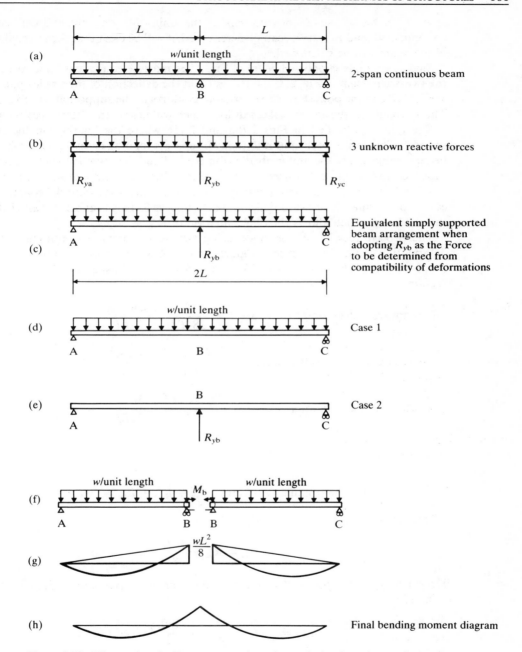

Figure 2.75   Illustrating the Force approach to the analysis of continuous beams

span $2L$ subjected to a combination of the uniformly distributed load and the concentrated load R_{yb}. If R_{yb} can be determined, it will be clear that the internal forces in the beam can be calculated.

Since the simply supported beam of Fig. 2.75c must be equivalent in all respects to the two-span beam of Fig. 2.75a, it follows that the deflection of the single-span beam of Fig. 2.75c at the point B must be zero—to conform to the support at B in Fig. 2.75a. This is, therefore, the additional condition which will permit the determination of R_{yb}.

The point is clarified in Figs 2.75d and 2.75e where the loading on the simply supported beam has been split into the two separate loading cases—one in which the beam is subjected to the uniformly distributed load and the other in which the beam is subjected to the upward concentrated force R_{yb}. For the combination of the cases 1 and 2 to be equivalent to the original two-span beam, the downward deflection at B in case 1 must equal the upward deflection at B in case 2. This is another way of stating the condition which allows the determination of R_{yb}.

The procedure will be illustrated using the results previously obtained for the deformation of a uniform simply supported beam. Acknowledging that the span of the simply supported beams in Figs 2.75d and 2.75e is $2L$, and using the expressions from Section 2.7:

$$\text{Downward deflection at B in case 1} = \frac{5w(2L)^4}{384EI} = \frac{5wL^4}{24EI}$$

$$\text{Upward deflection at B in case 2} = \frac{R_{yb}(2L)^3}{48EI} = \frac{R_{yb}L^3}{6EI}$$

$$\text{Since these must be equal:} \qquad \frac{R_{yb}L^3}{6EI} = \frac{5wL^4}{24EI}$$

$$R_{yb} = \frac{5wL}{4}$$

Using $\sum P_y$ for the original structure: $\quad R_{ya} + R_{yb} + R_{yc} = 2wL$

Since for symmetry: $\qquad\qquad R_{ya} = R_{yc}$

$$R_{ya} = R_{yc} = \frac{3wL}{8}$$

The single equation representing the bending moment at any section along the beam is therefore

$$M_x = \frac{3}{8}wLx - \frac{w}{2}x^2 + \frac{5wL}{4}[x - L]$$

The bending moment diagram could be obtained using this equation but it is preferable to calculate the bending moment at B directly and obtain the diagrams using the principle of superposition as discussed previously in relation to Fig. 2.48. Thus

$$M_b = \frac{3}{8} wL(L) - wL\left(\frac{L}{2}\right) = -\frac{wL^2}{8}$$

The two-span continuous beam can therefore be regarded as a combination of the two simply supported beams in Fig. 2.75f, in which moments of $wL^2/8$ are applied at the supports B. The corresponding bending moment diagrams are, therefore, as in Fig. 2.75g which combined give that in Fig. 2.75h.

Needless to say, any one of the three vertical reactions could have been chosen as the starting point for the analysis. For example, the two-span continuous beam could have been transposed to the statically determinate beam in Fig. 2.76a which is supported at B and C but cantilevers beyond B to carry the concentrated load R_{ya} in addition to the original distributed load. The condition permitting the calculation of R_{ya} is again that the downward deflection at A in Fig. 2.76b is equal to the upward deflection at A in Fig. 2.76c.

A further alternative would have been to transpose the two-span continuous beam to the two separate simply supported beams, each with the unknown hogging moments M_b acting at B. In this case, it is necessary to seek a deformation condition that will allow M_b to be determined.

The required deformation condition relates to the rotation of the beam at B. We know that for the original continuous beam there can be no rotation of the beam at B because of symmetry. It follows that for the structure in Fig. 2.76d to be equivalent to the original continuous beam, the anticlockwise rotation of the beam at B in Fig. 2.76e must equal the clockwise rotation of the beam at B in Fig. 2.76f. This condition will enable M_b to be determined.

For case 3 in Fig. 2.76e, an expression for the rotation was previously obtained in Section 2.7. Thus:

$$\text{Anticlockwise rotation at B in case 3} = \frac{wL^3}{24EI}$$

Case 4 in Fig. 2.76f was not previously solved and we will therefore calculate its rotation using the unit load method. The M_0 diagram corresponding to the load M_b in Fig. 2.76f is shown in Fig. 2.76g; the m_1 diagram corresponding to $X_1 = 1$ in Fig. 2.76h is shown in Fig. 2.76i.

$$\text{Coefficient for the product integral from Table 2.7} = \frac{1}{3}$$

$$\text{Clockwise rotation at B in case 4} = \frac{1}{EI} \times \frac{1}{3} \times L \times M_b \times 1$$

$$= \frac{M_b L}{3EI}$$

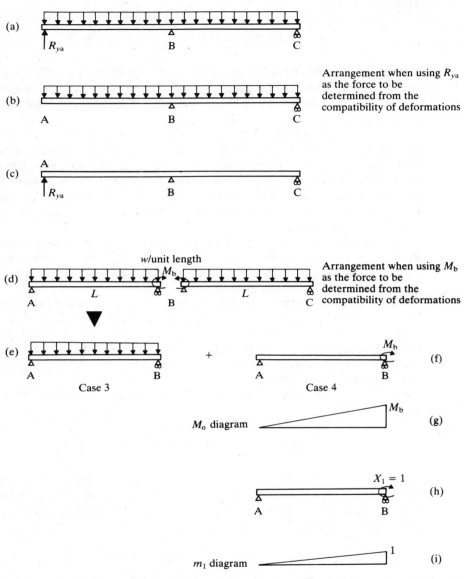

Figure 2.76   Alternative statically determinate structures for analysing the continuous beam using the Force approach

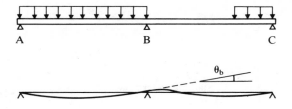

Rotation of AB at B = Rotation of BC at B = θ_b

Figure 2.77   Illustrating the deformation condition allowing the analysis of asymmetrical continuous beams

As the rotations must balance one another out:

$$\frac{M_b L}{3EI} = \frac{wL^3}{24EI}$$

$$M_b = \frac{wL^2}{8}$$

This is the same value as before and allows the bending moment diagram to be drawn.

This particular problem was simplified by the condition of symmetry and we were able to carry out the calculation of rotations without reference to the span BC. For non-symmetrical cases, there would be some rotation at B as in Fig. 2.77. Nevertheless, since the rotation of AB at B must be equal to the rotation of BC at B (as the beam is continuous), this deformation condition would still allow the value of M_b to be determined.

It will be of value to consider also pin-jointed frames. That in Fig. 2.78a will be recognizable as statically determinate, and this can be confirmed (see Section 2.1) from the equation

$$m + r = 2j$$

The pin-jointed frame in Fig. 2.78b is, however, statically indeterminate. The extra member prevents the calculation of the member forces using statical equilibrium alone. This will be apparent from the fact that three unknown member forces meet at the joint A but there are only two equations of equilibrium at the joint. Moreover, the inequality

$$m + r > 2j$$

confirms statical indeterminacy.

As with the continuous beam problem, analysis can commence by transposing the indeterminate structure into a statically determinate structure but with an unknown load acting. This is shown in Fig. 2.78c where the statically determinate frame is acted

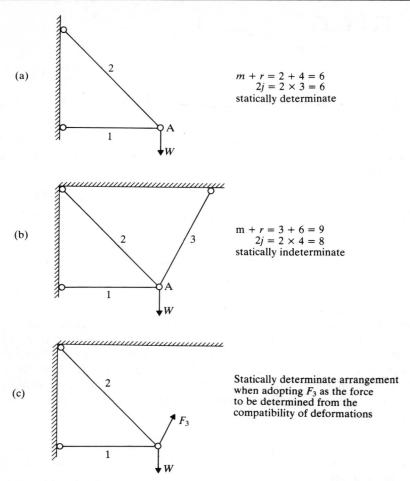

(a)

$$m + r = 2 + 4 = 6$$
$$2j = 2 \times 3 = 6$$
statically determinate

(b)

$$m + r = 3 + 6 = 9$$
$$2j = 2 \times 4 = 8$$
statically indeterminate

(c)

Statically determinate arrangement
when adopting F_3 as the force
to be determined from the
compatibility of deformations

Figure 2.78 Illustrating the Force approach to the analysis of statically indeterminate pin-jointed frames

upon by the known load W and the unknown force F_3, this latter being the force in member 3 of the original structure.

If F_3 can be determined from some known condition, so the forces in members 1 and 2 can be calculated. The deformation condition which allows the determination of F_3 is that the deflection at joint A of the statically determinate frame in Fig. 2.78c, under the effects of loads W and F_3, must be such that the extension of member 3 is compatible with the member force–extension relationship. This deformation condition allows an equation to be obtained from which F_3 can be determined.

An alternative approach to the analysis (see Fig. 2.79) is to consider initially the displacements of joint A (Δ_{xa} and Δ_{ya}) as the unknowns to be determined. The

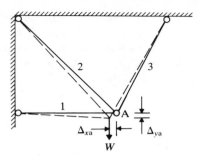

Figure 2.79   Illustrating the Displacement approach to the analysis of statically indeterminate frames

procedure is then identical to that outlined previously in the section on the calculation of displacements in statically determinate pin-jointed frames, namely:

- the determination of the extension of the members in terms of Δ_{xa} and Δ_{ya};
- the translation of these extensions to expressions for member forces;
- the development of the two equilibrium equations at joint A using $\sum P_x = 0$ and $\sum P_y = 0$;
- solving these two simultaneous equations to obtain Δ_{xa} and Δ_{ya};
- substituting these values of Δ in the earlier expressions to obtain the member forces.

It is of special importance to note that the displacement approach makes no distinction between statically determinate and statically indeterminate structures, and it will be recalled that this procedure was previously used to analyse statically determinate pin-jointed frames. The approach is equally applicable to structures involving bending, but these will not be considered at this stage.

To summarize then, there are basically two different approaches to the analysis of statically indeterminate structures:

1. In the *Force* approach (sometimes termed the Flexibility or the Compatibility approach), the statically indeterminate structure is transposed to a statically determinate structure but with a number of unknown *selected forces* acting in addition to the original load. Using the force–displacement relationships and equilibrium, the corresponding displacements are calculated in terms of the selected forces. A number of simultaneous equations can be set up by acknowledging compatibility of deformation. The equations enable the selected forces to be determined, from which the remaining forces and the displacements can be calculated.

2. In the *Displacement* approach (sometimes termed the Stiffness or the Equilibrium approach), certain *key displacements* are selected as unknowns. Using the force–displacement relationships and compatibility, the forces are calculated in terms of the key displacements. A number of simultaneous equations can be set up by considering equilibrium. The equations enable the key displacements to be determined from which the internal forces can be calculated.

Within each of these fundamental approaches, there are a number of different methods. With hand methods of analysing statically indeterminate structures, the easier approach will usually be that which has the fewer simultaneous equations to solve. With computer methods of analysis, this distinction is less important and the displacement approach is frequently the most advantageous.

Fundamentals of statically indeterminate structures

It will be apparent from the previous examples that the analysis of statically indeterminate structures requires the use of equations from

1. Equilibrium of forces.
2. Compatibility of deformations.
3. Force–displacement relationships of the members of the structure.

These same conditions are appropriate to statically determinate structures but whilst with such structures the calculations pertaining to forces and deformation can proceed independently, those for statically indeterminate structures are inter-related. It is, therefore, clear that the analyses of statically indeterminate structures are appreciably more complicated than those for statically determinate structures. It is then relevant to ask why these more complicated structures should be used. There are indeed both advantages and disadvantages in adopting statically indeterminate structures, and it will be of interest to students to itemize some of these prior to their further studies in this field.

Advantages

1. The statically indeterminate structure is considerably stiffer than the corresponding statically determinate structure.

   (This is illustrated in Fig. 2.80. The statically determinate, simply supported beam has a maximum deflection five times that of the comparable statically indeterminate, fixed-ended beam. The maximum deflection of the statically indeterminate portal frame is somewhat intermediate between these two extremes, depending on the stiffness of the columns.)
2. The maximum bending moments (or axial forces) in the statically indeterminate structure are generally lower than in the corresponding statically determinate structures.

   (This point is not actually confirmed by the continuous beam example in Fig. 2.75 where the maximum bending moment is identical to the maximum in the corresponding simply supported beams, namely $wL^2/8$. However, this is more the exception, and the situation illustrated in Fig. 2.71 is more typical.)

The implication of these advantages is that in design a smaller section can usually be adopted for the statically indeterminate structures, with corresponding economies in material.

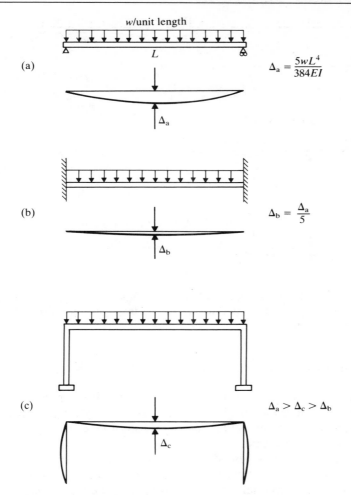

(a)

$$\Delta_a = \frac{5wL^4}{384EI}$$

(b)

$$\Delta_b = \frac{\Delta_a}{5}$$

(c)

$$\Delta_a > \Delta_c > \Delta_b$$

Figure 2.80   An advantage of statically indeterminate structures is their greater stiffness compared with corresponding determinate structures

Disadvantages

1. Any relative settlement of supports causes stresses in statically indeterminate structures but not in statically determinate structures.

   (This is illustrated in Fig. 2.81. The statically determinate bridge structure can undergo settlement of one of the supports without causing bending of the beams. On the other hand if the superstructure is continuous, bending of the beams is inevitable in order to accommodate any relative settlement. The resulting stresses are additional to those caused by the loads. This has a considerable effect in the design of structures in areas where subsidence is possible, mining areas for example.

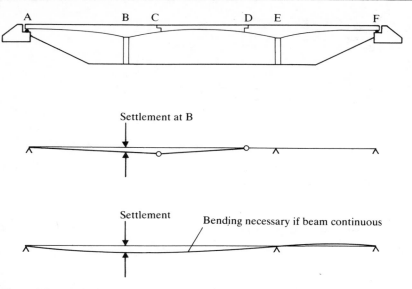

Figure 2.81   Relative settlement at supports can be accommodated by statically determinate structures without causing accompanying stresses but this may not be possible in statically indeterminate structures

The statically determinate structure would generally be preferred in such circumstances.)

2. The calculation of forces within a statically indeterminate structure due to an applied load necessitates the calculation of deformations, which in turn requires a knowledge of the dimensions of the cross-sections of the members, resulting in a trial-and-error design procedure.

(To commence the design of statically indeterminate structures, it is first necessary to select the size of the members *prior* to knowing the forces within the members. The subsequent analysis and knowledge of the forces may necessitate changes in the sections of the members, and therefore re-analysis and possible further changes in the sections until one converges on a suitable design. In contrast, the calculation of forces in a statically determinate structure caused by a particular load is independent of the size of members.)

There are also several other factors affecting the relative merits of statically determinate and statically indeterminate structures, but discussion on these is better left until the student has further experience of design. Notwithstanding, it must not be imagined that the structural engineer always has a simple choice in this matter. For example, the nature of reinforced concrete construction is such that rigid-joints between members are much easier to effect than are pin-joints, so that the choice of reinforced concrete as the structural material usually implies the adoption of statically indeterminate structures.

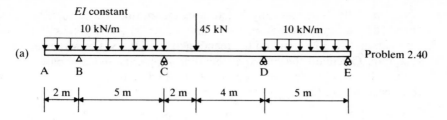

(a)   Problem 2.40

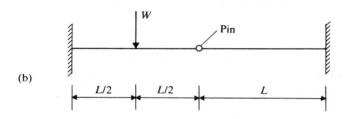

(b)

Problem 2.41

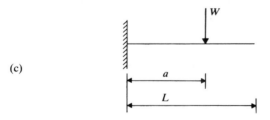

(c)

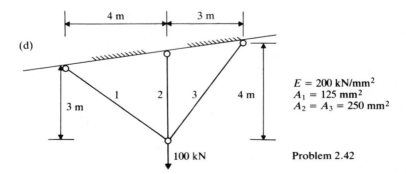

(d)

$E = 200$ kN/mm^2
$A_1 = 125$ mm^2
$A_2 = A_3 = 250$ mm^2

Problem 2.42

Figure 2.82   Beams used in Problems 2.40 to 2.42

Problem 2.40   The statically indeterminate, continuous beam ABCDE shown in Fig. 2.82a is supported at B, C, D and E. An analysis shows that the hogging moment at C is 30 kN m and the reaction at E is 20 kN. Draw the bending moment diagram. Calculate the mid-span bending moments and compare these with the values that would have obtained in a statically determinate system of beams if there had been no continuity at C and D.

Problem 2.41   The statically indeterminate beam in Fig. 2.82b has a uniform cross-section throughout. It can be considered as two cantilevers joined together by a pin connection. Given that the deflection at the end of the cantilever shown in Fig. 2.82c is

$$\frac{Wa^2}{2EI}\left(L - \frac{a}{3}\right)$$

use the principle of compatibility of deformations to calculate the vertical force transmitted by the pin.

Problem 2.42   Show that the frame in Fig. 2.82d is statically indeterminate. Obtain the forces in the three members in terms of the displacements of the loaded joint. Use the equilibrium at this joint to obtain the simultaneous equations involving the displacements and so obtain their values. Hence calculate the forces in the members. (If a computer and a suitable planar pin-jointed frame program are available to the student, use these to verify the solution.)

3. Structural analysis: mechanics of solids

The determination of the distribution of the internal forces in a structure caused by the loads applied to that structure was demonstrated, for simple cases, in Chapter 2 under the heading 'Mechanics of structures'. In the design of a structure it would be necessary to establish that the effects of these forces can satisfactorily be sustained at each and every part of the proposed structure.

Since the effects of an internal force acting on a particular section of a member can vary at different points in that section, it becomes necessary to translate that force into its effects at the various points—an analysis termed 'Mechanics of solids'. The intensities of the force and of the resulting deformation at a particular point are defined as the stress and strain respectively, and these terms were briefly introduced at an early stage in Chapter 1. It is the aim of this chapter to illustrate the calculation of the distribution of stress and strain across a section caused by a particular force acting on that section. For example, the distribution of direct stresses across a section of a beam due to a bending moment on that section; similarly, the distribution of shear stresses across a section of a beam by the shear force at that section, or the distribution of stresses and strains caused by a torsional moment on a circular shaft.

Since it is possible for several types of internal force to act on one particular cross-section (for example, a bending moment and a shear force will generally act together on a cross-section of a beam), it becomes necessary to be able to combine their effects at a particular point to give the 'worst' effect. This part of the problem, the determination of the *principal stresses* and *principal strains*, is another important aspect of the mechanics of solids demonstrated in this chapter. However, we will begin with the problems of determining the distribution of stress across a section as a result of a single internal force.

A. Stress distributions due to a single internal force

A beam is a structural member with the primary purpose of sustaining loads applied perpendicular to its longitudinal or axial direction. Generally, its dimension in this axial direction is significantly bigger than the dimensions of its cross-section. Figure 3.1a shows such a beam in the form of a cantilever AB subjected to an upward-acting

193

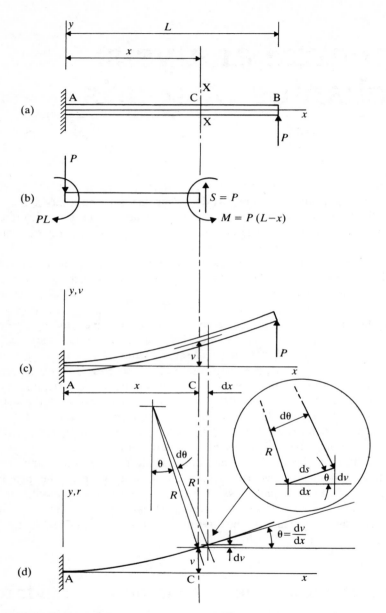

Figure 3.1   Showing the deflection of a cantilever under a point load

load P applied at the free end. The reactions at the fixed end A will be a downward-acting vertical load P and a clockwise moment PL. At a cross-section XX (Fig. 3.1b) distance x from A, the internal forces acting on the cross-section are the moment $M = P(L - x)$ and the shear force $S = P$ (see Chapter 2). These forces are sometimes referred to as stress resultants, because each represents the resultant of a distribution of stresses on the cross-section. In the case of M, it is a distribution of stresses which act in a direction normal to the cross-section, i.e., in the x direction, and are called normal, or direct, stresses. In the case of S, it is a distribution of stresses which act in a direction tangential to the cross-section, i.e., in the y direction, and are called shear stresses. Our task is to discover the nature of these distributions of normal stresses and shear stresses across any cross-section such as XX. The direct stresses associated with M and the shear stresses associated with S can be considered separately.

Also in this section, we will study the stress patterns which are generated in a circular shaft when a twisting moment, or torque, is applied to the shaft. The torque T gives rise to shear stresses acting throughout the thickness of the shaft, and their distribution proves to be a very simple one.

3.1 Longitudinal stresses in a beam due to a bending moment

The moment M at C in Fig. 3.1b causes the beam to bend. Therefore, it is called a bending moment. The accumulated effects of bending at all points along the beam result in the deflected shape shown in Fig. 3.1c. (The deflection of the beam is greatly exaggerated for the sake of clarity.) Two points about Fig. 3.1c are important. Firstly, the deflection is shown occurring in the plane xy, i.e., all points on the beam are moving vertically, in the same direction as P. For this to happen, the cross-sectional shape of the beam must be symmetrical about the y-axis. The plane xy is called the *plane of bending*. Secondly, the moment M will cause compressive x-direction stresses at the top of the beam cross-section and tensile x-direction stresses at the bottom.

The deflected beam in Fig. 3.1c shows a positive deflection v at C, that is, in the positive direction of y. In Fig. 3.1d, the deflected beam is represented by a single line and indicates that, at C, the deflection v is accompanied by a rotation θ, where θ is defined by the angle between the tangent at C of the deflected shape and the original (unloaded) direction of the beam. The angle θ will vary with x and its value at a particular point is referred to as the slope of the beam at that point. Because θ will be small, $\tan \theta = \theta = dv/dx$.

At any point, such as C, along the length of the beam, the deflected shape will have a radius R (Fig. 3.1d) which will also vary with x. If the deflection v of the beam and its slope θ are small then, to a first order of accuracy, the length of the deflected beam ds over the differential distance dx can be taken as equal to dx. Therefore, in Fig. 3.1d:

$$\frac{1}{R} = \frac{d\theta}{ds} \simeq \frac{d\theta}{dx} = \frac{d^2 v}{dx^2} \tag{3.1}$$

As pointed out in Chapter 2, R is called the *radius of curvature* and its reciprocal, namely d^2v/dx^2, is called the *curvature of the beam*. Like R, the curvature will, in general, vary with x.

An important point, which has been demonstrated by Fig. 3.1, is that, with the sign convention we have adopted, a positive moment M (as at C on Fig. 3.1b) causes positive curvature d^2v/dx^2, i.e., $\theta = dv/dx$ increases with increasing x. This will always be true, irrespective of the sign of dv/dx. (When dv/dx is negative, positive curvature means that dv/dx becomes a smaller negative value as x increases.)

An easy way to remember the relationship between moment and curvature is to note that positive M produces compression at positive y-values of the beam, and this corresponds to positive curvature.

To examine the distribution of normal (or direct) stresses on a cross-section subjected to bending moment M, we consider a situation in which M is constant as x varies, i.e., $dM/dx = 0$, often referred to as *pure bending* because, since the shear force $S = -dM/dx$, it follows that the shear is zero when M is constant.

Such a case is illustrated in Fig. 3.2. A simply supported beam AB of length L (Fig. 3.2a) is symmetrically loaded with two vertical loads, each equal to P, and deflects as shown in Fig. 3.2b. We want to examine the bending of the element of length δx situated between the loads at a distance x from A. The bending moment and shear force diagrams in Figs 3.2c and 3.2d demonstrate that the element is being subjected to pure bending, i.e., the bending moment, $M = Pa$, is constant across the length of the element which experiences zero shear force.

In Fig. 3.2g the element is reproduced to a larger scale, first in its undeformed state and then in the deformed state due to the action of the bending moment M. Notice that the deformed element has been drawn with the planes defining its length remaining plane, i.e., it has been assumed that cross-sectional planes remain plane after bending. This is a fundamentally important assumption. Also shown in Fig. 3.2g is the cross-sectional shape of the beam. This is arbitrary except for two important properties, namely, it is symmetrical about the y-axis and it remains constant along the entire length of the beam. A beam with a constant cross-section (of whatever shape) along its length is called *prismatic*.

Because of the bending moment M, the axis of the element bends to radius R (Fig. 3.2g). As a result, the longitudinal top fibres of the element become shorter and the bottom ones become longer. It follows that at some distance between the top and bottom, the longitudinal fibres do not change length but simply bend. Suppose this happens along the plane represented by the line cd along δx and the line NA on the cross-section. Thus, the length cd on the undeformed element equals the length of c'd' on the deformed element. Therefore, along c'd' there is zero longitudinal strain. This condition is true for all the longitudinal fibres at this level across the entire width of the cross-section, i.e., across NA, which is therefore known as the *neutral axis* of the cross-section.

On Fig. 3.2g, consider the element of length ab of dimensions $\delta x \times t \times \delta y$ situated at distance y from the neutral axis, NA.

Length of ab before bending $= ab = cd = c'd'$

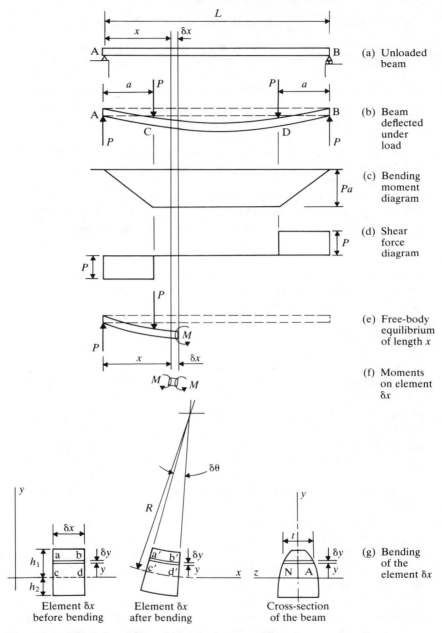

Figure 3.2   The deformation of a beam element subjected to pure bending

Length of the same element after bending = a'b'

Therefore longitudinal (x-direction) strain on the element is ε_x

where
$$\varepsilon_x = \frac{a'b' - ab}{ab} = \frac{a'b' - c'd'}{c'd'}$$

As
$$c'd' = R \times \delta\theta \quad \text{and} \quad a'b' = (R - y)\delta\theta$$

Then
$$\varepsilon_x = \frac{(R - y)\delta\theta - R\delta\theta}{R \times \delta\theta} = -\frac{y}{R} \tag{3.2}$$

The negative sign in Eq. (3.2) reflects the fact that, at positive y, the fibres experience negative strain when the curvature is positive, i.e., they become shorter because they are in compression.

Experimental evidence indicates that fibres experiencing tension or compression in a beam subjected to bending behave as they do in uniaxial tensile or compressive tests, i.e., they obey the uniaxial form of Hooke's Law. Therefore

$$\varepsilon_x = \frac{\sigma_x}{E} \tag{3.3}$$

where σ_x = the longitudinal stress in the beam element
and E = Young's Modulus

From Eqs (3.2) and (3.3):
$$\frac{\sigma_x}{E} = -\frac{y}{R}$$

Rearranging gives
$$\frac{\sigma_x}{y} = -\frac{E}{R} \quad \text{or} \quad \sigma_x = -\frac{E}{R}y \tag{3.4}$$

Equation (3.4) indicates that the longitudinal stress σ_x on any cross-section of the beam varies linearly with distance from the neutral axis NA, because E/R has a particular value at any cross-section. This linear variation of stress arises directly from the linear variation of strain across the section (from compressive strain at the top to tensile strain at the bottom) which, in turn, rests upon the assumption that plane sections remain plane after bending. The longitudinal stress at the neutral axis (where $y = 0$) is zero.

The distribution of σ_x across the cross-section occurs solely because of the presence of the bending moment M. There is no other force acting on the cross-section. Equilibrium at the cross-section enables us to find the position of the neutral axis NA, i.e., the values of h_1 and h_2 (Fig. 3.2g), and to establish a relationship between M and the curvature $1/R$.

Figure 3.3 summarizes the results of the analysis so far. The element of the beam of length δx (drawn in its undeformed state in Fig. 3.3b for convenience) shows the stresses σ_x acting on the element, compressive above the neutral axis and tensile below the neutral axis. This stress distribution exists because of M and is equivalent to M. The stress distribution is represented in a more conventional way in Fig. 3.3c and the

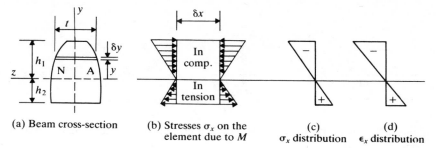

(a) Beam cross-section   (b) Stresses σ_x on the   (c)   (d)
                                element due to M σ_x distribution ϵ_x distribution

Figure 3.3   Stress and strain distributions on the cross-section of a beam caused by a bending moment

longitudinal strain (ε_x) distribution is shown in Fig. 3.3d. Stress and strain are both positive when tensile, and negative when compressive. (See also the section on sign convention in Chapter 1.)

The integration of the σ_x-distribution over the entire cross-sectional area of the beam will give the force on the cross-section in the longitudinal direction. This is zero, because no longitudinal force has been applied to the beam. Therefore

$$\int_{h_2}^{h_1} \sigma_x t \, dy = -\frac{E}{R} \int_{h_2}^{h_1} ty \, dy = 0$$

It follows that

$$\int_{h_2}^{h_1} ty \, dy = 0$$

because E/R cannot be zero. The quantity

$$\int_{h_2}^{h_1} ty \, dy$$

is the moment of the area of the cross-section about the line NA, the neutral axis. It is known as the *first moment of area* of the cross-section, to distinguish it from the *second* moment of area which was introduced in Chapter 2.

Thus, if A = area of the cross-section and $\bar{y}$ = distance of the centroid of A from NA

then

$$\int_{h_2}^{h_1} ty \, dy = A\bar{y} = 0 \quad \text{(from above)} \tag{3.5}$$

Since $A \neq 0$, then $\bar{y} = 0$

Therefore NA, *the neutral axis, passes through the centroid of area of the cross-section.*

The stress distribution σ_x is caused by M and is equivalent to M. Figure 3.4 shows

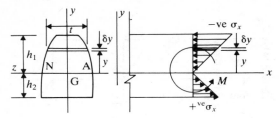

Figure 3.4   The equivalence of the bending moment and the corresponding stress distribution

M and its equivalent stress distribution acting on a positive face, i.e., negative (compressive) stress at positive y, and positive stress at negative y.

On the element of area $t\,dy$ at y from NA, the stress is negative. Therefore the normal force on the elements is $-\sigma_x t\,dy$. The moment about NA of the force on the element is δM, where

$$\delta M = -\sigma_x ty\,dy = \frac{E}{R}\,ty^2\,dy, \quad \text{from Eq. (3.4)}$$

therefore

$$M = \frac{E}{R}\int_{h_2}^{h_1} ty^2\,dy \tag{3.6}$$

The quantity

$$\int_{h_2}^{h_1} ty^2\,dy$$

is the moment of the first moment of area of the cross-section about the axis Gz and is therefore called *the second moment of area of the cross-section*. It is usually denoted by the symbol I. (I has a form similar to that of the moment of inertia, but with mass replaced by area. In fact, it is sometimes referred to, misleadingly, as the moment of inertia.)

Equation (3.6) can now be written as

$$M = \frac{E}{R}I \qquad \text{or} \qquad \frac{M}{I} = \frac{E}{R} \tag{3.7}$$

Combining Eq. (3.7) with Eq. (3.4), we get a most important double equation, namely

$$\frac{M}{I} = \frac{E}{R} = -\frac{\sigma_x}{y} \tag{3.8}$$

Because the second moment of area I is an important property of the beam cross-section, it is considered further in the sections which follow.

Second moment of area and the parallel axis theorem

Consider the cross-section shown in Fig. 3.5. Its shape is symmetrical about the y-axis, but otherwise is quite arbitrary. The centroid of the area is at point G.

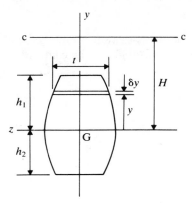

Figure 3.5   A cross-section of a beam symmetrical about the y-axis

Let the second moment of area of the cross-section about the axis Gz be I_z (referred to simply as I in Eqs (3.7) and (3.8)).

Then
$$I_z = \int_{h_2}^{h_1} ty^2 \, dy \tag{3.9}$$

where t is a known function of y.

We can equally well write an expression for the second moment of area of the cross-section about any line in the plane of the cross-section and parallel to Gz.

Thus, the second moment of area about the line cc, at any distance H from Gz, is given by I_c, where

$$I_c = \int_{h_2}^{h_1} t(H - y)^2 \, dy$$

$$= \int_{h_2}^{h_1} t(H^2 - 2Hy + y^2) \, dy$$

that is
$$I_c = \int_{h_2}^{h_1} tH^2 \, dy - 2 \int_{h_2}^{h_1} tHy \, dy + \int_{h_2}^{h_1} ty^2 \, dy \tag{3.10}$$

In Eq. (3.10):

$$\int_{h_2}^{h_1} tH^2 \, dy = AH^2 \text{ where } A = \text{area of the cross-section}$$

$$\int_{h_2}^{h_1} tHy \, \mathrm{d}y = \text{first moment of area of the cross-section about G}z. \text{ This must be zero because the } z\text{-axis passes through G, the centroid of the cross-section}$$

that is

$$\int_{h_2}^{h_1} tHy \, \mathrm{d}y = 0$$

and

$$\int_{h_2}^{h_1} ty^2 \, \mathrm{d}y = I_z$$

Therefore

$$I_c = AH^2 + I_z \tag{3.11}$$

Equation (3.11) expresses the *parallel axis theorem*. It says that the second moment of area of the cross-section about any line parallel to Gz is equal to the second moment of area about Gz plus the product of the cross-sectional area and the square of the distance of the line from Gz. Clearly, I_c has its minimum value when $H = 0$, i.e., when the line cc coincides with Gz. In other words, I_z is the minimum value of the second moment of area about all lines parallel to Gz.

If a similar analysis was carried out to investigate the second moment of area of the cross-section about all lines parallel to the axis Gy, it would be similarly found that I_y, the second moment of area about Gy, is the minimum value. The parallel axis theorem therefore applies also in the z direction.

For a cross-section such as that in Fig. 3.5, that is, one which is symmetrical about *either* Gy or Gz, these axes are called *principal axes* of the cross-section. (The location of the principal axes for the case of a cross-section which has no axis of symmetry will not be considered in this book. The principal axes are always at right-angles to each other.)

Note carefully that the locations of the principal axes of a cross-section depend entirely upon the shape of the cross-section itself and are not influenced by the forces applied to the cross section.

By contrast, the location of the neutral axis of a cross-section *is* dependent upon the force or forces applied to it. For example, when a bending moment is applied (in the absence of any other force) in the plane xy to the cross-section of Fig. 3.5, the neutral axis will coincide with the principal axis Gy. This is also true in the cases illustrated by Figs 2.54, 3.2, 3.3 and 3.4.

If, on the other hand, a moment is applied in the xz plane to the cross-section in Fig. 3.5 (again, in the absence of any other force), then the neutral axis will coincide with the principal axis Gy. In general, therefore, the location of the neutral axis is dependent upon the nature of the forces applied to the cross-section. This will be illustrated very clearly in Section 3.7 where the combination of a bending moment and an axial force is considered.

Examples of the calculation of second moments of area

In Section 2.7, the values of I_z (called simply I in Chapter 2) were found for rectangular and circular cross-sections respectively as $BD^3/12$ and $\pi R^4/4$, see Figs 3.6a and 3.6b. It follows that, for the rectangular cross-section in Fig. 3.6a, the second moment of area about Gy, i.e., $I_y = B^3D/12$ and, because of symmetry, for the circular cross-section in Fig. 3.6b, $I_y = I_z = \pi R^4/4$.

Figure 3.6c is a hollow, rectangular cross-section symmetrical about both Gy and Gz. From Eq. (3.9):

$$I_z = \int_{h_2}^{h_1} t y^2 \, dy$$

$$I_z = \int_{-d/2}^{d/2} 2\left(\frac{B}{2} - \frac{b}{2}\right) y^2 \, dy + \int_{d/2}^{D/2} By^2 \, dy + \int_{-D/2}^{-d/2} By^2 \, dy$$

which reduces to

$$I_z = \frac{BD^3}{12} - \frac{bd^3}{12} \tag{3.12}$$

Thus, I_z is the second moment of area of the outer rectangle minus the second moment of area of the inner (empty) rectangle. Similarly:

$$I_y = \frac{B^3D}{12} - \frac{b^3d}{12} \tag{3.13}$$

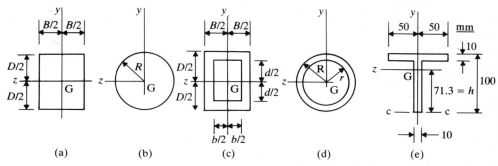

Figure 3.6   Various cross-sections used to demonstrate the calculation of second moments of area

Equations (3.12) and (3.13) demonstrate an important fact, namely, that second moments of area can be added and subtracted.

The second moment of area of the section in Fig. 3.6d, a hollow, circular shape, is given therefore by

$$I_z = I_y = \frac{\pi}{4}(R^4 - r^4) \tag{3.14}$$

EXAMPLE 3.1

The value of I_z for the T-section of Fig. 3.6e cannot be found until the centroid of area G is located. This is readily done by saying that the first moment of the total area about the line cc is equal to the sum of the first moments of area of the elements of the cross-section (the horizontal flange and the vertical web) about the same line. Thus, if G is located at distance $\bar{h}$ from the line cc:

$$[(100 \times 10) + (90 \times 10)]\bar{h} = (100 \times 10)95 + (90 \times 10)45$$

therefore $\qquad \bar{h} = 71.3$ mm

To find I_z for the cross-section, use is made of the parallel axis theorem (Eq. (3.11)) by finding the second moments of area of the flange and web about their own respective centroids, and then transferring these second moments of area to the Gz axis. Thus

$$I_z = \frac{100 \times 10^3}{12} + (100 \times 10)(95 - 71.3)^2 + \frac{10 \times 90^3}{12} + (90 \times 10)(71.3 - 45.0)^2 \text{ mm}^4$$

that is

$I_z = 1800 \times 10^3 \text{ mm}^4$

The value of I_y for the same cross-section is given by

$$I_y = \frac{10 \times 100^3}{12} + \frac{90 \times 10^3}{12} = 841 \times 10^3 \text{ mm}^4$$

To find I_z for the triangular cross-section in Fig. 3.7, for which the centroid of area G is obviously located at $D/3$ from the base, we can use either the expression

$$I_z = \int_{-D/3}^{2D/3} ty^2 \, dy \tag{3.15}$$

or, by making use of the parallel axis theorem

$$I_z = I_c - A\left(\frac{D}{3}\right)^2 \tag{3.16}$$

where $A = BD/2$, the area of the cross-section.

The second expression is probably simpler to use. If h is measured from the line cc,

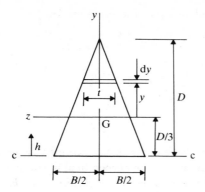

Figure 3.7   Example 3.1

then

$$t = \left(\frac{D - h}{D}\right)B$$

therefore, using Eq. (3.16):

$$I_z = \int_0^D \left(\frac{D - h}{D}\right)Bh^2 \, dh - \left(\frac{BD}{2}\right)\left(\frac{D}{3}\right)^2 = \frac{BD^2}{36} \qquad (3.17)$$

Problem 3.1   Show that, for the triangular cross-section of Fig. 3.7:

$$I_y = \frac{DB^3}{48}$$

Second moment of area and the perpendicular axis theorem

The cross-section shown in Fig. 3.8 has an arbitrary shape. It has an area equal to A and the centroid of A is located at the point G.

Consider an element of area δA, at distance r from G, which has coordinates z and y with respect to the axes Gz, Gy and coordinates $\bar{z}$ and $\bar{y}$ with respect to the axes G$\bar{z}$, G$\bar{y}$ which are located at an angle θ relative to Gz, Gy. Then

$$I_z = \sum (\delta A y^2); \qquad I_y = \sum (\delta A z^2) \qquad (3.18)$$

and
$$I_{\bar{z}} = \sum (\delta A \bar{y}^2); \qquad I_{\bar{y}} = \sum (\delta A \bar{z}^2) \qquad (3.19)$$

Let

$J =$ the second moment of area of the cross-section about the x-axis (which passes through G at right angles to the diagram), then

$$J = \sum (\delta A r^2) \qquad (3.20)$$

J is called the rotational or *polar second moment of area* about G.

$$J = \sum (\delta A r^2) = \sum \delta A(z^2 + y^2) = \sum (\delta A z^2) + \sum (\delta A y^2)$$

That is

$$J = I_z + I_y \tag{3.21}$$

Similarly, because $r^2 = \bar{z}^2 + \bar{y}^2$:

$$J = I_{\bar{z}} + I_{\bar{y}} \tag{3.22}$$

Thus, the polar second moment of area of the cross-section about G is equal to the sum of the second moments of area about any pair of perpendicular axes through G in the plane of the cross-section. This is known as the *perpendicular axis theorem*.

Also, because J, the polar second moment of area about Gx, is a constant for any particular cross-section, it follows that $(I_{\bar{z}} + I_{\bar{y}})$ is a constant, irrespective of the value of θ in Fig. 3.8.

Some important points about bending

1. $I_{\bar{z}}$ and $I_{\bar{y}}$ both change their values as θ varies (Fig. 3.8) but, because $(I_{\bar{z}} + I_{\bar{y}}) = J$ = constant, as $I_{\bar{z}}$ increases $I_{\bar{y}}$ must decrease and vice versa. It follows that when $I_{\bar{z}}$ reaches its maximum value, $I_{\bar{y}}$ must be a minimum. It can be shown that there is a unique value of θ when $I_{\bar{z}}$ is a maximum and $I_{\bar{y}}$ is a minimum. The corresponding axes are called *the principal axes of the cross-section*. They are known respectively as *the major principal axis* (about which the second moment of area is a maximum for any axis in the plane of the cross-section passing through G) and *the minor principal axis* (about which the second moment of area is a minimum for any axis in the plane of the cross-section passing through G).

   Thus, I_{max} occurs about the major principal axis and I_{min} occurs about the minor principal axis . Both principal axes pass through G and lie in the plane of the cross-section.

2. If a cross-section is symmetrical about at least one axis, then the axis of symmetry is a principal axis. The other principal axis passes through G at right angles to the

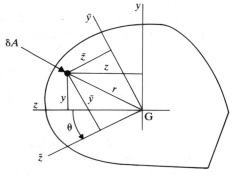

Figure 3.8   A cross-section of arbitrary shape

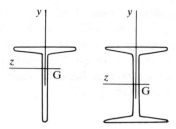

Figure 3.9 The positions of the principal axes of beam cross-sections when there is at least one axis of symmetry

axis of symmetry. In Fig. 3.9, the major principal axes are Gz and the minor principal axes are Gy for both cross-sections shown.

3. The locations of the principal axes and the corresponding maximum and minimum values of the second moment of area are functions of the area and shape of the cross-section. They are not related to the bending moment applied to the cross-section.

4. When a bending moment is applied to a beam cross-section in a plane which does not coincide with either of the two principal axes, the bending moment can always be resolved into the components about the principal axes.

5. The neutral axis of a cross-section coincides with a principal axis only when the bending moment is applied in the plane containing the other principal axis. Thus, although the locations of the principal axes of a cross-section never change, the position of the neutral axis depends upon the plane in which the bending moment is applied.

Section moduli of a cross-section

A section modulus Z is defined as

$$Z = \frac{\text{the second moment of area about a principal axis}}{\substack{\text{the greatest possible positive or negative distance from a} \\ \text{principal axis to a point on the cross-section}}}$$

Every cross-section therefore has four section moduli, two associated with each principal axis.

For example, the triangular cross-section of Fig. 3.7 has the following section moduli:

$$Z_{z1} = \frac{I_z}{\left(\frac{2}{3}D\right)} \qquad Z_{z2} = \frac{I_z}{\left(-\frac{D}{3}\right)}$$

$$Z_{y1} = \frac{I_y}{\left(\frac{B}{2}\right)} \qquad Z_{y2} = \frac{I_y}{\left(-\frac{B}{2}\right)}$$

$$(3.23)$$

The use of the section moduli can be illustrated by referring back to Eq. (3.8) which gives the bending stress as $\sigma_x = -My/I$. Thus, for bending about the z-axis (say), when $M = M_z$ and $I = I_z$, the maximum compressive stress will be $-M_z/Z_{z1}$ and the maximum tensile stress will be $-M_z/Z_{z2}$. To facilitate the quick calculation of these maximum compressive and tensile stresses in structural members, the values of section moduli are provided in the manuals of steelwork manufacturers.

Radii of gyration of a cross-section

If A is the area of a cross-section with a second moment of area I_z about the Gz principal axis, then the radius of gyration k_z about the axis Gz is defined by

$$I_z = Ak_z^2 \tag{3.24}$$

Similarly, the radius of gyration k_y about the Gy axis is defined by

$$I_y = Ak_y^2 \tag{3.25}$$

(Radii of gyration are not of immediate interest here, but they become very useful in the analysis of long columns.)

EXAMPLE 3.2

(a) For the beam cross-section shown in Fig. 3.10, find (i) the locations of the principal axis, (ii) the second moments of area I_z and I_y and (iii) the section moduli.

(b) If a bending moment of 1.1 kN m is applied to the cross-section in the xy plane (i.e., about the z-axis), find the distribution of normal stress (i.e., σ_x) on the cross-section.

(a) (i) The area of the cross-section $A = 732 \text{ mm}^2$. To locate the position of the centroid G, take moments of area about any line perpendicular to the z-axis. The

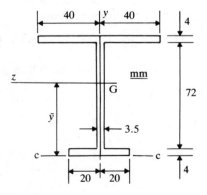

Figure 3.10 Example 3.2

line cc is convenient. Let G be situated at distance $\bar{y}$ above cc. Then, working in mm units:

moment of total area about cc = sum of moments of elements of area about cc

$$732\bar{y} = (80 \times 4 \times 78) + (72 \times 3.5 \times 40) + (40 \times 4 \times 2)$$

$$\bar{y} = 48.3 \text{ mm} = \text{location of G above cc}$$

The distance from Gz to the top fibre of the cross-section is therefore equal to 31.7 mm. Gy is an axis of symmetry. The principal axes are therefore Gy and Gz.
(ii) Using the parallel axis theorem to find the second moments of area of the top flange, the web and the bottom flange about Gz, then

$$I_z = \left[\left(\frac{80 \times 4^3}{12} \right) + (320 \times 29.7^2) \right] + \left[\left(\frac{3.5 \times 72^3}{12} \right) + (252 \times 8.3^2) \right]$$

$$+ \left[\left(\frac{40 \times 4^3}{12} \right) + (160 \times 46.3^2) \right] \text{mm}^4$$

that is
$$I_z = 752 \times 10^3 \text{ mm}^4$$

In the case of I_y, this is equal to the sum of the second moments of area about Gy of the three rectangles making up the total cross-section:

$$I_y = \left(\frac{4 \times 80^3}{12} \right) + \left(\frac{72 \times 3.5^3}{12} \right) + \left(\frac{4 \times 40^3}{12} \right) = 192.3 \times 10^3 \text{ mm}^4$$

(iii) The section moduli can now be calculated as follows:

$$Z_{z1} = \frac{752 \times 10^3}{31.7} \text{ mm}^3 = 23.7 \times 10^3 \text{ mm}^3$$

$$Z_{z2} = \frac{752 \times 10^3}{-48.3} \text{ mm}^3 = -15.6 \times 10^3 \text{ mm}^3$$

$$Z_{y1} = \frac{192 \times 10^3}{40} \text{ mm}^3 = 4.8 \times 10^3 \text{ mm}^3; \quad Z_{y2} = -Z_{y1} = -4.8 \times 10^3 \text{ mm}^3$$

(b) The stress distribution on the cross-section when a bending moment of 1.1 kN m ($= 1.1 \times 10^6$ N mm) is applied in the xy plane, i.e., $M_z = 1.1$ kN m, about the Gz axis, can be calculated from Eq. (3.8) (in which M is now called M_z), namely

$$\sigma_x = -M_z \frac{y}{I_z}$$

This distribution gives $\sigma_x = 0$ when $y = 0$ (i.e., at Gz, which is therefore the neutral axis when M_z is applied) and varies linearly from the minimum stress (i.e., the largest magnitude of the negative, or compressive, stress) at the top fibres of the cross-section (when $y = 31.7$ mm) to the maximum stress (i.e., the largest

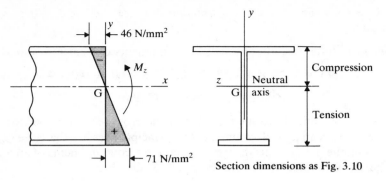

σ_x distribution due to $M_z = 1.1$ kN m

Figure 3.11   The distributions of normal stress, σ_x in Example 3.2

magnitude of the positive, tensile stress) at the bottom fibres of the cross-section (when $y = -48.3$ mm). Thus, in the top fibres:

$$\sigma_{x\,min} = -M_z \times \frac{31.7}{I_z} = -\frac{M_z}{Z_{z1}} = -\frac{1.1 \times 10^6}{23.7 \times 10^3} = \underline{-46 \text{ N/mm}^2}$$

and, in the bottom fibres:

$$\sigma_{x\,max} = -M_z \times \left(\frac{-48.3}{I_z}\right) = -\frac{M_z}{Z_{z2}} = \frac{1.1 \times 10^6}{15.6 \times 10^3} = \underline{71 \text{ N/mm}^2}$$

The corresponding stress distribution is shown in Fig. 3.11.

Problem 3.2   A universal beam (UB) has a doubly symmetrical cross-section of total depth over the flanges of 910 mm and a second moment of area about the major principal axis (i.e., I_z) of 3751×10^6 mm^4. The beam has a mass of 223 kg/m and is simply supported over a span of 9 m.

Calculate the maximum tensile and compressive bending stresses in the beam as a 400 kN wheel load moves slowly across the span.

3.2 Shear stresses in a beam due to a shear force

In general, the cross-section of a beam will be subjected to a bending moment and a shear force. Let us confine our attention here to cross-sections which are symmetrical about the y-axis and are subjected to a bending moment and a shear force which both act in the xy plane.

Such a case is shown in Fig. 3.12a, where the cantilever of length L and rectangular cross-section is subjected to an upward end load W.

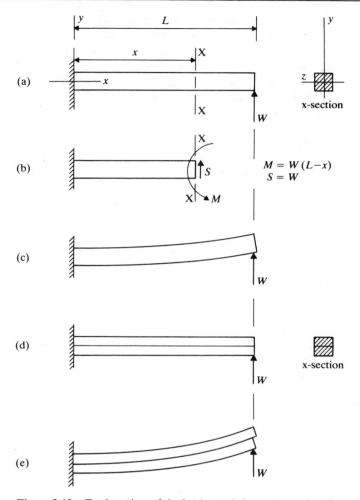

Figure 3.12   Explanation of the horizontal shear stresses in a beam

At the cross-section XX, distance x from A, the shear force S is equal to W and the bending moment M is equal to $W(L - x)$. (In the previous section, a moment about the z-axis was designated M_z but, because we are now considering only bending moments in the xy plane, we can drop the suffix.)

It is fairly clear that the shear force S must be the resultant of some sort of shear stress distribution acting vertically on the cross-section XX of the beam. What is not so obvious is that there is also a shear stress distribution acting horizontally, i.e., in the x direction, through the depth of the beam. That these horizontal shear stresses must exist is demonstrated by consideration of Figs 3.12c, 3.12d and 3.12e. Figure 3.12c shows the deflected shape of the cantilever due to the load W. Now suppose that the cantilever of Fig. 3.12a is replaced by two cantilevers, as shown in Fig. 3.12d, which lie one on top of the other and are not connected across the interface between them. Figure 3.12e shows how they will deflect under the load W. They will slide relative to each other along the interface. This happens because they act as two separate cantilevers. Thus, the lower face of the top cantilever is in tension and will therefore get longer, whilst the upper face of the bottom cantilever is in compression and gets shorter. This, of course, does not happen in the single, solid beam in Fig. 3.12c, and it follows that horizontal shear stresses must exist along the longitudinal section.

There is another way to demonstrate that these horizontal (x-direction) shear stresses exist, which also indicates their magnitude in relation to that of the vertical (y-direction) shear stress at any point through the depth of the cross-section. Consider the segment of beam of length dx in Fig. 3.13a. Suppose that it is part of a beam (such as the cantilever in Fig. 3.12a) along which the shear force S is constant.

Suppose the vertical shear stress on the element of height dy, situated at y from the x-axis, is τ. We can consider this to be constant over the differential distance dy. Vertical equilibrium of the element means that τ must act on both sides of the element as shown in Fig. 3.13a. It can be shown that equal and opposite shear stresses, also of magnitude τ, act along the top and bottom faces of the element as shown in Fig. 3.13b. These are known as *complementary shear stresses* and are dealt with in detail in Section 3.4.

The fact that, at any distance y from the x-axis, the vertical and horizontal shear stresses are equal in magnitude, remains true when there are normal (x-direction) stresses acting on the faces of the element, as there generally will be in a beam.

It is now necessary to find the values of the shear stresses through the depth of a beam when the shear force S applied to the cross-section is known.

Figure 3.14a shows the cross-section of a beam which is being subjected to bending and shear. The cross-section is symmetrical about the y-axis, but is otherwise of

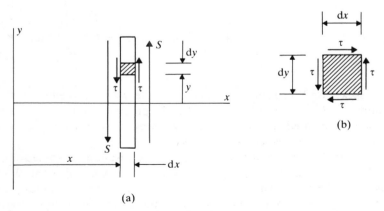

Figure 3.13   The shear stresses acting on an element in a short segment of a beam

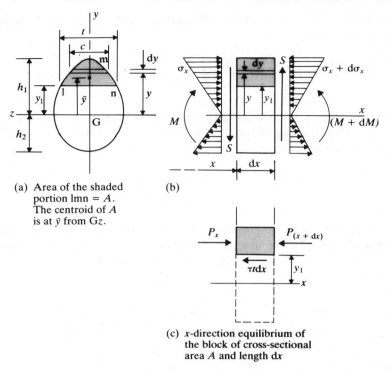

(a) Area of the shaded portion lmn = A. The centroid of A is at $\bar{y}$ from Gz.

(b)

(c) x-direction equilibrium of the block of cross-sectional area A and length dx

Figure 3.14 The determination of the shear stress acting at a particular level of the cross-section of a beam

arbitrary shape. The portion of the cross-section above $y = y_1$ has an area $\bar{A}$ and a width t at $y = y_1$. The element within $\bar{A}$, of width c and depth dy, has an area $d\bar{A}$.

At some distance x along the beam, the longitudinal element of length dx (Fig. 3.14b) is subjected to a shear force S. At x, the bending moment is M which changes to $(M + dM)$ over the distance dx, where $dM/dx = -S$ (see page 131). Notice that when S is positive, the change in bending moment over dx is negative.

The centroid of the cross-section is at G. Gz is a principal axis and is also the neutral axis. The bending stress at the left-hand end of the elemental strip of height dy at y from G, *acting in the direction shown in Fig. 3.14b*, is

$$\sigma_x = \frac{M}{I} y$$

Therefore, the force on the left-hand end of the elemental strip of cross-sectional area $t\, dy = d\bar{A}$, acting in the direction shown, is

$$\sigma_x\, d\bar{A} = \frac{M}{I} y\, d\bar{A}$$

It follows that the total force P_x on the left-hand face of the shaded block, acting in the direction shown in Fig. 3.14c, is

$$P_x = \int_{y_1}^{h_1} \frac{M}{I} y \, \mathrm{d}\bar{A}$$

where the integration is carried out over the area $\bar{A}$, bounded by lmn on Fig. 3.14a.

Similarly, the force acting on the right-hand face of the shaded block, *acting in the direction shown in Fig. 3.14c*, is

$$P_{(x+\mathrm{d}x)} = \int_{y_1}^{h_1} \left(\frac{M + \mathrm{d}M}{I}\right) y \, \mathrm{d}\bar{A}$$

These forces, together with the shear force $\tau t \, \mathrm{d}x$, *i.e., the horizontal shear stress τ (at y_1 from the x-axis) multiplied by the area of the bottom face of the shaded block*, must keep the shaded block in x-directional equilibrium. Therefore

$$P_x - P_{(x+\mathrm{d}x)} - \tau t \, \mathrm{d}x = 0$$

that is

$$\int_{y_1}^{h_1} \frac{M}{I} y \, \mathrm{d}\bar{A} - \int_{y_1}^{h_1} \left(\frac{M + \mathrm{d}M}{I}\right) y \, \mathrm{d}\bar{A} - \tau t \, \mathrm{d}x = 0$$

therefore

$$\tau = -\frac{\mathrm{d}M}{\mathrm{d}x} \frac{1}{It} \int_{y_1}^{h_1} y \, \mathrm{d}\bar{A} \tag{3.26}$$

In Eq. (3.26), the quantity $\int_{y_1}^{h_1} y \, \mathrm{d}\bar{A}$ is the moment about Gz (Fig. 3.14a) of the shaded area $\bar{A}$ above $y = y_1$. Suppose the centroid of this area is located at distance $\bar{y}$ from Gz (Fig. 3.14a). Then

$$\int_{y_1}^{h_1} y \, \mathrm{d}\bar{A} = \bar{A}\bar{y}$$

Also

$$\frac{\mathrm{d}M}{\mathrm{d}x} = -S$$

Therefore, Eq. (3.26) can be written as

$$\tau = \frac{S}{It} \bar{A}\bar{y} \tag{3.27}$$

Equation (3.27) gives the value of the shear stress (both horizontal *and vertical*) at a distance y_1 from Gz where the width of the beam is t, and $\bar{A}\bar{y}$ is the moment about Gz of the area of the cross-section lying between y_1 and the extremity of the cross-section at $y = h_1$.

The use of Eq. (3.27) will now be illustrated by its application to a rectangular cross-section and to an I-section beam.

Shear stress distribution in a beam of rectangular cross-section

The simply supported beam shown in Fig. 3.15a has the shear force diagram in Fig. 3.15b. The free-body diagram for the length between A and the section XX is drawn in Fig. 3.15c. At the section XX, there is a shear force S ($=4.8$ kN) and a bending moment M (with which we are not concerned here).

We will investigate the shear stress distribution through the depth of the beam at section XX if the beam has a rectangular cross-section of dimensions $b \times d$ as shown in Fig. 3.16a.

Using Eq. (3.27), we can write the expression for τ at any distance y_1 from the neutral axis Gz.

For the shaded area between $y = y_1$ and $y = d/2$:

$$\bar{A} = b\left(\frac{d}{2} - y_1\right), \qquad \bar{y} = \frac{1}{2}\left(\frac{d}{2} + y_1\right)$$

$$\bar{A}\bar{y} = \frac{b^2}{2}\left(\frac{d}{4} - y_1^2\right), \qquad I = \frac{bd^3}{12}$$

Therefore, from Eq. (3.27), the shear stress at y_1 from Gz is given by

$$\tau = \frac{S}{It}\,\bar{A}\bar{y} = \frac{S}{b}\frac{\dfrac{b}{2}\left(\dfrac{d^2}{4} - y_1^2\right)}{\dfrac{bd^3}{12}} = \frac{3}{2}\frac{S}{bd}\left[1 - \frac{2y_1^2}{d}\right] \tag{3.28}$$

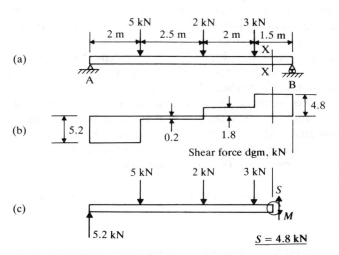

Figure 3.15 Illustrating the determination of the shear stress distribution at a cross-section of a beam

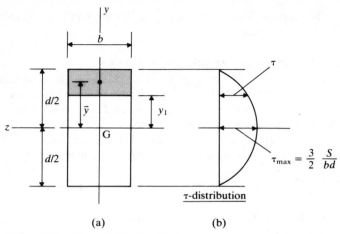

Figure 3.16   Shear stress distribution across a rectangular section

This is a parabolic variation as shown on Fig. 3.16b. The maximum value of τ occurs at the mid-height of the cross-section, i.e., when $y_1 = 0$, and is given by

$$\tau_{\text{max}} = \frac{3}{2} \frac{S}{bd} \tag{3.29}$$

Notice that τ_{max} is equal to $\frac{3}{2}$ times the mean shear stress S/bd.

Shear stress distribution in an I-section beam

It is evident from Fig. 3.16b that the shear stress at the top and bottom of the cross-section is zero. This is because the horizontal free surfaces at the top and bottom cannot carry shear stresses in their own planes. It follows that the vertical shear stress must also be zero at these extremities of the cross-section.

Looking now at the I-section in Fig. 3.17a, it is evident that the same argument must apply along the top and bottom extremities of the flanges. Such flanges are generally relatively thin, so the vertical shear stresses generated between the top and bottom faces (where they must be zero) are generally small enough to be neglected.

Nevertheless, the flanges do, of course, carry x-direction bending stresses (σ_x). The shear stresses associated with the changes in σ_x from one section of the beam to another (due to the change in bending moment between the two sections) act *horizontally* along the flanges with the complementary shear stress acting across the flange in the plane xy (Fig. 3.17b).

These stresses can be found by using Eq. (3.27), as illustrated in the following example.

EXAMPLE 3.3

A shear force of $S = 400$ kN is applied to the cross-section shown in Fig. 3.18a. S is applied vertically along the centre-line of the web, i.e., along Gy.

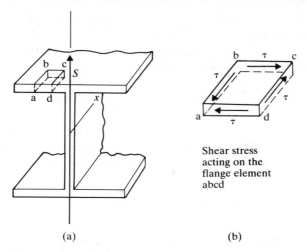

Shear stress
acting on the
flange element
abcd

(a)   (b)

Figure 3.17   The direction of the shear stresses in the flange of an I-beam

For the cross-section, the second moment of area is given by

$$I = \frac{1}{12}(300 \times 470^3 - 290 \times 430^3) = 674 \times 10^6 \text{ mm}^4$$

(Remember, I-values can be added and subtracted so the value above is found by subtracting the I-value of the 'empty' space on the cross-section from the I-value of the solid rectangular section of breadth 300 mm and depth 470 mm. However, care must be taken with accuracy when finding the relatively small difference between two large quantities, as in this case.)

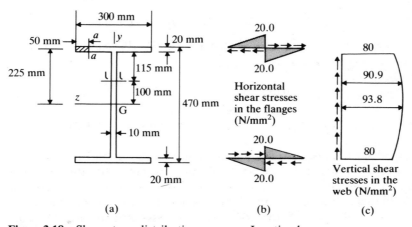

(a)   (b)   (c)

Figure 3.18   Shear stress distribution across an I-section beam

Consider the shear stress at the section aa on the top flange. In Eq. (3.27), the value of $\bar{A}$ in this case is the shaded area to the left of aa, i.e., $\bar{A} = (50 \times 20) = 1000 \text{ mm}^2$ and $\bar{y}$ is 225 mm (see Fig. 3.18a), and t, the flange thickness $= 20$ mm. Using Eq. (3.27) with these values and $S = 400 \text{ kN} = 400 \times 10^3 \text{ N}$, then

$$\tau_{aa} = \frac{400 \times 10^3}{(674 \times 10^6)20} (50 \times 20)(225) = 6.7 \text{ N/mm}^2$$

The growth of τ along the flange is linear with the length of the shaded area. It is therefore zero at the tip of the flange and becomes a maximum at the centre-line of the flange. This maximum value of the horizontal shear stress in the flange (at the y-axis) is therefore given by

$$\tau_y = \left(\frac{150}{50}\right) \times 6.7 \text{ N/mm}^2 = 21.0 \text{ N/mm}^2$$

The distribution of the horizontal shear stresses in all the flanges is shown in Fig. 3.18b.

The distribution of the shear stress in the web can now be calculated from Eq. (3.27) Thus, at section ll, 100 mm from Gz, in Eq. (3.28):

$$\bar{A} = (300 \times 20) + (115 \times 10) = 7150 \text{ mm}^2$$

$$\bar{y} = [(300 \times 20)225 + (115 \times 10)157.5]/7150 = 214.1 \text{ mm}$$

$$I = 674 \times 10^6 \text{ mm}^4$$

$$t = 10 \text{ mm}$$

Therefore $\quad \tau_{ll} = \dfrac{400 \times 10^3}{(674 \times 10^6)(10)} \times (7150)(2141) = 90.9 \text{ N/mm}^2$

The maximum shear stress will occur at G where $y_1 = 0$ and

$$\bar{A} = (300 \times 20) + (215 \times 10) = 8150 \text{ mm}^2$$

$$\bar{y} = [(300 \times 20)225 + (215 \times 10)107.5]/8150 = 194 \text{ mm}$$

$$I = 674 \times 10^6 \text{ mm}^4$$

$$t = 10 \text{ mm}$$

Therefore $\quad \tau_{max} = \dfrac{400 \times 10^3}{(674 \times 10^6)10} (8150)(194) = 93.8 \text{ N/mm}^2$

The total shear stress distribution on the web is shown on Fig. 3.18c.

It is common practice to assume, and with excellent justification, that the web of an I-section beam carries the entire vertical shear load S. For simplicity, it is usually assumed that it is uniformly distributed such that, at all points on the web, the vertical shear stress τ is equal to S divided by the web cross-sectional area. Thus, in this case, a uniform vertical shear stress of $(400 \times 10^3)/(430 \times 10) = 93 \text{ N/mm}^2$ would be assumed. Examination of the true distribution shown on Fig. 3.18c shows that such an assumption is very reasonable.

Problem 3.3 Find the shear stress distribution on a triangular cross-section such as that shown in Fig. 3.7 if $D = 90$ mm, $B = 60$ mm, and a vertical shear force of 150 kN acts along Gy.

3.3 Stress distributions in members of circular cross-section due to a torque

Torsion of a solid circular shaft

Figure 3.19 represents a solid circular shaft of length L and radius R. It is being subjected to equal and opposite torsional moments about its longitudinal axis, x. The torsional moment T is also called a twisting moment or a torque.

The questions we now address are, firstly, what is the stress pattern in the shaft which transmits T from one end to the other and, secondly, how does the shaft deform as a result of the presence of T and the consequent stresses in the shaft.

An appreciation of what is happening to the shaft is obtained by imagining it to be made up of a large number of thin circular discs all stuck together by glue spread across their flat surfaces. It is easy to see that when a torque is applied to a shaft constructed in this way, the discs will tend to rotate relative to one another and will be prevented from doing so by the glue. It is also possible to see that, should such a relative movement take place between one disc and its neighbour, it would be a rotation about the axis of the shaft, with points at some radius r moving tangentially. To resist such relative movement, the interface glue would therefore generate a shear stress acting tangentially at the interface between two discs and, because the relative tangential movement would increase with increasing distance from the longitudinal axis, one can deduce that the tangential shear stress required to prevent such movement would also increase with distance from the axis.

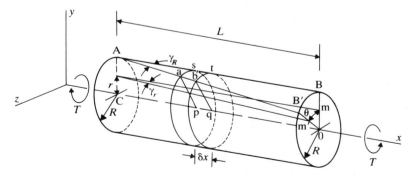

Figure 3.19 Deformation of a solid circular shaft when subjected to torsion

Returning now to Fig. 3.19, suppose that when T is applied at the right-hand end, the left-hand end does not move at all. It is like trying to remove an obstinate wood screw which will not budge; the screw resists the torque applied to it by generating an equal and opposite torque, but it remains fast in the wood. However, the shaft of the screwdriver is made of elastic material and it twists. In Fig. 3.19, a straight line AB drawn on the surface of the shaft parallel to the x-axis before T was applied becomes AB′ when T is applied, i.e., the shaft twists elastically through the angle θ over the length L. The line AB moves through the angle γ_R to AB′.

There are two important assumptions which are implicit in the behaviour of the shaft as depicted in Fig. 3.19.

The first is that cross-sections of the shaft, which were plane before T was applied, remain plane after the application of T. Thus, it is assumed that the radius OB moves to its new position OB′ and, in doing so, remains in the same plane at right angles to the x-axis.

The second assumption is that, during this movement, OB (and all radii) has remained straight.

To look more closely at what is happening in the shaft, consider the short length δx (Fig. 3.19) bounded by two cross-sections perpendicular to the x-axis. The top part of this disc is reproduced to a larger scale in Fig. 3.20a. When AB moved to AB′, the line st (which is on AB) moved to ab′, which is reproduced on Fig. 3.20a. The line ab′ has also moved through the angle γ_R. This is therefore the angle it makes with ab (Fig. 3.20a) which is parallel to the x-axis. An element abcd on the surface of the shaft has changed it shape to ab′c′d because of the application of T.

Similarly, at radius r, the element of original shape efgh has changed its shape to

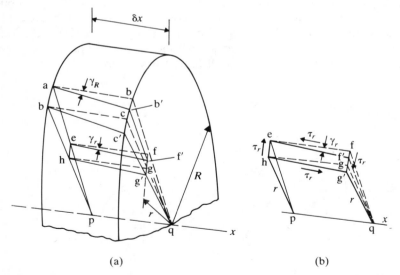

(a)   (b)

Figure 3.20   The deformation of an element of the shaft subjected to torsion

ef'g'h. The shear stress τ_r acting along the faces fg and he of the element arise directly from the presence of T. We saw in Section 3.2 (and will prove in Section 3.4) that such shear stresses are always accompanied by complementary shear stresses of equal magnitude acting on the other two faces of the elements, as shown in Fig. 3.20b.

Shear strain is defined as the change in a right angle on elements such as abcd and efgh. Therefore, the shear strain at radius R is γ_R and the shear strain at radius r is γ_r, both of which are indicated in Fig. 3.20.

The elastic relationship between shear stress and shear strain is

$$\frac{\text{shear stress}}{\text{shear strain}} = \text{elastic shear modulus}$$

The elastic shear modulus, denoted by G, is, like E, an empirical property of the material. Thus, in the shaft under torsion:

$$\frac{\tau_r}{\gamma_r} = \frac{\tau_R}{\gamma_R} = G \tag{3.30}$$

Referring to Fig. 3.19:

$$\text{the arc mm}' = r\theta = L\gamma_r \tag{3.31}$$

Replacing γ_r in Eq. (3.31) by τ_r/G from Eq. (3.30), then:

$$\tau_r = Gr\left(\frac{\theta}{L}\right) \tag{3.32}$$

i.e., the shear stress in the shaft is directly proportional to the distance r from the centre of the shaft. Note that (θ/L) is the twist per unit length of the shaft.

Equation (3.32) is illustrated in Fig. 3.21; the shear stress at the centre of the shaft is zero and has its maximum value τ_R at the outside of the shaft, given by

$$\tau_R = GR\left(\frac{\theta}{L}\right) \tag{3.33}$$

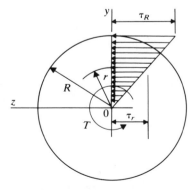

Figure 3.21   Distribution of shear stress across the section of a shaft

Consider now Fig. 3.22 which shows a circular annulus, within the cross-section, of width dr and radius r, carrying the shear stress τ_r.

On every short length ds of the annulus, the shear stress τ_r, when multiplied by the area on which it acts, produces a tangential force with a moment about O, the centre of the shaft. The total moment about O, produced by the shear stresses over the whole cross-section, must be equivalent to the torque T which caused the shear stresses.

Area of the annulus of radius r $= 2\pi r\, dr$

Tangential force acting on the annulus $= 2\pi r\tau_r\, dr$

Moment about O of this force $= (2\pi r\tau_r\, dr)r = 2\pi r^2\tau_r\, dr$

$$\therefore \text{ total moment about O} = T \qquad = \int_0^R 2\pi r^2\tau_r\, dr \qquad (3.34)$$

From Eq. (3.32):
$$\tau_r = Gr\left(\frac{\theta}{L}\right)$$

$\therefore$ in Eq. (3.34):
$$T = 2\pi G\left(\frac{\theta}{L}\right)\int_0^R r^3\, dr$$

i.e.
$$T = G\left(\frac{\theta}{L}\right)\frac{\pi R^4}{2} \qquad (3.35)$$

The quantity $\pi R^4/2$ is the polar second moment of area of the cross-section, J, as described in Section 3.1. Equation (3.35) can therefore be written as

$$T = G\left(\frac{\theta}{L}\right)J \qquad (3.36)$$

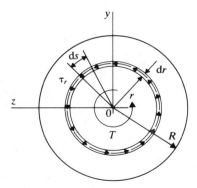

Figure 3.22 Illustrating the direction of the shear stresses on an annulus within the cross-section of a shaft

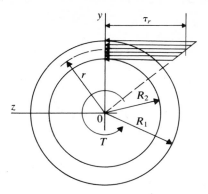

Figure 3.23 Distribution of shear stress on a section of a hollow thick-walled circular tube subjected to torsion

Combining Eq. (3.36) with Eq. (3.32), we get

$$\frac{T}{J} = \frac{\tau_r}{r} = G\left(\frac{\theta}{L}\right) \tag{3.37}$$

Torsion of a hollow (thick-walled) circular tube

A hollow shaft is a circular tube with an outside radius of R_1 and an inside radius of R_2, as shown in Fig. 3.23.

The arguments used to establish Eq. (3.37) are valid for this case, so there is no need for any further analysis. It follows that the shear stress distribution across the wall of the circular shaft is as shown on Fig. 3.23 where τ_r varies between $r = R_1$ and $r = R_2$ in accordance with Eq. (3.37).

The value of J for the hollow shaft is

$$J = \frac{\pi R_1^4}{2} - \frac{\pi R_2^4}{2} = \frac{\pi}{2}(R_1^4 - R_2^4)$$

or

$$J = \frac{\pi}{32}(D_1^4 - D_2^4) \tag{3.38}$$

where D_1 and D_2 are respectively the outside and inside diameters of the hollow shaft.

Torsion of a thin-walled circular tube

When the difference between R_1 and R_2 becomes small, such that t, the wall thickness of the tube, is of the order of $1/25$ of the mean radius R_m, see Fig. 3.24, where $R_m = (R_1 + R_2)/2$, then the variation of τ_r across t is small enough to be disregarded for most practical purposes. That is, the shear stress in the wall, τ, can then be regarded as constant across the wall thickness.

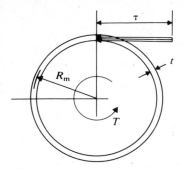

Figure 3.24 Shear stresses on a section of a thin-walled circular tube subjected to torsion

In this case, Eq. (3.37) becomes

$$\frac{T}{J} = \frac{\tau}{R_m} = G\left(\frac{\theta}{L}\right) \tag{3.39}$$

where

$$R_m = (R_1 + R_2)/2$$

$$t = R_1 - R_2 \tag{3.40}$$

and

$$J \simeq (2\pi R_m t)R_m^2 = 2\pi R_m^3 t$$

This approximate value of J is derived by taking the cross-sectional area of the wall, which is approximately $2\pi R_m t$, and multiplying it by R_m^2 to give its second moment of area about the x-axis.

Transmission of power through circular shafts

Probably the most common circumstance of a shaft being subjected to shear stresses due to torsion is the use of the shaft to transmit power, as in a motor car, a ship or innumerable machine applications. Generally, it is more efficient to use a hollow shaft rather than a solid one, because in the former case, a high shear stress is generated in all the material used (see Fig. 3.23), leading to a saving in material compared with a solid shaft to carry the same torque.

In a shaft the power transmitted = (torque × rotational speed) where the rotational speed is measured in radians per unit time, i.e.,

$$P = T\omega \tag{3.41}$$

The use of Eq. (3.41) is illustrated in Example 3.5 below.

EXAMPLE 3.4

A hollow shaft, of external diameter D_1 and internal diameter D_2, where $D_2 = 2D_1/3$, is required to transmit a torque of 125 kN m. Find the minimum value of D_1 if both the following conditions must be satisfied.

 (i) The maximum shear stress in the shaft must not exceed 60 N/mm² and
(ii) the twist of the tube over a length of 10 m must not exceed 1°.

Sketch the shear stress distribution across the wall of the tube. The shear modulus $G = 84 \times 10^3$ N/mm^2.

(i) $\tau_{max} \not> 60$ N/mm^2. Use $T/J = \tau_r/r$ from Eq. (3.37).

$$T = 125 \times 10^6 \text{ N mm}, \quad r = R = D_1/2, \quad J = \frac{\pi}{32}(D_1^4 - D_2^4), \quad D_2 = 2D_1/3$$

$$J = \frac{\pi}{32}\left[D_1^4 - \left(\frac{2}{3}\right)^4 D_1^4 \right] = 0.078 D_1^4$$

Therefore, using Eq. (3.37):

$$\frac{125 \times 10^6}{0.078 D_1^4} = 60 \times \frac{2}{D_1}, \text{ which gives } \underline{D_1 = 237 \text{ mm}}$$

(ii) The angle of twist, $\theta \not> 1°$ when $L = 10$ m. Use $T/J = G(\theta/L)$ from Eq. (3.37). Converting θ to radians, then $\theta \not> (1 \times \pi/180)$ radians. Therefore

$$\frac{125 \times 10^6}{0.078 D_1^4} = 84 \times 10^3 \left(\frac{\pi}{180} \times \frac{1}{10 \times 10^3} \right), \text{ which gives } \underline{D_1 = 324 \text{ mm}}$$

Therefore, the minimum allowable value of $\underline{D_1 = 324 \text{ mm}}$ if neither condition is to be violated.

Putting $D_1 = 324$ mm, then $J = 0.078(324)^4$ mm$^4 = 8.6 \times 10^8$ mm^4:

$$\frac{D_2}{2} = \frac{1}{2} \times \frac{2}{3} \times 324 = 108 \text{ mm}$$

Therefore at $r = D_2/2$ $\tau_r = \dfrac{125 \times 10^6}{8.6 \times 10^8} \times 108 = \underline{15.7 \text{ N/mm}^2}$

and at $r = D_1/2$, $\tau_r = 15.7 \times \dfrac{324}{2 \times 108} = \underline{23.7 \text{ N/mm}^2}$

The shear stress distribution is shown on Fig. 3.25.

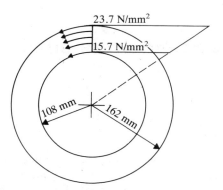

Figure 3.25   Example 3.4

EXAMPLE 3.5

A solid shaft is to transmit 500 kW at 600 r.p.m. Find the minimum required shaft diameter if the maximum shear stress in the shaft must not exceed 40 N/mm².

$$P = \text{power transmitted} = (\text{torque}) \times (\text{rotational speed}) = T \times \omega$$

$$\tau_{max} \not> 40 \text{ N}, \quad P = 500 \text{ kW}, \quad 1 \text{ kW} = 1 \text{ kN m/s}, \quad \omega = \frac{600}{60} \times 2\pi \text{ radians/sec}$$

From Eq. (3.41): $T = \dfrac{P}{\omega} = \dfrac{500 \times 10^6}{10 \times 2\pi} = 7.96 \times 10^6 \text{ N mm}$

$$J = \frac{\pi D^4}{32} \quad \text{where } D = \text{required diameter}$$

∴ from Eq. (3.37):

$$\frac{7.96 \times 10^6}{\pi D^4 / 32} = \frac{40}{D/2} \quad \text{from which } \underline{D = 100.5 \text{ mm}}$$

Problem 3.4   Two thin-walled circular tubes, of the same length but of different mean diameters, are assembled concentrically. The ends of both tubes are fixed to rigid circular plates, one at each end (Fig. 3.26). The outside tube is steel, of mean diameter $2d$, wall thickness t, and shear modulus G_s. The inside tube is copper, of mean diameter d, wall thickness t, and shear modulus G_c. If a torque T is applied to this assembly, find in terms of G_s and G_c:

(i) the ratio of the shear stress in the steel tube to that in the copper tube,
(ii) the ratio of the torque transmitted by the steel tube to that transmitted by the copper tube.

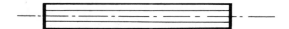

Figure 3.26   Problem 3.4

B. Complex stresses and strains and stress–strain relationships

3.4 The analysis of stress

Definition of a stress

Consider the solid, three-dimensional body shown in Fig. 32.7a. It has any general shape, is subjected to a set of external forces P_1 to P_6, and is in equilibrium, i.e., it is stationary. It follows that the forms P_1 to P_6 are also in equilibrium. Because the

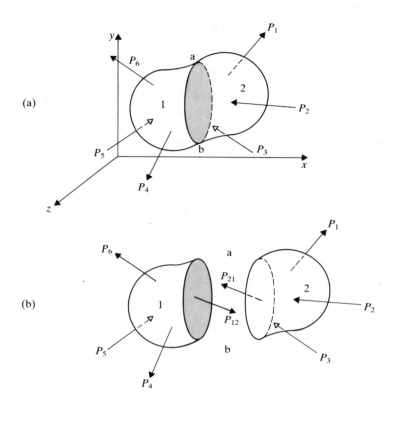

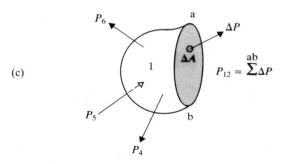

Figure 3.27   Forces within a body

points of application of the forces are separated physically, they must communicate with one another through the body itself. The body is transmitting the forces and, as a result, is said to be stressed, or in a state of stress. It is necessary to clarify exactly what this means.

Imagine a plane ab which divides the body into two parts, part 1 to the left of the plane and part 2 to the right. There are internal forces acting across the plane ab, and these forces hold together parts 1 and 2 of the body. Suppose the resultant of the forces acting on part 1 from part 2 is P_{12} and the resultant of the forces acting on part 2 from part 1 is P_{21}.

If, therefore, the body is cut into its two parts by slicing along the plane ab and these resultant (internal) forces are applied to the two parts as external forces, the two parts will each remain in equilibrium as shown in Fig. 3.27b. Also, equilibrium across the plane ab clearly demands that

$$P_{12} + P_{21} = 0 \tag{3.42}$$

Figures 3.27a and 3.27b simply demonstrate a fundamental law of mechanics which says that if a body is in equilibrium under the action of external forces, then any arbitrary portion of the body must be in equilibrium under the action of the external and internal forces acting on that portion. Thus, in Fig. 32.7b, part 1 is in equilibrium under the actions of P_4, P_5, P_6 (which are external forces) and P_{12} (which is an internal force).

Let us concentrate on part 1 of the body in Fig. 3.27b and, in particular, on the exposed face ab. The force P_{12} is the resultant of the forces distributed across the surface but it does not tell us anything about that distribution. Suppose that on any small area, ΔP acts. The resultant force P_{12} is the addition of all such small forces across the face ab, that is

$$P_{12} = \overset{ab}{\Sigma} \Delta P \tag{3.43}$$

Now considering only ΔP acting on ΔA, the stress at this point is defined by f, where

$$f = \underset{\Delta A \to 0}{\text{limit}} \frac{\Delta P}{\Delta A} \tag{3.44}$$

and acts in the same direction as ΔP.

The force ΔP acting on ΔP can be resolved into two components, one component acting normal to the surface ab and the other acting tangential to the surface. Let these two components by ΔP_N and ΔP_T respectively. Then

$$\text{the normal (or } direct\text{) stress at the point, } \sigma = \underset{\Delta A \to 0}{\text{limit}} \frac{\Delta P_N}{\Delta A} \tag{3.45}$$

and

$$\text{the tangential (or } shear\text{) stress at the point, } \tau = \underset{\Delta A \to 0}{\text{limit}} \frac{\Delta P_T}{\Delta A} \tag{3.46}$$

These two 'types' of stress, namely, direct stress and shear stress, are the only types of stress used in solid mechanics and structural analysis. In simple terms, σ is the direct force per unit area and τ is the shear force per unit area.

The general state of uniform stress on a three-dimensional element

We now want to examine the full range of stresses which might act at a point within a stressed body. Suppose that the point is within the three-dimensional rectangular element shown in Fig. 3.28 and, because the element is of differential size, $dx \times dy \times dz$, we will neglect any variation of stresses across its surface and assume that they are equal to the stresses at the point itself. The differential element is therefore said to be in a state of uniform stress.

The total stress condition is shown in Fig. 3.28. At first sight, this is a formidable picture so it is a good idea to build it up step by step. You will find that it is very helpful to copy the diagram as it is described.

The positive directions of all the stresses acting on the element are defined by Fig. 3.28 and the sign convention is consistent with that used for forces in Section 1.5. Accordingly, the positive faces of the element (i.e., those faces from which the outgoing normals are in the positive directions of the axes) are bfgc, aefb and abcd. As for forces, stresses which act in a positive direction on a positive face are said to be positive.

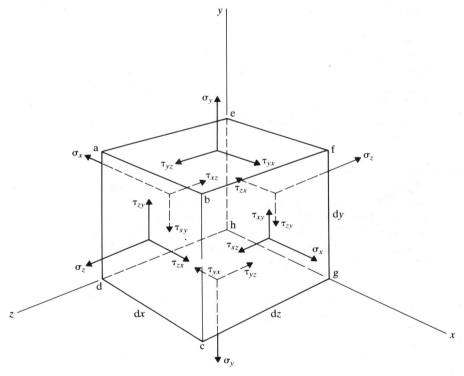

Figure 3.28   General state of stress on a 3-dimensional element

Consider only the positive x-face bfcg for the moment. There are three stresses acting on this face: a normal stress σ_x and the two directional components of the shear stress, namely, τ_{xy} (on the positive x-face in the direction of y, hence the suffix xy) and τ_{xz} (on the positive x-face in the direction of z).

These three stresses σ_x, τ_{xy} and τ_{xz} fully describe the stress condition on the face bfgc because there is no other possible stress component which cannot be incorporated within them.

Similarly, the stresses on the positive y-face and on the positive z-face can be drawn in as shown. Check the suffices on the shear stresses in accordance with the rule described for τ_{xy} and τ_{xz}.

The stresses acting on the negative faces of the element (i.e., those faces opposite the positive faces) follow from our assumption of uniform stress. If, therefore, there is a uniform tensile stress σ_x acting throughout the element in the x direction, then the stress on the negative x-face (which has the same area as the positive x-face) must be equal and opposite. So it is with all the other stresses including, of course, the shear stresses.

Thus, although the element has six faces, with three stresses on each face, there are only nine different stresses involved. These nine stresses fully define the uniform state of stress of the element and they can be conveniently written in the form of a *stress array*, F, where

$$F = \begin{bmatrix} \sigma_x & \tau_{xy} & \tau_{xz} \\ \tau_{yx} & \sigma_y & \tau_{yz} \\ \tau_{zx} & \tau_{zy} & \sigma_z \end{bmatrix} \tag{3.47}$$

The elements of F define the stress state of the element with respect to the axes x, y and z. This is easy to see if you imagine each stress in the array to have a numerical value, but we will come back to this shortly. In the meantime, it is possible to simplify the expression for F.

Complementary shear stress

The stress array F suggests that there are six independent values of shear stress. In fact, there are only three independent values of shear stress which, together with the three normal stresses, fully define the stress state of the element. This is easily demonstrated by reference to Fig. 3.29 and by writing the rotational equilibrium equation for the element in each of its three orthogonal planes.

In Fig. 3.29a, taking moments about d, gives

$$(\tau_{xy}\,dy\,dz)\,dx - (\tau_{yz}\,dx\,dz)\,dy = 0 \text{ hence } \tau_{xy} = \tau_{yx}$$

In Fig. 3.29b, taking moments about g, gives

$$(\tau_{yz}\,dx\,dz)\,dy - (\tau_{zy}\,dx\,dy)\,dz = 0 \text{ hence } \tau_{yz} = \tau_{zy}$$

In Fig. 3.29c, taking moments about e, gives

$$(\tau_{zx}\,dx\,dy)\,dz - (\tau_{xz}\,dy\,dz)\,dx = 0 \text{ hence } \tau_{zx} = \tau_{xz}$$

$$\left.\vphantom{\begin{matrix}1\\1\\1\\1\\1\\1\end{matrix}}\right\} \tag{3.48}$$

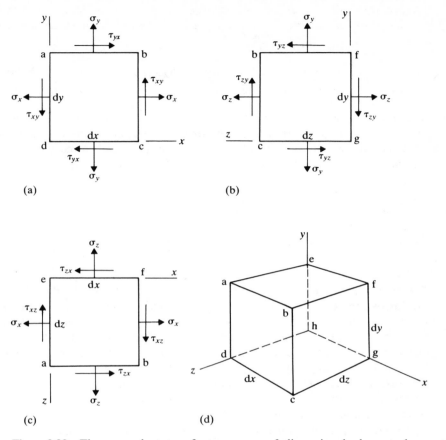

Figure 3.29 The general state of stress on a 3-dimensional element demonstrated in 2-dimensional diagrams

These equal-magnitude pairs of shear stresses, each pair acting in the same plane along adjacent faces of the element, are called *complementary shear stresses* and were first introduced in Section 3.2. It follows that the stress array, Eq. (3.47), is symmetrical about the leading diagonal.

EXAMPLE 3.6
To illustrate (i) the symmetry of the stress array (because of complementary shear stresses) and (ii) the superposition of stress arrays.

A solid body is acted upon separately by two sets of external forces. At a certain point in the body, the stress state, measured with respect to a set of orthogonal axes xyz is given by the stress array F_1 when the first set of forces acts. The second set of forces produces the stress array F_2 at the same point and with respect to the same set of axes. What is the stress state at the point if both sets of forces act simultaneously?

$$F_1 = \begin{bmatrix} 30 & 10 & 20 \\ 10 & -60 & 0 \\ 20 & 0 & 150 \end{bmatrix} \text{N/mm}^2; \qquad F_2 = \begin{bmatrix} 0 & 50 & 50 \\ 50 & 0 & -50 \\ 50 & -50 & -10 \end{bmatrix} \text{N/mm}^2$$

Solution   Any stress component is proportional to the corresponding component of force. Therefore, if several force components are applied, the corresponding total stress component is the sum of the stresses due to each force component acting separately. It follows that, when both sets of forces act together, the resulting stress state is:

$$F_1 + F_2 = \begin{bmatrix} 30 + 0 & 10 + 50 & 20 + 50 \\ 10 + 50 & -60 + 0 & 0 - 50 \\ 20 + 50 & 0 - 50 & 150 - 10 \end{bmatrix} \text{N/mm}^2 = \begin{bmatrix} 30 & 60 & 70 \\ 60 & -60 & -50 \\ 70 & -50 & 140 \end{bmatrix} \text{N/mm}^2$$

Change of stress axes

If a body is subjected to a set of external forces, then, provided these forces remain unchanged, the stress state of any particular point within the body must also remain unchanged. But the definition of this stress state at the point, expressed in terms of the numbers entered into the stress array, is dependent upon the orientation of the element at the point in question, that is, the orientation of the chosen orthogonal axes. If, therefore, the values in the stress array are known for a given set of axes, it follows (because the stress state remains unchanged) that the values in the stress array corresponding to any other set of axes can be deduced. To demonstrate this fact in three dimensions is rather protracted, so we will confine attention to an element which is stressed only in two dimensions, i.e., in the plane xy. This is known as a state of *plane stress* and reduces the stress state to that described by Fig. 3.29a, the stresses in Fig. 3.29b and Fig. 3.29c all now being zero, except σ_x and σ_y.

The analysis of plane stress

The plane stress condition is illustrated in Fig. 3.30. Note that there is no stress on the faces abcd and efgh, and that τ_{xz} and τ_{yz} do not exist. The stress array is therefore

$$F = \begin{bmatrix} \sigma_x & \tau_{xy} & 0 \\ \tau_{yx} & \sigma_y & 0 \\ 0 & 0 & 0 \end{bmatrix} \text{with respect to the axes } xyz \qquad (3.49)$$

(Remember that $\tau_{yx} = \tau_{xy}$.)

Let us now consider how we might define this state of stress with respect to any other orthogonal axes $\bar{x}\bar{y}$ which are orientated at an angle θ to xy, where θ is measured positive anticlockwise. To do this, the wedge aekjdh indicated in Fig. 3.30 and isolated in Fig. 3.31 is examined.

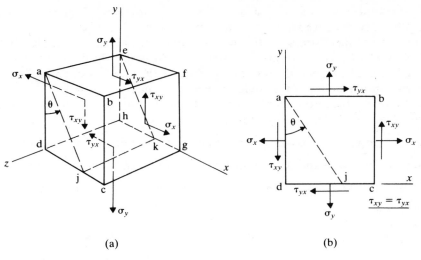

(a)   (b)

Figure 3.30   An element in a state of plane stress in the x–y plane

This wedge is a portion of the element previously considered. It is therefore in equilibrium under the actions of the forces on its faces, i.e., referring to Fig. 3.30b, the forces on the faces ad, dj and ja. We already know about the stresses on the faces ad and dj. The normal to the face ja (Fig. 3.30b) lies at an angle θ from the direction of x. Let the normal and shear stresses acting on the face aj of the wedge be $\sigma_{\bar{x}}$ and $\tau_{\bar{x}\bar{y}}$ respectively.

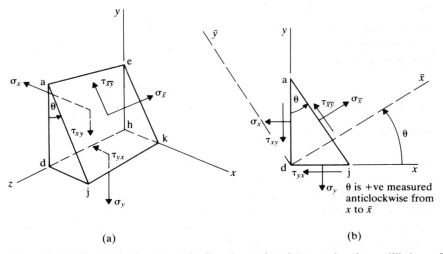

(a)   (b)

Figure 3.31   To obtain the stresses in directions other than x and y, the equilibrium of a wedge-shaped portion of the element is considered

We want to find that the values of $\sigma_{\bar{x}}$ and $\tau_{\bar{x}\bar{y}}$ expressed in terms of σ_x, σ_y, τ_{xy} and θ so that, provided we know the values of σ_x, σ_y and τ_{xy}, we can find the values of the normal and shear stresses acting on any plane orientated at any angle θ to the x-axis.

Consider the equilibrium of the wedge aekjdh, first in the direction of the normal to the face aj and then in the direction parallel to the face aj. For convenience, let the z-direction dimension of the wedge be equal to unity, i.e., jk = 1.

Resolving the forces acting on the wedge in the direction of the normal to aj, i.e., in the direction $\bar{x}$ (Fig. 3.31b) gives

$$\sigma_{\bar{x}}\text{aj} - (\sigma_x\text{ad}) \cos \theta - (\sigma_y\text{dj}) \sin \theta - (\tau_{xy}\text{ad}) \sin \theta - (\tau_{yx}\text{dj}) \cos \theta = 0$$

But ad $=$ aj cos θ and dj $=$ aj sin θ and $\tau_{yx} = \tau_{xy}$. Therefore

$$\sigma_{\bar{x}}\text{aj} - \sigma_x\text{aj} \cos^2 \theta - \sigma_y\text{aj} \sin^2 \theta - \tau_{xy}\text{aj} \sin \theta \cos \theta - \tau_{xy}\text{aj} \sin \theta \cos \theta = 0$$

Therefore

$$\sigma_{\bar{x}} = \sigma_x \cos^2 \theta + \sigma_y \sin^2 \theta + 2\tau_{xy} \sin \theta \cos \theta \tag{3.50}$$

Alternatively, from Eq. (3.50), putting $\cos^2 \theta = \dfrac{1}{2}(1 + \cos 2\theta)$ and

$$\sin^2 \theta = \frac{1}{2}(1 - \cos 2\theta)$$

$$\sigma_{\bar{x}} = \sigma_x \frac{1}{2}(1 + \cos 2\theta) + \sigma_y \frac{1}{2}(1 - \cos 2\theta) + \tau_{xy} \sin 2\theta$$

that is

$$\sigma_{\bar{x}} = \frac{1}{2}(\sigma_x + \sigma_y) + \frac{1}{2}(\sigma_x - \sigma_y) \cos 2\theta + \tau_{xy} \sin 2\theta \tag{3.51}$$

Equation (3.51) is an important result.

Resolving the forces acting on the wedge in the direction parallel to aj, i.e., in the direction $\bar{y}$, gives

$$\tau_{\bar{x}\bar{y}}\text{aj} + (\sigma_x\text{ad}) \sin \theta - (\sigma_y\text{dj}) \cos \theta - (\tau_{xy}\text{ad}) \cos \theta + (\tau_{yx}\text{dj}) \sin \theta = 0$$

that is $\tau_{\bar{x}\bar{y}} + \sigma_x \sin \theta \cos \theta - \sigma_y \sin \theta \cos \theta - \tau_{xy} \cos^2 \theta + \tau_{xy} \sin^2 \theta = 0$

therefore $\tau_{\bar{x}\bar{y}} = -(\sigma_x - \sigma_y) \sin \theta \cos \theta + \tau_{xy}(\cos^2 \theta - \sin^2 \theta) \tag{3.52}$

or $\tau_{\bar{x}\bar{y}} = -\dfrac{1}{2}(\sigma_x - \sigma_y) \sin 2\theta + \tau_{xy} \cos 2\theta \tag{3.53}$

Equation (3.53) is also an important result.

Equations (3.51) and (3.53) have fulfilled the desired objective, namely, to express the stresses on face aj, whose normal is located at any angle θ anticlockwise from the x-axis, in terms of σ_x, σ_y, τ_{xy} and θ.

Now let us see what this means. Figure 3.32 reproduces the axes xy; the normal to aj is the axis $\bar{x}$, lying at θ anticlockwise from x, and the axis $\bar{y}$ is at right angles to $\bar{x}$. As

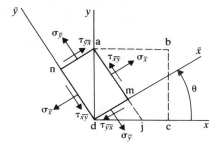

Figure 3.32   The stress system in the x and y directions is equivalent to a different stress system acting in oblique orthogonal directions

shown in Fig. 3.32, we can construct a new element amdn in the axes $\bar{x}\bar{y}$. The fact that this element is a different shape from the original element, and is smaller, is of no consequence because both elements are differentially small in the plane xy and both therefore represent the stress state at the same point.

From Eqs (3.51) and (3.53) we know the stresses $\sigma_{\bar{x}}$ and $\tau_{\bar{x}\bar{y}}$ on the face am. $\tau_{\bar{x}\bar{y}}$ is shown acting in the positive direction, namely, in the positive direction of $\bar{y}$ on the positive $\bar{x}$-face of the element amdn. The directions of the other shear stresses are therefore as shown and, because of complementary shear stress, $\tau_{\bar{y}\bar{x}} = \tau_{\bar{x}\bar{y}}$. If $\tau_{\bar{x}\bar{y}}$ should be negative, then it would act along the face am in the opposite direction to that shown in Fig. 3.32, and the other shear stresses would also act in directions opposite to those shown.

(The values of $\sigma_{\bar{y}}$ and $\tau_{\bar{y}\bar{x}}$, which act on the face na of the element in Fig. 3.32, can be found by replacing θ by $(\theta + 90°)$ in Eqs (3.51) and (3.53) respectively, because this is equivalent to rotating the $\bar{x}$-axis through a further $90°$. $\tau_{\bar{x}\bar{y}}$ in Fig. 3.32 would then become $\tau_{\bar{x}\bar{y}}$ (at $\theta + 90°$) and would have the same magnitude as $\tau_{\bar{x}\bar{y}}$ at θ, but with the opposite sign.)

The stress array, $\bar{F}$, for the element amdn in Fig. 3.32 is, therefore

$$\bar{F} = \begin{bmatrix} \sigma_{\bar{x}} & \tau_{\bar{x}\bar{y}} & 0 \\ \tau_{\bar{y}\bar{x}} & \sigma_{\bar{y}} & 0 \\ 0 & 0 & 0 \end{bmatrix} \text{ with respect to the axes } \bar{x}\bar{y}z \qquad (3.54)$$

The stress arrays F (Eq. (3.48)) and $\bar{F}$ (Eq. (3.54)) both describe the same state of stress at the same point, but they give combinations of normal and shear stress acting on planes (for example, bc and aj) passing through the point at different orientations. This is illustrated in Example 3.7.

EXAMPLE 3.7

At a point in a steel plate, illustrated by Fig. 3.33a, the state of stress is defined by F,

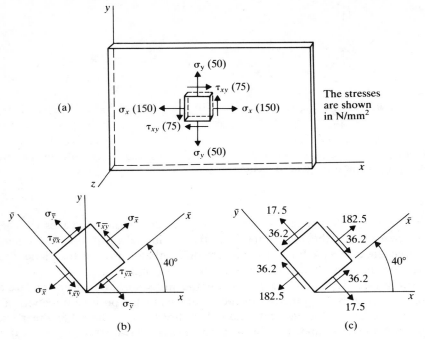

Figure 3.33 Example 3.7

where

$$F = \begin{bmatrix} 150 & 75 & 0 \\ 75 & 50 & 0 \\ 0 & 0 & 0 \end{bmatrix} \text{ N/mm}^2 \text{ with respect to the axes } xyz$$

Find the stress array for the same point with respect to the axes $\bar{x}$, $\bar{y}$, z where $\bar{x}$ lies at 40° anticlockwise from x. Sketch the element in the axes $\bar{x}$, $\bar{y}$ and the stresses acting on its faces.

The stress values given in the array F are shown in Fig. 3.33a, namely

$$\sigma_x = 150 \text{ N/mm}^2, \quad \sigma_y = 50 \text{ N/mm}^2, \quad \tau_{xy} = \tau_{yx} = 75 \text{ N/mm}^2, \quad \theta = +40°$$

The stresses which must be evaluated are illustrated on Fig. 3.33b.

From Eq. (3.51), putting $2\theta = 80°$:

$$\sigma_{\bar{x}} = \frac{1}{2}(150 + 50) + \frac{1}{2}(150 - 50)\cos 80 + 75 \sin 80 = 182.5 \text{ N/mm}^2$$

and, from Eq. (3.53):

$$\tau_{\bar{x}\bar{y}} = -\frac{1}{2}(150 - 50)\sin 80 + 75 \cos 80 = -36.2 \text{ N/mm}^2$$

In order to find $\sigma_{\bar{y}}$, put $\theta = (40 + 90)°$ in Eq. (3.51), that is

$$2\theta = 2(40 + 90)° = 260°$$

therefore

$$\sigma_{\bar{y}} = \frac{1}{2}(150 + 50) + \frac{1}{2}(150 - 50)\cos 260 + 75 \sin 260 = 17.5 \text{ N/mm}^2$$

Notice that $\tau_{\bar{x}\bar{y}}$ is negative. This means that the shear stress on the positive $\bar{x}$-face of the element in the axes $\bar{x}\bar{y}$ is acting in the negative direction of $\bar{y}$, and it follows that the shear stresses on the other three faces are also negative. They therefore act as shown in Fig. 3.33c.

(If θ is put equal to $(40 + 90)° = 130°$ in Eq. (3.53), then $\tau_{\bar{x}\bar{y}}$ would equal $+36.2 \text{ N/mm}^2$. That is, the shear stress normal to the $\bar{x}$-axis, located at $130°$ anticlockwise from the x-axis, would act in the positive direction on the positive $\bar{x}$-face. It is evident from Fig. 3.33c that this is correct.)

The stress array illustrated in Fig. 3.33c is therefore

$$\bar{F} = \begin{bmatrix} 182.5 & -36.2 & 0 \\ -36.2 & 17.5 & 0 \\ 0 & 0 & 0 \end{bmatrix} \text{N/mm}^2 \text{ with respect to the axes } \bar{x}\bar{y}z \text{ at } 40° \text{ anticlockwise from the axes } xy$$

Principal planes, principal stresses and principal axes

It should now be clear that the state of stress at a point can be described in an infinite number of ways, depending upon the directions in which the stresses are evaluated, These different stress values in different directions at the same point are, of course, real and, for design purposes, it is important to know about them. For example, in Fig. 3.33, known tensile stresses of 150 N/mm^2 and 50 N/mm^2 were applied to the plate in the x and y directions respectively, but it would be unsatisfactory to design the plate to carry a maximum tensile stress of 150 N/mm^2 because, at $40°$ anticlockwise from x, there is a bigger tensile stress of 182.5 N/mm^2 and there is no reason to think, at the moment, that this is the maximum value. Therefore, it is necessary to investigate the variations of σ_x and $\tau_{\bar{x}\bar{y}}$ as θ changes from 0 to $180°$. (There is no necessity to investigate these variations from 0 to $360°$ because, if we know the stress conditions on one face then we also know those on the opposite face.) This can be done by using Eqs (3.51) and (3.53), which are repeated here for convenience:

$$\sigma_{\bar{x}} = \frac{1}{2}(\sigma_x + \sigma_y) + \frac{1}{2}(\sigma_x - \sigma_y)\cos 2\theta + \tau_{xy}\sin 2\theta \tag{3.55}$$

$$\tau_{\bar{x}\bar{y}} = -\frac{1}{2}(\sigma_x - \sigma_y)\sin 2\theta + \tau_{xy}\cos 2\theta \tag{3.56}$$

To see what happens to the values of $\sigma_{\bar{x}}$ and $\tau_{\bar{x}\bar{y}}$ when θ is varied, the element used in Example 3.7 is reproduced in Fig. 3.34a. Figures 3.34b and 3.34c show the variation of $\sigma_{\bar{x}}$ and $\tau_{\bar{x}\bar{y}}$ with θ (from 0 to $180°$) according to Eqs (3.55) and (3.56) respectively. Both

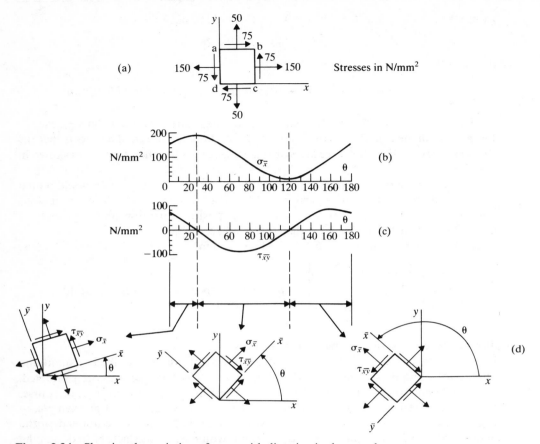

Figure 3.34　Showing the variation of stress with direction in the x–y plane

stresses move through a complete cycle as θ changes from 0 to 180°, i.e., the stresses on face bc (Fig. 3.34a), at $\theta = 0$, are reproduced at $\theta = 180°$, that is, on face da. Both $\sigma_{\bar{x}}$ and $\tau_{\bar{x}\bar{y}}$ pass through maximum and minimum values. In this case, $\sigma_{\bar{x}}$ is never zero and remains positive (tensile) at every value of θ, but this will not always be the case. The value of $\tau_{\bar{x}\bar{y}}$ passes through zero twice and *this will always be the case*. Notice that the zero values of $\tau_{\bar{x}\bar{y}}$ coincide (*as they always will*) with the maximum and minimum values of $\sigma_{\bar{x}}$. In Fig. 3.34d, the directions of $\sigma_{\bar{x}}$ and $\tau_{\bar{x}\bar{y}}$ are drawn on the face of the element being investigated, as θ moves through values corresponding to the change in sign of $\tau_{\bar{x}\bar{y}}$ from positive to negative and back to positive.

We are now in a position to define *principal planes*, *principal stresses* and *principal axes*.

A *principal plane* is a plane along which the shear stress is zero. The normal stresses acting on principal planes are called *principal stresses*.

The directions of principal stresses are called *principal axes*. It follows that the principal axes are normal to the principal planes.

The locations of the principal planes are found by putting $\tau_{\bar{x}\bar{y}} = 0$ in Eq. (3.56). The principal planes therefore occur when

$$0 = -\frac{1}{2}(\sigma_x - \sigma_y) \sin 2\theta + \tau_{xy} \cos 2\theta$$

i.e., when
$$\tan 2\theta = \frac{2\tau_{xy}}{(\sigma_x - \sigma_y)} \tag{3.57}$$

Equation (3.57) is satisfied by two values of 2θ between 0 and 360°, and these two values will differ by 180°. It is, therefore, satisfied by two values of θ between 0 and 180° which will differ by 90°. These two values of θ will locate the principal planes. Thus, there are two principal planes at right angles to each other whose angular locations are defined, with respect to the x-axis, by the roots of Eq. (3.57).

If Eq. (3.55) is differentiated with respect to θ and the result put to zero, the consequent values of θ will give the angular locations of the stationary (maximum and minimum) values of $\sigma_{\bar{x}}$. Thus

$$\frac{d\bar{\sigma}_x}{d\theta} = -(\sigma_x - \sigma_y) \sin 2\theta + 2\tau_{xy} \cos 2\theta = 0$$

i.e.,
$$\tan 2\theta = \frac{2\tau_{xy}}{(\sigma_x - \sigma_y)} \tag{3.58}$$

Equation (3.58) therefore locates the maximum and minimum values of $\sigma_{\bar{x}}$. It is identical with Eq. (3.57). Therefore, the stationary (maximum and minimum) values of $\sigma_{\bar{x}}$ (located by Eq. (3.58)) act on the principal planes (located by Eq. (3.57)). It follows that the maximum and minimum values of $\sigma_{\bar{x}}$ are the principal stresses and their directions define the principal axes.

Note that, although we now know that the maximum and minimum values of $\sigma_{\bar{x}}$ (the principal stresses) act on the principal planes, at this stage we do not know which principal stress (maximum or minimum) acts on which principal plane. This will be discussed later.

The values of the principal stresses

It has now been determined that the principal stresses are the maximum and minimum values of $\sigma_{\bar{x}}$ and that they act on the principal planes, which are defined as planes along which the shear stress is zero.

Let the principal stresses be σ_1 and σ_2. The values of σ_1 and σ_2 can be found by substituting into Eq. (3.55) those values of 2θ between 0 and 360° which satisfy Eq. (3.57). However, it is useful to express σ_1 and σ_2 directly in terms of σ_x, σ_y and τ_{xy} only. In order to do this, it is convenient to derive first the expressions for $\sin 2\theta$ and $\cos 2\theta$ which satisfy Eq. (3.57) and then substitute these expressions into Eq. (3.55) to define the values of σ_1 and σ_2.

From standard trigonometrical relations:

$$\cos 2\theta = 1/\sec 2\theta = 1/(1 + \tan^2 2\theta)^{1/2}$$

and

$$\sin 2\theta = \tan 2\theta \cos 2\theta = \tan 2\theta/(1 + \tan^2 2\theta)^{1/2} \qquad (3.59)$$

Substituting the value of $\tan 2\theta$ from Eq. (3.57) into Eq. (3.59), then, at the principal planes:

$$\cos 2\theta = (\sigma_x - \sigma_y)/[(\sigma_x - \sigma_y)^2 + 4\tau_{xy}^2]^{1/2}$$

and

$$\sin 2\theta = 2\tau_{xy}/[(\sigma_x - \sigma_y)^2 + 4\tau_{xy}^2]^{1/2} \qquad (3.60)$$

Putting the values of $\cos 2\theta$ and $\sin 2\theta$ from Eq. (3.60) into Eq. (3.55), the principal stresses σ_1 and σ_2 are therefore given by

$$\sigma_{1,2} = \frac{1}{2}(\sigma_x + \sigma_y) \pm \frac{1}{2}(\sigma_x - \sigma_y)\frac{(\sigma_x - \sigma_y)}{\sqrt{((\sigma_x - \sigma_y)^2 + 4\tau_{xy}^2)}}$$

$$\pm \tau_{xy}\frac{2\tau_{xy}}{\sqrt{((\sigma_x - \sigma_y)^2 + 4\tau_{xy}^2)}}$$

$$= \frac{1}{2}(\sigma_x + \sigma_y) \pm \frac{1}{2}\left[\frac{(\sigma_x - \sigma_y)^2 + 4\tau_{xy}^2}{\sqrt{((\sigma_x - \sigma_y)^2 + 4\tau_{xy}^2)}}\right]$$

$$= \frac{1}{2}(\sigma_x + \sigma_y) \pm \sqrt{\left(\left(\frac{\sigma_x - \sigma_y}{2}\right)^2 + \tau_{xy}^2\right)}$$

Therefore

$$\sigma_1 = \frac{1}{2}(\sigma_x + \sigma_x) + \sqrt{\left(\left(\frac{\sigma_x - \sigma_y}{2}\right)^2 + \tau_{xy}^2\right)}$$

$$(3.61)$$

and

$$\sigma_2 = \frac{1}{2}(\sigma_x + \sigma_y) - \sqrt{\left(\left(\frac{\sigma_x - \sigma_y}{2}\right)^2 + \tau_{xy}^2\right)}$$

Equations (3.55), (3.56), (3.57) and (3.61) are important.

Figure 3.35 represents an element subjected to principal stresses. The correspond-

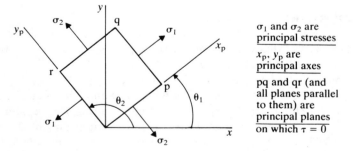

σ_1 and σ_2 are principal stresses

x_p, y_p are principal axes

pq and qr (and all planes parallel to them) are principal planes on which $\tau = 0$

Figure 3.35 On principal planes, the shear stress is zero

ing principal stress array (for plane stress) is therefore

$$F_p = \begin{bmatrix} \sigma_1 & 0 & 0 \\ 0 & \sigma_2 & 0 \\ 0 & 0 & 0 \end{bmatrix} \begin{array}{l} \text{with respect to principal axes } x_p y_p \\ \text{located at } \theta_1 \text{ from the axes } xy \end{array}$$

Which principal stress acts on which principal plane?

At the moment we have located the principal planes by the two values of 2θ between 0 and 360° which satisfy Eq. (3.57), i.e., by the two values of θ between 0 and 180° which satisfy Eq. (3.57).

From Eq. (3.61) we know the two values of the principal stresses: σ_1, the maximum principal stress, and σ_2, the minimum principal stress.

We do not yet know which of the two values of θ identifies the plane acted upon by σ_1 or which identifies the plane acted upon by σ_2.

To find out, we must find the sign of the second derivative, $d^2\sigma_{\bar{x}}/d\theta^2$ of Eq. (3.55) when the two solutions of Eq. (3.57) are substituted into this second derivative. If the sign is negative then we have located σ_1, the maximum principal stress (because $d\sigma_{\bar{x}}/d\theta$ is reducing with θ and therefore $\sigma_{\bar{x}}$ is a maximum). If the sign is positive, then we have located σ_2, the minimum principal stress. The second derivative of Eq. (3.55) is

$$\frac{d^2\sigma_{\bar{x}}}{d\theta^2} = -2(\sigma_x - \sigma_y)\cos 2\theta - 4\tau_{xy}\sin 2\theta \tag{3.62}$$

Location and magnitude of the maximum and minimum shear stresses

Equation (3.56) gives the value of $\tau_{\bar{x}\bar{y}}$ acting on any plane such as aj in Fig. 3.31b, that is, normal to the direction of the axis $\bar{x}$. The stationary values of $\tau_{\bar{x}\bar{y}}$ will therefore be located by equating to zero the derivative of $\tau_{\bar{x}\bar{y}}$ with respect to θ.

From Eq. (3.56):

$$\frac{d\tau_{\bar{x}\bar{y}}}{d\theta} = -(\sigma_x - \sigma_y)\cos 2\theta - 2\tau_{xy}\sin 2\theta = 0 \tag{3.63}$$

at maximum and minimum values of $\tau_{\bar{x}\bar{y}}$, which therefore occur when

$$\tan 2\theta = -\frac{(\sigma_x - \sigma_y)}{2\tau_{xy}} \tag{3.64}$$

Equation (3.64) is satisfied by two values of 2θ, between 0 and 360°, which differ by 180°, that is, by two values of θ between 0 and 180°, which differ by 90° and locate the planes of maximum and minimum shear stress.

The right-hand side of Eq. (3.64) is the negative reciprocal of the right-hand side of Eq. (3.57), which located the principal planes. It follows that the values of 2θ which satisfy Eq. (3.64) are 90° different from those which satisfy Eq. (3.57). Therefore, the planes of maximum and minimum shear stress lie at 45° to the principal planes and at 45° to the principal stresses. (Refer again to Figs 3.34b and 3.34c.)

The values of the maximum and minimum shear stresses can be found, in terms of σ_x, σ_y and τ_{xy}, by substituting into Eq. (3.56) the values of 2θ which satisfy Eq. (3.64).

Recalling Eq. (3.59) and substituting for tan 2θ from Eq. (3.64) then, at the locations of the maximum and minimum shear stresses:

$$\cos 2\theta = 2\tau_{xy}/[(\sigma_x - \sigma_y)^2 + 4\tau_{xy}^2]^{1/2}$$

and

$$\sin 2\theta = -(\sigma_x - \sigma_y)/[(\sigma_x - \sigma_y)^2 + 4\tau_{xy}^2]^{1/2}$$

(3.65)

Putting these values of cos 2θ and sin 2θ into Eq. (3.56):

$$\tau_{\bar{x}\bar{y}\,\substack{max \\ min}} = \pm \left\{ \frac{(\sigma_x - \sigma_y)^2}{2\sqrt{((\sigma_x - \sigma_y)^2 + 4\tau_{xy}^2)}} + \frac{2\tau_{xy}}{\sqrt{((\sigma_x - \sigma_y)^2 + 4\tau_{xy}^2)}} \right\}$$

that is

$$\tau_{\bar{x}\bar{y}\,\substack{max \\ min}} = \pm \sqrt{\left(\left[\frac{\sigma_x - \sigma_y}{2} \right]^2 + \tau_{xy}^2 \right)}$$

(3.66)

Clearly, the maximum and minimum values of $\tau_{\bar{x}\bar{y}}$ are equal in magnitude, opposite in sign and 90° apart. (In fact, it is often the habit to refer simply to the 'maximum' shear stress because of this equality of magnitude, and frequently it is more important to know the maximum magnitude of the shear stress than whether it is positive or negative.)

Having found the stationary values of $\tau_{\bar{x}\bar{y}}$ (Eq. (3.66)) and the planes upon which they act (Eq. (3.64)), there remains the problem of identifying which shear stress ($\tau_{\bar{x}\bar{y}max}$ or $\tau_{\bar{x}\bar{y}min}$) acts on which plane. This can be resolved by finding the sign of the second derivative of Eq. (3.56), i.e., of

$$\frac{d^2 \tau_{\bar{x}\bar{y}}}{d\theta^2} = 2(\sigma_x - \sigma_y) \sin 2\theta - 4\tau_{xy} \cos 2\theta$$

(3.67)

If Eq. (3.67) is negative, then the corresponding value of θ locates the normal, $\bar{x}$ (with respect to x) to the plane along which acts the maximum shear stress, $\tau_{\bar{x}\bar{y}max}$, and, if it is positive, the minimum value $\tau_{\bar{x}\bar{y}min}$ has been located.

The maximum shear stress in terms of the principal stresses

In view of Eq. (3.66), Eq. (3.61) can be written as

$$\sigma_1 = \frac{1}{2}(\sigma_x + \sigma_y) + \tau_{\bar{x}\bar{y}max}$$

and

$$\sigma_2 = \frac{1}{2}(\sigma_x + \sigma_y) - \tau_{\bar{x}\bar{y}max}$$

that is

$$2\tau_{\bar{x}\bar{y}max} = \sigma_1 - \sigma_2$$

therefore

$$\tau_{\bar{x}\bar{y}max} = \frac{1}{2}(\sigma_1 - \sigma_2)$$

(3.68)

Equation (3.68) is an important result; the maximum shear stress in the plane xy is equal to half the difference between the principal stresses in the plane xy.

The values of the normal (or direct) stresses on planes of maximum and minimum shear stress

The direct stress acting on any plane normal to the direction θ is given by Eq. (3.55). On planes of maximum and minimum shear stress, the value of 2θ is given by Eq. (3.64), i.e.

$$\tan 2\theta = -\frac{(\sigma_x - \sigma_y)}{2\tau_{xy}}$$

From Eq. (3.59):

$$\cos 2\theta = \frac{1}{(1 + \tan^2 2\theta)^{1/2}}$$

therefore

$$\sin 2\theta = \frac{\tan 2\theta}{(1 + \tan^2 2\theta)^{1/2}} = -\frac{(\sigma_x - \sigma_y)}{2\tau_{xy}} \times \frac{1}{(1 + \tan^2 2\theta)^{1/2}}$$

at planes of maximum and minimum shear stress.
Substituting these values of $\cos 2\theta$ and $\sin 2\theta$ into Eq. (3.55) gives

$$\sigma_{\bar{x}} = \frac{1}{2}(\sigma_x + \sigma_y) \tag{3.69}$$

Equation (3.69) gives the value of the direct stress on the plane of maximum shear stress *and* on the plane of minimum shear stress, i.e., the direct stresses on these two planes are always of the same sign and of equal magnitude, given by the mean of σ_x and σ_y.

The stress arrays which define the states of stress on the planes of maximum and minimum shear stress will be called the maximum and minimum shear stress arrays and will be denoted by F_{T1} and F_{T2}.

Mohr's circle of stress

All the relationships developed in the previous analysis of stress can be represented very simply, and to great advantage, by Mohr's stress circle construction. So far, attention has been confined to plane stress (i.e., in the xy plane with all z-direction stresses equal to zero) and, for this condition, Mohr's circle is ideal. In fact, once having become familiar with Mohr's circle and satisfied that it is a true representation of a state of plane stress, *it can be used exclusively and there is no necessity to remember the equations so far presented* in this section.

Consider again the stress state defined by a stress array F and suppose that we know the values of σ_x, σ_y and τ_{xy}, i.e.

$$F = \begin{bmatrix} \sigma_x & \tau_{xy} & 0 \\ \tau_{yx} & \sigma_y & 0 \\ 0 & 0 & 0 \end{bmatrix} \text{ with respect to the axes } xyz$$

F describes the stress state of the element abcd on Fig. 3.36a. The z dimension of the element is not shown in Fig. 3.36a. Imagine it is an element from a flat plate with a relatively small dimension in the z direction. This is a good example, because plates are frequently subjected to plane stress.

To draw Mohr's circle of stress, first construct the axes σ horizontally and $\tau_{\bar{x}\bar{y}}$ *positive downward*, as shown in Fig. 3.36b. (When numerical values are assigned to the stresses, the axes would, of course, be marked with the appropriate units.)

Consider the stresses acting on bc, the positive x-face of the element in Fig. 3.36a. At $\theta = 0$, the axes $\bar{x}\bar{y}$ coincide with xy, so that on bc, the positive $\bar{x}$-face, $\sigma = \sigma_{\bar{x}} = \sigma_x$ and $\tau_{\bar{x}\bar{y}} = \tau_{xy}$. Plot the values σ_x and τ_{xy} as coordinates of the point R. If τ_{xy} is *positive*, its value is plotted downward.

Now imagine the axes $\bar{x}\bar{y}$ rotated through 90° anticlockwise from xy, i.e., $\theta = 90°$, so that ab now becomes the positive $\bar{x}$-face of the element and carries stresses $\sigma = \sigma_y$ and $\tau_{\bar{x}\bar{y}} = -\tau_{yx} = -\tau_{xy}$. (At $\theta = 90°$, $\sigma_{\bar{x}}$ is now equal to σ_y.) Plot the values σ_y and $-\tau_{xy}$ as the coordinates of the point S. R and S will be equally distant from the σ-axis, R at $+\tau_{xy}$ and S at $-\tau_{xy}$.

Draw a circle on RS as diameter. This is Mohr's circle of stress for the stress state defined by F, illustrated in Fig. 3.36a. It will be shown that the stresses on any plane orientated at any value of θ to xy within the element in Fig. 3.36a, can be represented by the coordinates of a point on the circle. The centre of the circle is always on the σ-axis. The circle should be thought of as a hoop and we will be interested only in points on the hoop. The space inside and outside the hoop has no significance at this stage.

The simplicity of constructing the Mohr's circle of stress is emphasized by the following instruction: plot the point R at σ_x, τ_{xy} (below the σ-axis if τ_{xy} is positive, above the σ-axis if τ_{xy} is negative). Then plot S at σ_y, $-\tau_{xy}$, where S is always on the opposite side of the σ-axis from R and at equal distance from the σ-axis. Draw the circle on RS as diameter.

Some properties of the stress circle follow from its construction. Let the radius be r, the centre be C, and L and M the intercepts on the σ-axis of perpendiculars from R and S respectively. Then:

$$\text{MC} = \text{CL} = \left(\frac{\sigma_x - \sigma_y}{2}\right)$$

$$\text{OC} = \left(\frac{\sigma_x + \sigma_y}{2}\right) \tag{3.70}$$

and

$$r = \sqrt{\left(\left[\frac{\sigma_x - \sigma_y}{2}\right]^2 + \tau_{xy}^2\right)}$$

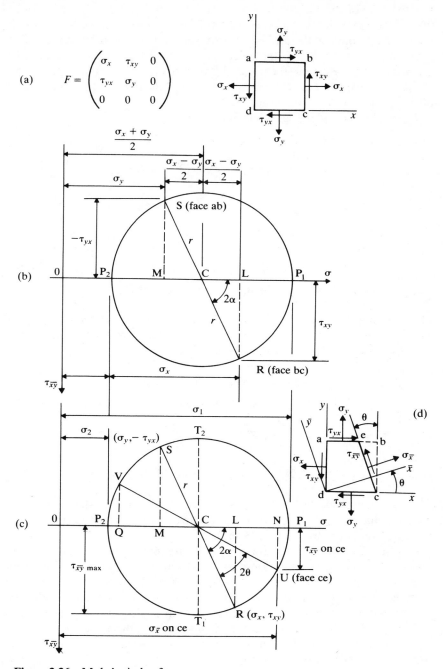

Figure 3.36 Mohr's circle of stress

For the sake of clarity, the circle is reproduced in Fig. 3.36c. In Fig. 3.36c, move to a point U which is situated on the circle at 2θ anticlockwise from R. Locate the point V diametrically opposite from U and drop perpendiculars to N and Q.

What we want to show is that, just as the point R represents the stresses (σ_x and τ_{xy}) on face bc of the element, the point U represents the stresses on a plane or face of the element located at θ (*not* 2θ) anticlockwise from the plane bc, i.e., the plane ce (and all planes parallel to ce) shown in Fig. 3.36d.

To do this, consider Figs 3.36c and 3.36b:

$$ON = OC + CN$$

$$= \frac{1}{2}(\sigma_x + \sigma_y) + r\cos(2\alpha - 2\theta)$$

$$= \frac{1}{2}(\sigma_x + \sigma_y) + r(\cos 2\alpha \cos 2\theta + \sin 2\alpha \sin 2\theta)$$

$$= \frac{1}{2}(\sigma_x + \sigma_y) + r\frac{CL}{r}\cos 2\theta + r\frac{LR}{r}\sin 2\theta$$

i.e.
$$ON = \frac{1}{2}(\sigma_x + \sigma_y) + \frac{1}{2}(\sigma_x - \sigma_y)\cos 2\theta + \tau_{xy}\sin 2\theta \qquad (3.71)$$

The right-hand side of Eq. (3.71) is identical with the right-hand side of Eq. (3.55) which defines $\sigma_{\bar{x}}$, the direct stress on the plane on the element located at θ anticlockwise from bc (see Fig. 3.36d).

On Fig. 3.36c:

$$UN = r\sin(2\alpha - 2\theta)$$

$$= r(\sin 2\alpha \cos 2\theta - \cos 2\alpha \sin 2\theta)$$

$$= r\frac{LR}{r}\cos 2\theta - r\frac{CL}{r}\sin 2\theta$$

i.e.
$$UN = -\frac{1}{2}(\sigma_x - \sigma_y)\sin 2\theta + \tau_{xy}\cos 2\theta \qquad (3.72)$$

The right-hand side of Eq. (3.72) is identical with the right-hand side of Eq. (3.56), which defines $\tau_{\bar{x}\bar{y}}$, the shear stress on the plane on the element located at θ anticlockwise from bc (see Fig. 3.36d).

ON ($=\sigma_{\bar{x}}$) and UN ($=\tau_{\bar{x}\bar{y}}$) are the coordinates of the point U in the σ and $\tau_{\bar{x}\bar{y}}$ axes. It follows that the stress condition on a face located at θ anticlockwise on the element from bc is represented by the coordinates of a point (U) located at 2θ anticlockwise from R on the Mohr's circle. (The rotational movement θ on the element and the corresponding rotation 2θ on the circle will always be in the same direction.)

In Fig. 3.36c, the point V, located at $180°$ from U, therefore has coordinates which equal the combination of direct stress and shear stress acting on the face located at $90°$ to ec (Fig. 3.36d) on the element.

These properties of Mohr's circle of stress will now be illustrated in Example 3.8.

EXAMPLE 3.8

Repeat Example 3.7, using Mohr's circle of stress to find the required answers.
The state of stress of the element is defined by

$$F = \begin{bmatrix} \sigma_x & \tau_{xy} & 0 \\ \tau_{xy} & \sigma_y & 0 \\ 0 & 0 & 0 \end{bmatrix} = \begin{bmatrix} 150 & 75 & 0 \\ 75 & 50 & 0 \\ 0 & 0 & 0 \end{bmatrix} \text{N/mm}^2 \text{ with respect to the axes } xyz$$

Find the stress array $\bar{F}$ with respect to axes $\bar{x}$, $\bar{y}$ located 40° anticlockwise from xy.

The Mohr's circle is constructed on Fig. 3.37 by plotting R and S to represent the stresses on faces bc and ab of the element and drawing a circle on RS as diameter. (Note that the shear stress on face ab would be negative if ab was the positive x-face, i.e., when $\theta = 90°$, hence $\tau_{\bar{x}\bar{y}}$ for this face is plotted at S.) To investigate the stresses on the plane mn at 40° anticlockwise from x on the element, locate the point U at $(2 \times 40°) = 80°$ anticlockwise from R on the circle. As we move from R to U, the sign of $\tau_{\bar{x}\bar{y}}$ changes from positive to negative on passing through zero at the σ-axis.
On the circle:

$$2\alpha = \tan^{-1}\frac{75}{50} = 1.5$$

$$\therefore \ 2\alpha = 56.31°$$

$$\therefore \ \text{angle UCN} = (80 - 56.31)° = 23.69°$$

Also,
$$OC = \frac{1}{2}(\sigma_x + \sigma_y) = \frac{1}{2}(150 + 50) = 100 \text{ N/mm}^2$$

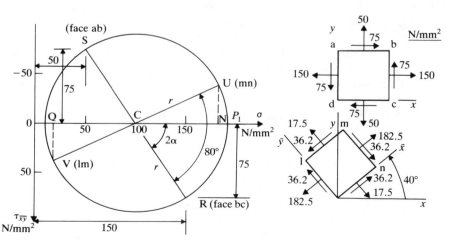

Figure 3.37   Example 3.8

and
$$r = \sqrt{\left(\left[\frac{\sigma_x - \sigma_y}{2}\right]^2 + \tau_{xy}^2\right)} = \sqrt{(50^2 + 75^2)} = 90.14 \text{ N/mm}^2$$

$\therefore$ at U, $\sigma_{\bar{x}} = 100 + 90.14 \cos 23.69 = 182.5 \text{ N/mm}^2$

and $\tau_{\bar{x}\bar{y}} = -90.14 \sin 23.69 = -36.2 \text{ N/mm}^2$

and at V, $\sigma_{\bar{y}} = 100 - 90.14 \cos 23.69 = 17.5 \text{ N/mm}^2$

Thus,

$$\bar{F} = \begin{bmatrix} 182.5 & -36.2 & 0 \\ -36.2 & 17.5 & 0 \\ 0 & 0 & 0 \end{bmatrix} \text{N/mm}^2 \text{ with respect to axes at } 40° \text{ anticlockwise from } xy$$

(It might be wondered why, in the stress array, the shear stress on the face lm is -36.2 N/mm^2 instead of $+36.2 \text{ N/mm}^2$, because the point V is certainly at a positive value of $\tau_{\bar{x}\bar{y}}$. The explanation is that, if the axes $\bar{x}\bar{y}$ were rotated through a further 90° from the position shown on the element in Fig. 3.37, then lm would become the positive $\bar{x}$-face and $\tau_{\bar{x}\bar{y}}$ would be positive. However, the stress array is written with respect to $\bar{x}\bar{y}$ at 40° anticlockwise from xy in which case, as can be seen on Fig. 3.37, lm is the positive $\bar{y}$-face. The simplest procedure is to determine the value of $\tau_{\bar{x}\bar{y}}$ on mn and then simply draw in the complementary shear stress on the face lm.)

In this example, notice again that the value of $\tau_{\bar{x}\bar{y}}$ has changed sign as the $\bar{x}$-axis moved from x to 40° anticlockwise from x. On the circle, this is represented by moving along the circle from R to U and, as can be seen, $\tau_{\bar{x}\bar{y}}$ will move from $+75 \text{ N/mm}^2$ (at R), to zero at P_1, and then become negative above the horizontal axis.

Principal stresses, their magnitudes and locations from Mohr's stress circle

It is important to emphasize that the stress circle represents the state of plane stress at a point in the material. Every diametrically opposite pair of points on the circle defines this stress state by combinations of $\sigma_{\bar{x}}$, $\tau_{\bar{x}\bar{y}}$ and $\sigma_{\bar{y}}$ orientated at a particular angular location from the reference axes x and y.

Referring again to Fig. 3.36, the principal stresses σ_1 and σ_2 (i.e., maximum and minimum values of σ) are obviously represented by the points P_1 and P_2 and occur on principal planes (i.e., planes upon which $\tau_{\bar{x}\bar{y}}$ is zero).

From Fig. 3.36:

$$\sigma_1 = OP_1 = OC + r = \frac{1}{2}(\sigma_x + \sigma_y) + \sqrt{\left(\left[\frac{\sigma_x - \sigma_y}{2}\right]^2 + \tau_{xy}^2\right)}$$

(3.73)

and $$\sigma_2 = OP_2 = OC - r = \frac{1}{2}(\sigma_x + \sigma_y) - \sqrt{\left(\left[\frac{\sigma_x - \sigma_y}{2}\right]^2 + \tau_{xy}^2\right)}$$

Equations (3.73) are identical to Eqs (3.61).

Also, σ_1 is located on the circle at $2\theta = 2\alpha$ anticlockwise from the direction of x (represented by point R), i.e., at

$$\tan 2\theta = \frac{2\tau_{xy}}{(\sigma_x - \sigma_y)} \tag{3.74}$$

Equation (3.74) is identical to Eqs (3.57) and (3.58) which were previously derived to locate the principal stresses and the zero shear stress.

The maximum and minimum shear stresses and their locations from Mohr's stress circle

Whereas previously, when using equations only, care had to be taken to identify the principal stresses and the maximum and minimum shear stresses with the planes upon which they acted, the pictorial nature of the stress circle removes this difficulty. For example, it is clear that, on the Mohr's circle in Fig. 3.36c, the maximum principal stress (σ_1 at P_1) occurs at 2α anticlockwise from R and not at $(2\alpha + 180°)$ anticlockwise from R. It therefore occurs at α anticlockwise from the direction of x on the element and not at $(\alpha + 90°)$ from the direction of x.

Similarly, from Fig. 3.36c, the maximum value of $\tau_{\bar{x}\bar{y}}$, i.e., $\tau_{\bar{x}\bar{y}max}$, is represented by the point T_1 and the minimum value $\tau_{\bar{x}\bar{y}min}$ by the point T_2.

Thus

$$\tau_{\substack{\bar{x}\bar{y}max \\ min}} = \pm r = \pm \sqrt{\left(\left[\frac{\sigma_x - \sigma_y}{2}\right]^2 + \tau_{xy}^2\right)} \tag{3.75}$$

which is identical to Eq. (3.66). Also, from the circle:

$$\tau_{\bar{x}\bar{y}max} = r = \frac{1}{2}(\sigma_1 - \sigma_2) \tag{3.76}$$

which is identical to Eq. (3.68).

Remembering that θ is measured positively anticlockwise from the x direction on the element (Fig. 3.36d) and 2θ from the corresponding point R on the circle, the points T_1 and T_2 are located on Fig. 3.36c by

$$\tan 2\theta = -\frac{(\sigma_x - \sigma_y)}{2\tau_{xy}} \tag{3.77}$$

which is identical to Eq. (3.64), but now there is no doubt which root locates T_1 ($\tau_{\bar{x}\bar{y}max}$) and which locates T_2 ($\tau_{\bar{x}\bar{y}min}$).

EXAMPLE 3.9
At a point in a flat plate lying in the xy plane, the stress state with respect to the orthogonal axes xyz is given by

$$F = \begin{bmatrix} -100 & -50 & 0 \\ -50 & 30 & 0 \\ 0 & 0 & 0 \end{bmatrix} \text{N/mm}^2 \text{ with respect to } xyz$$

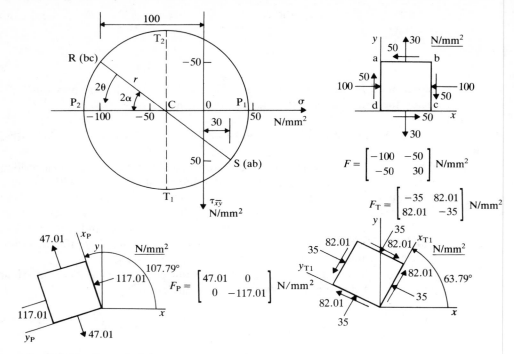

Figure 3.38   Example 3.9

Use Mohr's circle of stress to find the principal stress array F_p and the maximum shear stress array F_{T_1}. Sketch the elements corresponding to F_p and F_{T_1}.

From the given stress array, $\sigma_x = -100 \text{ N/mm}^2$, $\sigma_y = 30 \text{ N/mm}^2$, $\tau_{xy} = -50 \text{ N/mm}^2$. The stressed element in the axes xy is shown on Fig. 3.38. The point R (-100, -50) represents the stresses on the face bc and should always be plotted first (remember that τ_{xy} is plotted upward when it is negative). S represents the stresses on the face ab, where $\sigma = \sigma_y$ and $\tau_{\bar{x}\bar{y}} = 50 \text{ N/mm}^2$, remembering again that, if ab were the positive $\bar{x}$-face, then $\tau_{\bar{x}\bar{y}}$ is positive.

Referring to Fig. 3.38:

$$OC = \frac{1}{2}(\sigma_x + \sigma_y) = \frac{1}{2}(-100 + 30) = -35 \text{ N/mm}^2$$

$$r = \sqrt{\left(\left[\frac{\sigma_x + \sigma_y}{2}\right]^2 + \tau_{xy}^2\right)} = \sqrt{\left(\left[\frac{-100 - 30}{2}\right]^2 + (-50)^2\right)}$$

$$= 82 \text{ N/mm}^2$$

$$2\alpha = \tan^{-1}\frac{50}{65} = 0.769, \qquad 2\alpha = 37.57°$$

$$\sigma_1 = OC + r = -35 + 82.0 = 47 \text{ N/mm}^2$$

$$\sigma_2 = OC - r = -35 - 82.0 = -117 \text{ N/mm}^2$$

$$\tau_{\bar{x}\bar{y}\max} = r = 82 \text{ N/mm}^2$$

P_1 is located at $(180 + 2\alpha) = 216.6°$ anticlockwise from R

$\therefore$ σ_1 is located at $215.6°/2 = 107.8°$ anticlockwise from the x-direction

T_1 is located at $(90 + 2\alpha) = 127.6°$ anticlockwise from R

$\therefore$ $\tau_{\bar{x}\bar{y}\max}$ is located at $127.6°/2 = 63.8°$ anticlockwise from the x-direction

When $\qquad \tau_{\bar{x}\bar{y}} = \tau_{\bar{x}\bar{y}\max}, \qquad \sigma_{\bar{x}} = \sigma_{\bar{y}} = OC = -35 \text{ N/mm}^2$

Thus

$$F_p = \begin{bmatrix} 47.0 & 0 & 0 \\ 0 & -117.0 & 0 \\ 0 & 0 & 0 \end{bmatrix} \text{N/mm}^2 \text{ at } 107.8° \text{ anticlockwise from } xy$$

(or at $72.2°$ clockwise from xy, if this description is preferred)

and

$$F_{T_1} = \begin{bmatrix} -35.0 & 82.0 & 0 \\ 82.0 & -35.0 & 0 \\ 0 & 0 & 0 \end{bmatrix} \text{N/mm}^2 \text{ at } 63.8° \text{ anticlockwise from } xy$$

The elements carrying F_p and F_{T_1} are illustrated in Fig. 3.38.

Problem 3.5   For the state of stress defined in Example 3.7 (Figs 3.33 and 3.34), use Mohr's circle of stress to find the locations of the principal planes, the directions of the principal axes and the values of the principal stresses. Sketch the element bounded by the principal planes and write the principal stress array.

Problem 3.6   For the state of stress defined in Example 3.7 (Figs 3.33 and 3.34), use Mohr's circle of stress to find the magnitudes and directions of the maximum and minimum shear stresses and the direct stresses acting on the maximum shear planes. Write the maximum and minimum shear stress arrays F_{T_1} and F_{T_2}, and sketch the element bounded by the maximum and minimum shear planes, showing all the stresses acting on the elements.

Special cases of plane stress

1. *Uniform axial stress* (tension or compression)
Figure 3.39 shows a flat bar subjected to a tensile load P. Over the central section of the length, where the cross-sectional area is A, it may be assumed that the load is uniformly distributed across the cross-section with a tensile stress $\sigma_x = P/A$. The

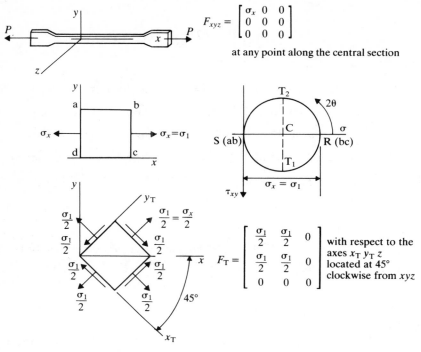

Figure 3.39   Stresses within a member subjected to an external axial force only

stresses σ_y, τ_{xy} and all the z-direction stresses are zero. The stress array, with respect to x, y, z is therefore

$$F = \begin{bmatrix} \sigma_x & 0 & 0 \\ 0 & 0 & 0 \\ 0 & 0 & 0 \end{bmatrix} = F_p = \begin{bmatrix} \sigma_1 & 0 & 0 \\ 0 & 0 & 0 \\ 0 & 0 & 0 \end{bmatrix}$$

This is a principal stress array.

The stress circle for this case is drawn in Fig. 3.39 and the magnitude of the shear stress reaches its minimum value ($\tau_{\bar{x}\bar{y}min}$) at point T_2 which is $2\theta = 90°$ anticlockwise from R and its maximum value ($\tau_{\bar{x}\bar{y}max}$) at T_1 which is $2\theta = 90°$ clockwise from R. Thus, all planes at 45°, both clockwise and anticlockwise, from the direction of x carry a tensile stress equal to $\sigma_x/2$ and a shear stress of the same magnitude.

(If, instead of a flat bar, the tensile specimen is a round bar and one imagines a rotation of the y and z axes about the x-axis, then it becomes clear that all planes at 45° to the x-axis will carry this same combination of tensile stress and shear stress. It is often observed that mild steel round bars, subjected to tension, eventually fracture along these planes of maximum shear stress to form cup and cone shaped failure surfaces.)

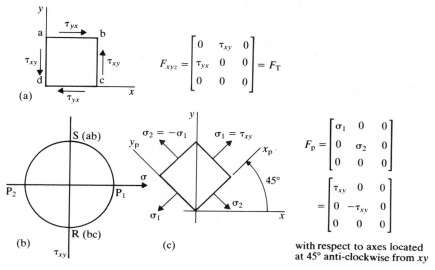

Figure 3.40   Stresses within an element subjected to shear stresses only in the x and y directions

2. Pure shear stress in the plane xy

Figure 3.40a shows an element of plate in the xy plane subjected to shear stress τ_{xy}, with $\sigma_x = \sigma_y = \sigma_z = 0$. In this case, the stress circle, Fig. 3.40b, indicates that the maximum and minimum principal stresses σ_1 and σ_2 (respresented by points P_1 and P_2) act on planes located at 45° from the direction of x (represented by point R) and are both of magnitude equal to τ_{xy}, σ_1 being tensile and σ_2 compressive. The element carrying the principal stresses is shown in Fig. 3.40c.

Three-dimensional stress

At the beginning of Section 3.4, the general state of three-dimensional uniform stress on an element was briefly considered and illustrated in Fig. 3.28. For any given state of such three-dimensional stress, it transpires that there is always a set of three orthogonal planes upon which all the shear stresses are simultaneously zero, i.e., there are three principal planes at right angles to one another. Figure 3.41 shows this element bounded by principal planes upon which, by definition, the direct stresses σ_1, σ_2 and σ_3 are acting in the directions of the three principal axes x_p, y_p and z_p. It is possible to draw a set of three Mohr's circles of stress to represent this three-dimensional state of stress.

Suppose, for the sake of illustration, that $\sigma_1 > \sigma_2 > \sigma_3$ and consider first the stresses acting in the plane xy, which contains the axes x_p, y_p and the principal stresses σ_1 and σ_2. Let this be plane ①, see Fig. 3.41.

The stress circle for stresses acting in plane ① is represented by the circle ① in Fig. 3.42, defined by the principal stresses σ_1 and σ_2 at points P_1 and P_2. Similarly, the

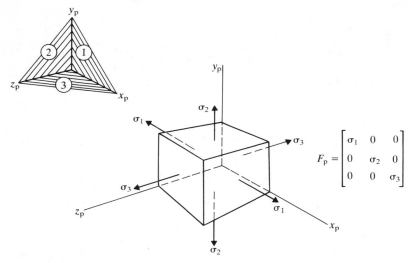

Figure 3.41   A 3-dimensional element subjected to principal stresses in the 3 orthogonal directions

stresses acting on plane ② are represented by circle ②, defined by the principal stresses σ_2 and σ_3 (P_2 and P_3) and those in plane ③ by circle 3, defined by σ_1 and σ_3 at P_1 and P_3.

Thus, circle ① gives combinations of σ and τ on faces of the element parallel to z_p, circle ② gives combinations of σ and τ on faces parallel to x_p, and circle ② gives combinations of σ and τ on faces parallel to y_p. This is a limited picture, because there is an infinite number of sets of orthogonal planes within the element as the axes are rotated about their origin with three degrees of rotational freedom. The combinations of direct stresses and shear stresses acting on such planes are, in fact, defined by points within the shaded area in Fig. 3.42, but it is beyond our present ambition to investigate this situation any further.

However, one very important point does become clear from Fig. 3.42, namely, that the maximum shear stress in the element is the radius of circle ③ and acts in the plane ③. This can be stated as follows.

In a three-dimensional state of stress, where the principal stresses are σ_1, σ_2 and σ_3 such that $\sigma_1 > \sigma_2 > \sigma_3$, then

$$\tau_{\max} = \frac{1}{2}(\sigma_1 - \sigma_3) \tag{3.78}$$

and occurs in the same plane as σ_1 and σ_3. Alternatively, one can simply say that the maximum shear stress is half the difference between the maximum and minimum principal stresses.

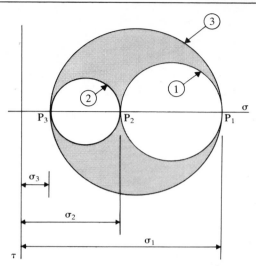

Figure 3.42 Mohr's circles of stress for the 3-dimensional element in Fig. 3.41

3.5 The analysis of strain

When considering stress, it was helpful to think of the total stress as consisting of a stress σ acting normal to the plane and a shear stress τ acting along the plane.

Similarly, with strain we will analyse the total strain condition in terms of a direct strain ε and a shear strain γ. Before defining ε and γ, there is a distinction which must be made between the displacement of an element and the distortion or deformation of the element. An element in a structure will usually be both displaced and deformed. The strain of an element is associated only with its deformation and not with its bodily displacement.

Let us dwell on this distinction by referring to Fig. 3.43 which shows a cantilever beam subjected to vertical and horizontal loads. The cantilever will deflect and the element abcd in the web will experience a bodily displacement in the plane xy together with a change of shape, or *deformation*. This change of shape or deformation is defined in terms of the *strains* on the element and is directly associated with the stresses acting upon it. (The *bodily displacement* of the element is the result of the accumulated strains on all the other elements between itself and the built-in end of the cantilever.) So, strain is a measure of deformation or change of shape (due to the stresses acting on the element) and not of bodily displacement.

Returning now to direct and shear strain, the direct strain ε is familiar and was defined in Chapter 1 as the change in length per unit length. Thus, for the element of unstressed dimensions $a \times b$ in Fig. 3.44, the application of σ_x alone will cause a change in length Δa and a strain in the x-direction, $\varepsilon_x = \Delta a/a$(Fig. 3.44a). Similarly, $\varepsilon_y = \Delta b/b$ if σ_y is applied alone (Fig. 3.44b). (It will be noticed that the element also changes its dimension in the direction perpendicular to the applied stress. This is called the Poisson's ratio effect and we will deal with it later in this chapter.)

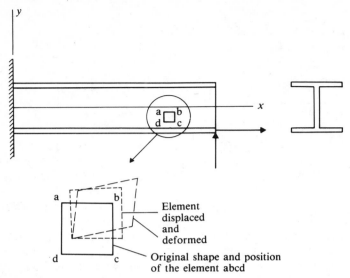

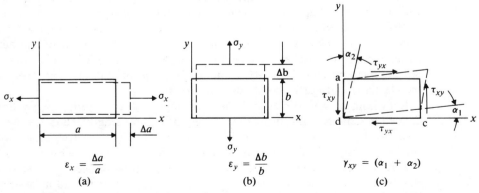

Figure 3.43 Illustrating the displacement and deformation of an element in the web of an I-beam

Also shown in Fig. 3.44c is the element subjected to shear stress alone and the resulting deformation or shear strain. Shear strain is a measure of angular distortion and is defined as the angular change (measured in radians) of a right angle. Thus, in Fig. 3.44c, the original right angle adc changes by $(\alpha_1 + \alpha_2)$. Therefore, the shear strain, $\gamma_{xy} = (\alpha_1 + \alpha_2)$.

Signs of direct strains and shear strains

Tensile direct strain will be designated as positive and compressive strain negative. Both ε_x and ε_y as drawn in Fig. 3.44 are positive.

Figure 3.44 Direct and shear strains of an element demonstrated in separate diagrams

Shear strains will be called positive if the right angle of the unstrained element lying in the axes xy is *reduced*. Thus, the shear strain in Fig. 3.44 is a positive shear strain. If the angle of the element lying in the xy axes is increased, the shear strain will be negative. This definition of positive shear strain is consistent with the definition of positive shear stress, as can be seen from Fig. 3.44c.

Strains in terms of displacements

The distinction has been made above between the *bodily* displacement of an element and its deformation due to strain. But, of course, the two are associated, because the displacement continues to change within the dimensions of the element, and it is the rate at which this displacement occurs which is defined as the strain.

To deduce the relationships between strains and displacements, consider the element abcd, of unstrained dimensions dx, dy, in Fig. 3.45.

Suppose the element has suffered a *bodily* displacement from its original position in the xy plane. This displacement is defined by the movements u_d and v_d in the x and y directions respectively, and by a rotation ϕ. The displacements u and v continue to change with x and y, and these changes cause the distortion of the element to its final shape $a_1b_1c_1d$. Let the change of u with increasing x, i.e., $\partial u/\partial x$, the change of u with increasing y, i.e., $\partial u/\partial y$, the change of v with increasing x, i.e., $\partial v/\partial x$, and change of v with increasing y, i.e., $\partial v/\partial y$, all be constant over the dimensions dx and dy of the element.

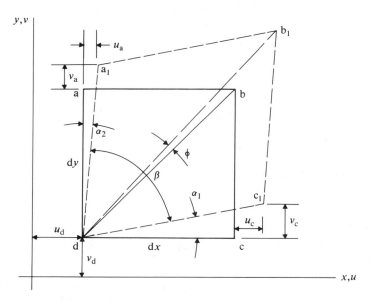

Figure 3.45   Displacement and deformation of an element

Then, on Fig. 3.45:
$$u_c = \frac{\partial u}{\partial x}\, dx, \qquad v_c = \frac{\partial v}{\partial x}\, dx$$

$$u_a = \frac{\partial u}{\partial y}\, dy, \qquad v_a = \frac{\partial v}{\partial y}\, dy$$

Also,
$$(dc_1)^2 = (dx + u_c)^2 + v_c^2 = \left\{dx\left(1 + \frac{\partial u}{\partial x}\right)\right\}^2 + \left(\frac{\partial v}{\partial x}\, dx\right)^2$$

$$= (dx)^2\left\{\left(1 + \frac{\partial u}{\partial x}\right)^2 + \left(\frac{\partial v}{\partial x}\right)^2\right\}$$

$$\therefore \; dc_1 = dx\left\{1 + 2\,\frac{\partial u}{\partial x} + \left(\frac{\partial u}{\partial x}\right)^2 + \left(\frac{\partial v}{\partial x}\right)^2\right\}^{1/2} \tag{3.79}$$

Expanding the right-hand side of Eq. (3.79) gives

$$dc_1 = dx\left\{1 + \frac{\partial u}{\partial x} + \frac{1}{2}\left(\frac{\partial u}{\partial x}\right)^2 + \frac{1}{2}\left(\frac{\partial v}{\partial x}\right)^2 + \cdots\right\} \tag{3.80}$$

The rates of displacement with distance, $\partial u/\partial x$ and $\partial v/\partial x$, will generally be very small compared with unity. Therefore, for small deformations, $(\partial u/\partial x)^2$ and $(\partial v/\partial x)^2$ can be ignored and Eq. (3.80) becomes

$$dc_1 = dx\left(1 + \frac{\partial u}{\partial x}\right) \tag{3.81}$$

The x-direction strain is therefore

$$\varepsilon_x = \frac{dc_1 - dc}{dc} = \frac{dx\left(1 + \dfrac{\partial u}{\partial x}\right) - dx}{dx}$$

that is
$$\varepsilon_x = \frac{\partial u}{\partial x} \tag{3.82}$$

Similarly, the y-direction strain is found to be

$$\varepsilon_y = \frac{\delta v}{\delta y} \tag{3.83}$$

The shear strain on the element is defined by

$$\gamma_{xy} = \left(\frac{\pi}{2} - \beta\right) = (\alpha_1 + \alpha_2) \tag{3.84}$$

From Fig. 3.45:
$$\tan \alpha_1 = \frac{v_c}{dx + u_c} = \frac{(\partial v/\partial x)\, dx}{dx + (\partial u/\partial x)\, dx} = \frac{\partial v/\partial x}{1 + \partial u/\partial x}$$

and
$$\tan \alpha_2 = \frac{u_a}{dy + v_a} = \frac{(\partial u/\partial y)\, dy}{dy + (\partial v/\partial y)\, dy} = \frac{\partial u/\partial y}{1 + \partial v/\partial y}$$

For small angles, $\tan \alpha_1 \simeq \alpha_1$, $\tan \alpha_2 \simeq \alpha_2$ and, assuming that $(\partial u/\partial x) \ll 1$ and $(\partial v/\partial y) \ll 1$, then

$$\gamma_{xy} = \frac{\partial v}{\partial x} + \frac{\partial u}{\partial y} \tag{3.85}$$

In summary, the strains on the element in the plane xy are completely defined, in terms of the displacements, by Eqs (3.82), (3.83) and (3.85).

An important point to note is that the direct strains, ε_x and ε_y, are independent of the shear strain γ_{xy} and vice versa.

Strain array for an element subjected to plane stress

When considering the element subjected to plane stress, we found that, because of complementary shear stress, $\tau_{xy} = \tau_{yx}$, i.e., there was only a single value of shear stress on the element. Similarly, it is clear from the above discussion that the shear strain on the element in the xy plane is fully defined by γ_{xy}. (Alternatively, the shear strain could be equally well designated as γ_{yx} which would also be positive if it described a reduction in the angle adc on the element in Fig. 3.45. Thus $\gamma_{xy} = \gamma_{xy}$.)

An element subjected to the stress array defined by

$$F = \begin{bmatrix} \sigma_x & \tau_{xy} & 0 \\ \tau_{xy} & \sigma_y & 0 \\ 0 & 0 & 0 \end{bmatrix} \text{ with respect to the axes } xyz \tag{3.86}$$

experiences a state of strain defined by the strain array

$$S = \begin{bmatrix} \varepsilon_x & \tfrac{1}{2}\gamma_{xy} & 0 \\ \tfrac{1}{2}\gamma_{xy} & \varepsilon_y & 0 \\ 0 & 0 & \varepsilon_z \end{bmatrix} \text{ with respect to the axes } xyz \tag{3.87}$$

There are two points about Eq. (3.87) which call for comment. Firstly, it contains $\tfrac{1}{2}\gamma_{xy}$ (rather than γ_{xy}) and $\tfrac{1}{2}\gamma_{yx}$ in the positions corresponding to τ_{xy} and τ_{yx} in Eq. (3.86). This proves to be convenient and the reason will become clear later.

Secondly, Eq. (3.87) contains ε_z. This is because the stresses σ_x and σ_y both cause strain in the z direction (just as they cause strains in the y and x directions respectively on Fig. 3.44). A state of plane stress does not, therefore, produce a state of plane strain. The strain γ_{xy}, however, does not cause shear strain in any other plane, hence the zero elements in the S-array of Eq. (3.87).

Our analysis will be concerned predominantly with the strains in the xy plane, so that, although ε_z exists, we will largely ignore its presence.

Transformation of the strain axes

Figure 3.46a shows the undeformed shape, abcd, and the strained shape, $a_1b_1c_1d$, of an element dx, dy lying in the plane xy. The deformed shape shown is greatly exaggerated, of course. In reality, we are talking about *small* deformations so that, for

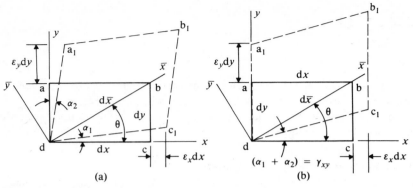

(a)   (b)

Figure 3.46   Deformation of an element subjected to both direct and shear strains

example, the length ac_1 can be regarded as equal to its projection on the x-axis. It is convenient for analytical purposes to rotate the deformed element $a_1b_1c_1d$ through α_2, so that it rests back on the y-axis, as shown in Fig. 3.46b. The shape of $a_1b_1c_1d$ is identical in Figs 3.46a and 3.46b. The shear strain, $\gamma_{xy} = (\alpha_1 + \alpha_2)$, is now the angle c_1dc on Fig. 3.46b. The direct strains are

$$\varepsilon_x = \frac{dc_1 - dc}{dc} \quad \text{and} \quad \varepsilon_y = \frac{da_1 - da}{da}$$

Let us suppose that ε_x, ε_y, ε_z and γ_{xy} for the element in Fig. 3.46b are all known. The state of strain with respect to the axes, xyz, namely

$$S = \begin{bmatrix} \varepsilon_x & \tfrac{1}{2}\gamma_{xy} & 0 \\ \tfrac{1}{2}\gamma_{xy} & \varepsilon_y & 0 \\ 0 & 0 & \varepsilon_z \end{bmatrix}$$

is therefore known.

We want to describe the same state with respect to the axes $\bar{x}\bar{y}z$, where $\bar{x}\bar{y}$ are axes located at an angle θ (measured positive anticlockwise) from xy and, of course, in the same plane. That is, we want to find

$$\bar{S} = \begin{bmatrix} \varepsilon_{\bar{x}} & \tfrac{1}{2}\gamma_{\bar{x}\bar{y}} & 0 \\ \tfrac{1}{2}\gamma_{\bar{x}\bar{y}} & \varepsilon_{\bar{y}} & 0 \\ 0 & 0 & \varepsilon_z \end{bmatrix}$$

with respect to the axes $\bar{x}\bar{y}z$ and to express $\varepsilon_{\bar{x}}$, $\varepsilon_{\bar{y}}$ and $\gamma_{\bar{x}\bar{y}}$ in terms of ε_x, ε_y, γ_{xy} and θ. The value of ε_z will remain the same; it is not affected by rotation of the element in the xy plane.

To do this, consider the direct strains and the shear strain separately. As we have seen, from Eqs (3.82), (3.83) and (3.85), direct strains and shear strains do not affect each other to a first order of accuracy.

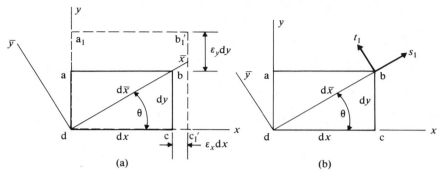

Figure 3.47   Deformations of an element subjected to direct strains only in the x and y directions

The effect of the direct strains ε_x and ε_y

Figure 3.47 isolates the change of shape of the element (see Fig. 3.46) due to ε_x and ε_y alone, i.e., it only shows the effects of the changes in lengths in the x and y directions and ignores the change in shape due to shear.

In Fig. 3.47a, the movement of the point b through the distance bb$'_1$ has components $\varepsilon_x \, dx$ and $\varepsilon_y \, dy$ in the x and y directions respectively. This same movement, i.e., the distance, bb$'_1$, can be represented equally well by components parallel to the $\bar{x}$ and $\bar{y}$ axes. These are shown in Fig. 3.47b and designated s_1 and t_1 respectively.

The relationships between these two ways of defining the distance bb$'_1$, are, from simple geometry:

$$s_1 = (\varepsilon_x \, dx) \cos \theta + (\varepsilon_y \, dy) \sin \theta \tag{3.88}$$

$$t_1 = -(\varepsilon_x \, dx) \sin \theta + (\varepsilon_y \, dy) \cos \theta \tag{3.89}$$

Putting $dx = d\bar{x} \cos \theta$; $dy = d\bar{x} \sin \theta$, then Eqs (3.88) and (3.89) can be written as:

$$s_1 = d\bar{x}(\varepsilon_x \cos^2 \theta + \varepsilon_y \sin^2 \theta) \tag{3.90}$$

$$t_1 = -d\bar{x}(\varepsilon_x - \varepsilon_y) \sin \theta \cos \theta \tag{3.91}$$

We will come back to Eqs (3.90) and (3.91).

The effect of the shear strain γ_{xy}

Figure 3.48 isolates the change of shape of the element (see Fig. 3.46) due to the shear strain γ_{xy} alone, i.e., it ignores the changes in the lengths in the x and y directions due to ε_x and ε_y.

In Fig. 3.48a, the movement of the point b through the distance bb$''_1$ is equal to $\gamma_{xy} \, dx$, which has components s_2 and t_2 parallel to the $\bar{x}$ and $\bar{y}$ axes, respectively, where

$$s_2 = \gamma_{xy} \, dx \sin \theta = \gamma_{xy} \, d\bar{x} \sin \theta \cos \theta \tag{3.92}$$

$$t_2 = \gamma_{xy} \, dx \cos \theta = \gamma_{xy} \, d\bar{x} \cos^2 \theta \tag{3.93}$$

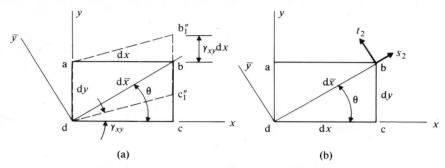

Figure 3.48   Deformations of an element subjected only to shear stress in the x and y directions

The combined effect of the direct strains and the shear strain

If the two effects which have been considered separately above are now combined, the total movement of the point b on the element in directions parallel to the $\bar{x}$ and $\bar{y}$ axes respectively are, from Eqs (3.90), (3.91), (3.92) and (3.93):

$$s_\theta = s_1 + s_2 = d\bar{x}(\varepsilon_x \cos^2 \theta + \varepsilon_y \sin^2 \theta + \gamma_{xy} \sin \theta \cos \theta) \tag{3.94}$$

$$t_\theta = t_1 + t_2 = d\bar{x}\{-(\varepsilon_x - \varepsilon_y) \sin \theta \cos \theta + \gamma_{xy} \cos^2 \theta\} \tag{3.95}$$

Recalling that

$$\sin^2 \theta = \frac{1}{2}(1 - \cos 2\theta); \qquad \cos^2 \theta = \frac{1}{2}(1 + \cos 2\theta); \qquad \sin \theta \cos \theta = \frac{1}{2}\sin 2\theta$$

then Eqs (3.94) and (3.95) can be written in terms of 2θ as

$$s_\theta = d\bar{x}\left\{\varepsilon_x \frac{1}{2}(1 + \cos 2\theta) + \varepsilon_y \frac{1}{2}(1 - \cos 2\theta) + \gamma_{xy} \frac{1}{2}\sin 2\theta\right\}$$

that is $$s_\theta = d\bar{x}\left\{\frac{1}{2}(\varepsilon_x + \varepsilon_y) + \frac{1}{2}(\varepsilon_x - \varepsilon_y)\cos 2\theta + \frac{\gamma_{xy}}{2}\sin 2\theta\right\} \tag{3.96}$$

and similarly

$$t_\theta = d\bar{x}\left\{-\frac{1}{2}(\varepsilon_x - \varepsilon_y)\sin 2\theta + \frac{\gamma_{xy}}{2}(1 + \cos 2\theta)\right\} \tag{3.97}$$

In summary, s_θ and t_θ, given by Eqs (3.96) and (3.97), are the components of the movement of point b, along and perpendicular to $d\bar{x}$, due to the strains ε_x, ε_y and γ_{xy} to which the element is subjected.

The direct strain ε at any angle θ from the direction of the x-axis

In Eq. (3.96), s_θ is the change in length of the line of original length $d\bar{x}$. Therefore, the direct strain in the xy plane at any angle θ from the direction of the x-axis is

$$\varepsilon_{\bar{x}} = \frac{s_\theta}{d\bar{x}}$$

that is

$$\varepsilon_{\bar{x}} = \frac{1}{2}(\varepsilon_x + \varepsilon_y) + \frac{1}{2}(\varepsilon_x - \varepsilon_y)\cos 2\theta + \frac{\gamma_{xy}}{2}\sin 2\theta \tag{3.98}$$

The shear strain $\gamma_{\bar{x}\bar{y}}$ on a rectangular element orientated at any angle θ from the x-axis

Because shear strain is defined as the change in a right angle, we must obviously examine the changes in directions of two of its sides. Thus, in Fig. 3.49, the sides bd and ld are at right angles to each other when the element bdlm is unstressed.

The application of stress causes the side bd to move through an angle ϕ_θ where

$$\phi_\theta = \frac{t_\theta}{d_{\bar{x}}} \quad \text{and, therefore, from Eq. (3.97)}$$

$$\phi_\theta = -\frac{1}{2}(\varepsilon_x - \varepsilon_y)\sin 2\theta + \frac{\gamma_{xy}}{2}(1 + \cos 2\theta) \tag{3.99}$$

At the same time, the side ld moves through the angle $\phi_{(\theta + \pi/2)}$ where

$$\phi_{(\theta + \pi/2)} = \frac{t_{(\theta + \pi/2)}}{d_{\bar{y}}}$$

and $t_{(\theta + \pi/2)}$ is found from Eq. (3.97) by replacing θ with $(\theta + \pi/2)$, and $d\bar{x}$ by $d\bar{y}$, i.e.

$$\phi_{(\theta + \pi/2)} = \frac{1}{2}(\varepsilon_x - \varepsilon_y)\sin 2\theta + \frac{\gamma_{xy}}{2}(1 - \cos 2\theta) \tag{3.100}$$

It follows that the original right angle ldb has been reduced by the angle $\phi_\theta - \phi_{(\theta + \pi/2)}$

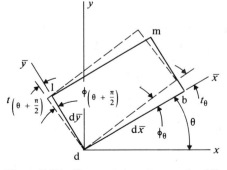

Figure 3.49   Shear strains at an angle oblique to the x and y directions

which, by definition, is therefore the shear strain on the element bdlm lying in the $\bar{x}\bar{y}$ axes. Therefore

$$\gamma_{\bar{x}\bar{y}} = \phi - \phi_{(\theta + \pi/2)}$$

i.e. $$\gamma_{\bar{x}\bar{y}} = \left\{ -\frac{1}{2}(\varepsilon_x - \varepsilon_y) \sin 2\theta + \frac{\gamma_{xy}}{2}(1 + \cos 2\theta) \right\} - \left\{ \frac{1}{2}(\varepsilon_x - \varepsilon_y) \sin 2\theta \right.$$

$$\left. + \frac{\gamma_{xy}}{2}(1 - \cos 2\theta) \right\}$$

i.e. $$\frac{\gamma_{\bar{x}\bar{y}}}{2} = -\frac{1}{2}(\varepsilon_x - \varepsilon_y) \sin 2\theta + \frac{\gamma_{xy}}{2} \cos 2\theta \tag{3.101}$$

Equations (3.98) and (3.101) are important. They are expressing the direct and shear strains at any angle θ to the direction of the x-axis in terms of the strains ε_x, ε_y, γ_{xy} and 2θ, and are equivalent to Eqs (3.55) and (3.56) previously derived for the direct and shear stresses $\sigma_{\bar{x}}$ and $\tau_{\bar{x}\bar{y}}$. Examine these two pairs of equations side by side and note their similarity.

Mohr's circle of strain

Comparison of the two pairs of Eqs (3.55) and (3.56) with (3.98) and (3.101) shows that the latter pair is identical with the former if $\sigma_{\bar{x}}$, $\tau_{\bar{x}\bar{y}}$, σ_x, σ_y and τ_{xy} are replaced by $\varepsilon_{\bar{x}}$, $\frac{1}{2}\gamma_{\bar{x}\bar{y}}$, ε_x, ε_y and $\frac{1}{2}\gamma_{xy}$ respectively.

Two things immediately follow from this comparison and from the study of stress on an element, namely:

(i) if the state of strain is defined at a point and with respect to the axes xyz by the strain array S, where

$$S = \begin{bmatrix} \varepsilon_x & \frac{1}{2}\gamma_{xy} & 0 \\ \frac{1}{2}\gamma_{xy} & \varepsilon_y & 0 \\ 0 & 0 & \varepsilon_z \end{bmatrix} \tag{3.102}$$

then the same state of strain can be restated with respect to $\bar{x}\bar{y}z$ by $\bar{S}$ where

$$\bar{S} = \begin{bmatrix} \varepsilon_{\bar{x}} & \frac{1}{2}\gamma_{\bar{x}\bar{y}} & 0 \\ \frac{1}{2}\gamma_{\bar{x}\bar{y}} & \varepsilon_{\bar{y}} & 0 \\ 0 & 0 & \varepsilon_z \end{bmatrix} \tag{3.103}$$

in which $\varepsilon_{\bar{x}}$ or $\varepsilon_{\bar{y}}$ are given by putting the appropriate values of θ (corresponding to the directions of $\bar{x}$ or $\bar{y}$ relative to x), in Eq. (3.98) and $\frac{1}{2}\gamma_{\bar{x}\bar{y}}$ is given by putting θ in Eq. (3.101), i.e., the angle of $\bar{x}$ relative to x, and

(ii) the state of strain at the point can be represented in the xy plane by a circle drawn in the axes $\varepsilon_{\bar{x}}$ and $\gamma_{\bar{x}\bar{y}}/2$.

To draw Mohr's circle of strain to represent the state of strain defined by S in Eq. (3.102) (the elements of the array all being known), first construct the axes ε and $\frac{1}{2}\gamma_{\bar{x}\bar{y}}$ as shown in Fig. 3.50d. Just as we plotted $\tau_{\bar{x}\bar{y}}$ downward when positive, so also $\frac{1}{2}\gamma_{\bar{x}\bar{y}}$ will be plotted positive downward.

Represent the direct strain and shear strain combination associated with the x direction (see Figs 3.50b and 3.50c) by the point R $\left(\varepsilon_x, \frac{1}{2}\gamma_{xy}\right)$ as shown in Fig. 3.50d. Represent the direct strain and shear strain associated with the y direction by the point S $\left(\varepsilon_y, -\frac{1}{2}\gamma_{xy}\right)$. The reason for this negative sign of $\frac{1}{2}\gamma_{xy}$ when plotting S is because the direction y is equivalent to the direction x rotated through 90° anticlockwise. That is, when $\theta = 90°$, the direction of $\bar{x}$ coincides with the direction of y and therefore $\frac{1}{2}\gamma_{\bar{x}\bar{y}} = -\frac{1}{2}\gamma_{xy}$. It is similar to the reasoning and procedure used when constructing the Mohr's circle of stress.

If γ_{xy} is positive, R is plotted below the ε-axis. The point S is always plotted on the opposite side of the ε-axis from R.

Draw a circle on RS as diameter. This is Mohr's circle of strain for the xy plane and points on the circle represent combinations of $\varepsilon_{\bar{x}}$ and $\frac{1}{2}\gamma_{\bar{x}\bar{y}}$ in every direction at the point where the state of strain is defined by the strain array S.

Referring to Fig. 3.50d, the following properties of the strain circle follow from its construction.

The centre of the circle, C, will always line on the ε-axis.

$$MC = CL = \frac{(\varepsilon_x - \varepsilon_y)}{2}$$

$$OC = \frac{(\varepsilon_x + \varepsilon_y)}{2} \tag{3.104}$$

and the radius,
$$r = \sqrt{\left(\left[\frac{\varepsilon_x - \varepsilon_y}{2}\right]^2 + \left[\frac{\gamma_{xy}}{2}\right]^2\right)}$$

The circle is reproduced in Fig. 3.50e. Let the points U and V be located on the circle at 2θ anticlockwise from R and S respectively.

By direct comparison with the reasoning which led to Eqs (3.71) and (3.72) for $\sigma_{\bar{x}}$ and $\tau_{\bar{x}\bar{y}}$ (relating to the stress circle), consideration of the geometry of Fig. 3.53e leads

$$S = \begin{bmatrix} \varepsilon_x & \tfrac{1}{2}\,\gamma_{xy} & 0 \\ \tfrac{1}{2}\,\gamma_{xy} & \varepsilon_y & 0 \\ 0 & 0 & \varepsilon_z \end{bmatrix}$$

with respect to xyz

(a)

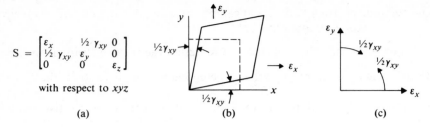

(b)

(c)

The state of strain S can be represented by (b) or, more conveniently, by (c)

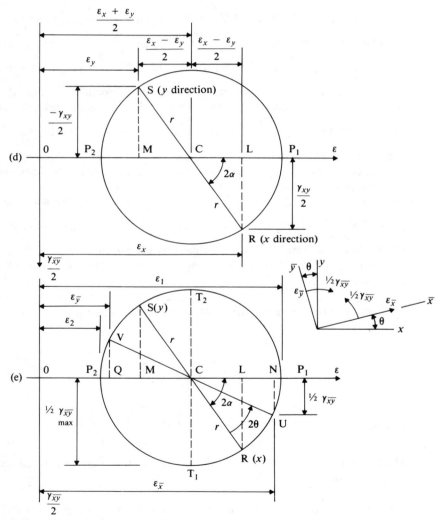

Figure 3.50   Mohr's circle of strain

to the following results. At the point U:

$$ON = \frac{1}{2}(\varepsilon_x + \varepsilon_y) + \frac{1}{2}(\varepsilon_x - \varepsilon_y)\cos 2\theta + \frac{\gamma_{xy}}{2}\sin 2\theta \qquad (3.105)$$

$$UN = -\frac{1}{2}(\varepsilon_x - \varepsilon_y)\sin 2\theta + \frac{\gamma_{xy}}{2}\cos 2\theta \qquad (3.106)$$

By comparison with Eqs (3.98) and (3.101), it is seen that Eqs (3.105) and (3.106) are the expressions for $\varepsilon_{\bar{x}}$ and $\frac{1}{2}\gamma_{\bar{x}\bar{y}}$ respectively in the direction located on the element at θ anticlockwise from the direction of the x-axis.

It follows that a rotational movement of 2θ on the strain circle represents a rotational movement of θ *in the same direction* on the element. If U represents the direct and shear strain combinations in the direction at θ anticlockwise from x on the element (i.e., in the direction $\bar{x}$), then the point V represents the strain combination at $(\theta + 90)°$ anticlockwise from x, i.e., in the direction $\bar{y}$.

Principal strains and principal planes from Mohr's strain circle

A principal plane was previously defined as a plane along which the shear stress is zero and it was shown that such planes, separated by 90°, carry the maximum and minimum values of direct or normal stresses, namely, the principal stresses. The principal planes so defined also experience zero shear strain and corresponding principal (maximum and minimum) direct or normal strains.

The principal strains are designated ε_1 and ε_2 and, as illustrated in Fig. 3.50e, are represented by the points P_1 and P_2 on the strain circle. Remember that the strain circle drawn in Fig. 3.50 represents the state of strain in the xy plane, so that ε_1 and ε_2 are the maximum and minimum direct strains in that plane.

From Fig. 3.50e, it is clear that the maximum principal strain (at P_1) occurs at angle α anticlockwise from the direction of x on the element and the minimum principal strain occurs at 90° to the maximum principal strain.

The principal strain array in the xy plane is therefore

$$S_p = \begin{bmatrix} \varepsilon_1 & 0 & 0 \\ 0 & \varepsilon_2 & 0 \\ 0 & 0 & \varepsilon_z \end{bmatrix} \text{ at } \alpha \text{ anticlockwise from the axes } xy \qquad (3.107)$$

Maximum and minimum shear strains from Mohr's strain circle

These are located in Fig. 3.50e at the points T_1 and T_2. Therefore, they occur on planes separated from the principal planes by 45° and it follows that planes of maximum and minimum shear strain coincide with planes of maximum and minimum shear stress and that on such planes, the direct strains are equal to OC, where

$$OC = \left(\frac{\varepsilon_x + \varepsilon_y}{2}\right) = \left(\frac{\varepsilon_1 + \varepsilon_2}{2}\right) \qquad (3.108)$$

The maximum shear strain array in the plane xy is therefore

$$S_{T_1} = \begin{bmatrix} \varepsilon_{\bar{x}} & \frac{1}{2}\gamma_{\bar{x}\bar{y}\,max} & 0 \\ \frac{1}{2}\gamma_{\bar{x}\bar{y}\,max} & \varepsilon_{\bar{y}} & 0 \\ 0 & 0 & \varepsilon_z \end{bmatrix}$$

i.e. $$S_{T_1} = \begin{bmatrix} \frac{1}{2}(\varepsilon_x + \varepsilon_y) & \frac{1}{2}\gamma_{\bar{x}\bar{y}\,max} & 0 \\ \frac{1}{2}\gamma_{\bar{x}\bar{y}\,max} & \frac{1}{2}(\varepsilon_x + \varepsilon_y) & 0 \\ 0 & 0 & \varepsilon_z \end{bmatrix} \qquad (3.109)$$

EXAMPLE 3.10

At a point on the surface of a steel plate lying in the xy plane, the strain array with respect to the axes xyz is given by S, where

$$S = \begin{bmatrix} 700 & 300 & 0 \\ 300 & 200 & 0 \\ 0 & 0 & \varepsilon_z \end{bmatrix} \times 10^{-6}$$

Use Mohr's circle of strain to find the directions of the principal strains. Write the principal strain array and the maximum shear strain array in the plane xy. Sketch an element subjected to the maximum shear strain, indicating how it deforms.

The strain array S and its simple representation in the axes xy are shown in Figs 3.51a and 3.51b respectively.

The circle to respresent the state of strain in the xy plane is constructed (see Fig. 3.51c) by plotting the point R $(\varepsilon_x = 700 \times 10^{-6}, \frac{1}{2}\gamma_{xy} = 300 \times 10^{-6})$ to represent the strain combination in the x direction and the point S $\left(\varepsilon_y = 200 \times 10^{-6}, -\frac{1}{2}\gamma_{yx} = \right.$ $-300 \times 10^{-6})$ to represent the strain combination in the y direction.

From the resulting circle drawn on the diameter RS:

In Fig. 5.54c,

$$\widehat{OC} = \left(\frac{\varepsilon_x + \varepsilon_y}{2}\right) = \frac{1}{2}(700 + 200) \times 10^{-6} = 450 \times 10^{-6}$$

$$LC = MC = \left(\frac{\varepsilon_x - \varepsilon_y}{2}\right) = \frac{1}{2}(700 - 200) \times 10^{-6} = 250 \times 10^{-6}$$

$$\therefore r = \{\sqrt{(300^2 + 250^2)}\} \times 10^{-6} = 390.5 \times 10^{-6}$$

$$\therefore \varepsilon_1 = (450 + 390.5) \times 10^{-6} \quad \text{i.e., } \varepsilon_1 = 840.5 \times 10^{-6}$$

and $$\varepsilon_2 = (450 - 390.5) \times 10^{-6} = 59.5 \times 10^{-6}$$

$$2\alpha = \tan^{-1}\frac{300}{250} = 50.20°, \qquad \therefore \alpha = 25.10°$$

$$S = \begin{bmatrix} 700 & 300 & 0 \\ 300 & 200 & 0 \\ 0 & 0 & \varepsilon_z \end{bmatrix} \times 10^{-6}$$

(a)

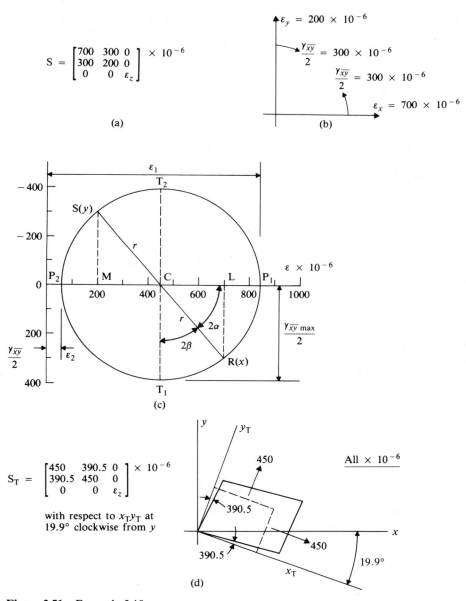

(b)

(c)

$$S_T = \begin{bmatrix} 450 & 390.5 & 0 \\ 390.5 & 450 & 0 \\ 0 & 0 & \varepsilon_z \end{bmatrix} \times 10^{-6}$$

with respect to $x_T y_T$ at 19.9° clockwise from y

(d)

Figure 3.51   Example 3.10

Thus, the principal strain array in the xy plane is S_p, where

$$S_p = \begin{bmatrix} 840.5 & 0 & 0 \\ 0 & 59.5 & 0 \\ 0 & 0 & \varepsilon_z \end{bmatrix} \times 10^{-6} \quad \begin{array}{l} \text{with respect to axes } x_p y_p \\ \text{located at } 25.1° \text{ anticlockwise} \\ \text{from } xy \end{array}$$

From above,

$$\frac{\gamma_{\bar{x}\bar{y}\,max}}{2} = r = 390 \times 10^{-6}$$

and

$$\beta = (45 - \alpha)° = 19.9° \text{ at point } T_1$$

Also, at T_1,

$$\varepsilon_{\bar{x}} = \varepsilon_{\bar{y}} = OC = 450 \times 10^{-6}$$

Thus the maximum shear strain array in the xy plane is S_T, where

$$S_{T_1} = \begin{bmatrix} 450. & 390.5 & 0 \\ 390.5 & 450 & 0 \\ 0 & 0 & \varepsilon_z \end{bmatrix} \times 10^{-6} \quad \begin{array}{l} \text{with respect to axes } x_T y_T \\ \text{located at } 19.9° \text{ clockwise} \\ \text{from } xy \end{array}$$

The deformed element described by S_{T_1} is shown in Fig. 3.51d.

EXAMPLE 3.11

A rectangular plate ABCD (Fig. 3.52) has dimensions AB = 1000 mm, BC = 600 mm when unstressed. In-plane loads are applied which cause the dimensions to become 1003.2 mm and 600.5 mm, and the angle ADC to change from 90° to 89.8°. Find the principal strain array in the plane (xy) of the plate and sketch the shape of an element subjected to maximum shear strain.

From Fig. 3.52

$$\varepsilon_x = (1003.2 - 1000)/1000 = 3.20 \times 10^{-3}$$

$$\varepsilon_y = (600.5 - 600)/600 = 0.83 \times 10^{-3}$$

$$\gamma_{xy} = (90.0 - 89.8)\pi/180 = 3.49 \times 10^{-3}$$

($+$ve because the angle in the xy becomes smaller)

The strain array with respect to the xyz axes is therefore

$$S = \begin{bmatrix} 3.20 & 1.75 & 0 \\ 1.75 & 0.83 & 0 \\ 0 & 0 & \varepsilon_z \end{bmatrix} \times 10^{-3}$$

The corresponding strain circle is drawn on Fig. 3.52b, based upon RS as diameter.

$$OC = (3.20 + 0.83)\frac{10^{-3}}{2} = 2.02 \times 10^{-3}$$

$$CM = LC = (3.20 - 0.83)\frac{10^{-3}}{2} = 1.19 \times 10^{-3}$$

$$\therefore r = 10^{-3}\sqrt{\{1.19^2 + 1.75^2\}} = 2.12 \times 10^{-3}$$

$$2\alpha = \tan^{-1}\frac{RM}{CM} = \tan^{-1}\frac{1.75}{1.19} = 55.8°, \qquad \therefore \alpha = 27.9°$$

$$\varepsilon_1 = OC + r = (2.02 + 2.12) \times 10^{-3} = 4.14 \times 10^{-3}$$

$$\varepsilon_2 = OC - r = (2.02 - 2.12) \times 10^{-3} = -0.10 \times 10^{-3}$$

The principal strain array in the plane xy is therefore

$$S_p = \begin{bmatrix} 4.14 & 0 & 0 \\ 0 & -0.10 & 0 \\ 0 & 0 & \varepsilon_z \end{bmatrix} \times 10^{-3}$$

with respect to $x_p y_p$ located 27.9° anticlockwise from the direction of the x axis

$\gamma_{\bar{x}\bar{y}\,max} = 2r = 4.24 \times 10^{-3}$ at $\frac{1}{2}(90 - 55.8)°$, i.e., 17.1° clockwise from xy. The element experiencing the maximum shear strain in the xy plane is shown in Fig. 3.52c.

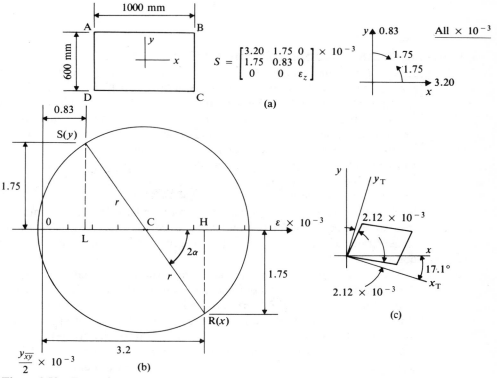

Figure 3.52 Example 3.11

Problem 3.7   Each of the strain arrays below represents the state of strain at a point on a plate lying in the xy plane. The strains are with respect to the xy axes. In each case, find the directions of the principal strains, the principal strain array and the maximum shear strain in the plane of the plate.

In each case, find the maximum shear strain array and sketch the deformed shape of the element suffering maximum shear strain.

$$S_1 = \begin{bmatrix} 400 & 200 & 0 \\ 200 & -200 & 0 \\ 0 & 0 & \varepsilon_z \end{bmatrix} \times 10^{-6} \qquad S_2 = \begin{bmatrix} 0 & -200 & 0 \\ -200 & 400 & 0 \\ 0 & 0 & \varepsilon_z \end{bmatrix} \times 10^{-6}$$

$$S_3 = \begin{bmatrix} 200 & 0 & 0 \\ 0 & 200 & 0 \\ 0 & 0 & \varepsilon_z \end{bmatrix} \times 10^{-6} \qquad S_4 = \begin{bmatrix} -200 & 200 & 0 \\ 200 & -600 & 0 \\ 0 & 0 & \varepsilon_z \end{bmatrix} \times 10^{-6}$$

Problem 3.8   A rectangle is drawn on a flat sheet of thin rubber. The sheet is then subjected to strains $\varepsilon_x = 3 \times 10^{-3}$, $\varepsilon_y = -2 \times 10^{-3}$, $\gamma_{xy} = 2 \times 10^{-3}$. If the angles of the rectangle do not change when these strains are applied, at what orientation to the xy axes was the rectangle drawn?

3.6 Stress–strain relations (or constitutive relations)

In structural analysis and design, it is necessary to know how structural materials respond to the application of stress. In structures which are statically indeterminate, it is impossible to analyse the structure without a knowledge of the stress–strain relations, because the distribution of the applied loads through the structure is dependent upon the compatible deformations of its various parts and these deformations are, in turn, dependent upon the strain-response of the material to stress. For statically determinate structures, although the stresses can be determined without a knowledge of the material stress–strain relationship, the calculation of deflections requires this knowledge. Furthermore, for all structures, it is necessary to know the level of stress which can occur before the material fails (fractures) completely, becomes incapable of carrying further increase in stress without experiencing unacceptably large strain or begins to creep (i.e., exhibits continuously increasing strain at a constant stress).

The relationship between stress and strain for any material is an empirical one; it is determined only by experiment and observation. It cannot be deduced theoretically.

For the large majority of structural analyses, the particular stress–strain relationship used is the simplest, namely, the uniaxial stress–strain characteristic, which is found by applying either a tensile force or a compressive force to a specimen of the material and measuring the corresponding strains in the direction of the force. In such tests, it is assumed that the applied force is distributed as a uniform stress on the cross-section of the specimen at the point where the strain is measured.

Metals are generally subjected to both tensile and compressive testing, but other materials, such as concrete, masonry and brick are usually tested only in compression (because their tensile strength is small and tensile testing is difficult). Timber is subjected to compression, shear and bending tests.

In determining and stating the stress–strain relationship for a material, there are certain precautions to be observed. It must be known whether the material is *homogeneous*, i.e., whether it has the same composition and physical properties throughout its mass, so that the specimen tested is representative of the material as a whole. Steel is generally regarded as a homogeneous material but, in spite of that, it is a common experience, especially when using steel in research projects, to test a number of specimens from different locations in the batch and to find small differences in the stress–strain relationships of the specimens. Most mechanical testing is directional as far as the material is concerned. For example, a tensile specimen is cut or machined from the material in a particular direction. An *isotropic* material can be defined as having the same physical properties in all directions. If these properties change with direction, the material is *anisotropic*. Timber is an obvious anisotropic material because of its grain structure.

Chapter 4 deals in some detail with the stress–strain relationships for the two most important materials used in structural engineering, namely, steel and concrete.

Stress–strain relations for elastic, isotropic materials

At this point, we confine our attention to only part of the total picture of the relationships which exist between stress and strain. The material is assumed to be elastic, i.e., the imposition of an increment of stress causes an increment of strain which totally disappears when the increment of stress is removed. In an isotropic material, the same increment of stress will cause the same increment of strain in any direction at a particular point in the material. We will also assume that the material is homogeneous, i.e., that the strain response to stress is the same throughout the mass of the material.

Poisson's ratio

Poisson's ratio is an empirically-determined material constant which relates the strain in all directions at right angles to the direction of the applied stress to the strain in the same direction as the applied stress.

Figure 3.53 shows the (exaggerated) change in shape of a round bar when subjected to a tensile stress σ_x. The bar gets longer and thinner. Therefore, it experiences a strain

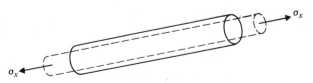

Figure 3.53 Illustrating the deformation of a round bar subjected to an axial tensile stress. It becomes longer but thinner

(equal to its change in length divided by its original length) in the direction of the stress, and a lateral strain equal to the change in diameter divided by the original diameter. It does not matter which diameter we choose because all diameters will experience the same change. The laterial strain is, therefore, the same in all directions perpendicular to the direction of the stress.

The strain in the lateral direction will always have the opposite sign to that of the strain in the direction of the stress. In this case (Fig. 3.53), the length of the bar increases, therefore the strain in the direction of σ_x is positive, whereas in the lateral direction the strain is negative. Notice that in any lateral direction, the bar experiences *a strain without any stress* in the direction of that strain.

Poisson's ratio is defined as v, where

$$v = -\left(\frac{\text{strain in any direction perpendicular to that of the applied stress}}{\text{strain in the direction of the applied stress}}\right)$$

Because the numerator and the denominator of this ratio will always be of opposite sign, it follows that v is always a positive quantity. If the specimen in Fig. 3.53 had been subjected to a compressive stress, the signs of both strains would be reversed.

A note on plane stress and plane strain

The condition of *plane stress* (in the xy plane) is defined by the stress array:

$$F = \begin{bmatrix} \sigma_x & \tau_{xy} & 0 \\ \tau_{yx} & \sigma_y & 0 \\ 0 & 0 & 0 \end{bmatrix}$$

The absence of direct stress in the z direction ($\sigma_z = 0$) will generally mean that there will exist a strain in the z direction, i.e., $\varepsilon_z \neq 0$, because of Poisson's ratio. The corresponding strain array will therefore be

$$S = \begin{bmatrix} \varepsilon_x & \frac{1}{2}\gamma_{xy} & 0 \\ \frac{1}{2}\gamma_{xy} & \varepsilon_y & 0 \\ 0 & 0 & \varepsilon_z \end{bmatrix}$$

The condition of *plane strain* (in the xy plane) is defined by the strain array:

$$S = \begin{bmatrix} \varepsilon_x & \frac{1}{2}\gamma_{xy} & 0 \\ \frac{1}{2}\gamma_{xy} & \varepsilon_y & 0 \\ 0 & 0 & 0 \end{bmatrix}$$

In this case, the absence of direct strain in the z direction ($\varepsilon_z = 0$) will generally mean that a stress will exist in the z direction, i.e., $\sigma_z \neq 0$. This stress will be necessary in order to prevent any strain in the z direction. The corresponding stress array will therefore be

$$F = \begin{bmatrix} \sigma_x & \gamma_{xy} & 0 \\ \gamma_{xy} & \sigma_y & 0 \\ 0 & 0 & \sigma_z \end{bmatrix}$$

The transformation equations derived for plane stress were based entirely upon equilibrium in the xy plane. Therefore, they remain valid if σ_z is not zero and this includes the plane strain condition. It follows that Mohr's circle of stress, which represents the transformation equations for stress, also remains true in the xy plane when $\sigma_z \neq 0$.

The analysis of strain was also conducted for the case of plane stress, when it was recognized that, for this condition, ε_z would generally exist (Eqs (3.86) and (3.87)). Again, the strain transformation equations in the xy plane (and their representation by the Mohr's circle of strain) did not involve strains in the z direction. It follows, therefore, that the analysis, and the circle, remain valid whatever the value of ε_z, which includes the case of plain strain.

Mohr's circles of stress and strain are therefore both valid for both plane stress and plane strain conditions.

Stress–strain relations for direct stress and strain

Let us now consider what happens to an element of material within the bar shown in Fig. 3.53 when the stress σ_x is applied. The element and its distorted shape are shown in Fig. 3.54.

The element of original dimensions, dx, dy, dz, will change shape as a result of the

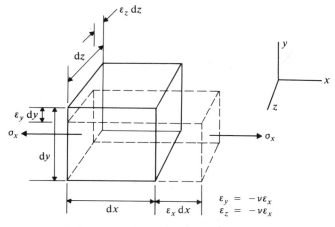

Figure 3.54   Deformation of an element of the bar in Fig. 3.53

application of σ_x, to the deformed element of dimensions $(dx + \varepsilon_x\,dx)$, $(dy + \varepsilon_y\,dy)$, $(dz + \varepsilon_z\,dz)$, where, because of the Poissons's ratio effect,

$$\varepsilon_y = -\nu\varepsilon_x$$

$$\varepsilon_z = -\nu\varepsilon_x$$

Because
$$\varepsilon_x = \frac{\sigma_x}{E}$$

then
$$\varepsilon_y = \varepsilon_z = -\nu\frac{\sigma_x}{E} \tag{3.110}$$

where E is Young's modulus.

The stress array for the element in Fig. 3.54 is F_1, where

$$F_1 = \begin{bmatrix} \sigma_x & 0 & 0 \\ 0 & 0 & 0 \\ 0 & 0 & 0 \end{bmatrix}$$

From Eq. (3.110) it follows that the stress array F_1 causes the strain array S_1, where

$$S_1 = \begin{bmatrix} \varepsilon_x & 0 & 0 \\ 0 & -\nu\varepsilon_x & 0 \\ 0 & 0 & -\nu\varepsilon_x \end{bmatrix} = \frac{1}{E}\begin{bmatrix} \sigma_x & 0 & 0 \\ 0 & -\nu\sigma_x & 0 \\ 0 & 0 & -\nu\sigma_x \end{bmatrix} \tag{3.111}$$

Now suppose that, instead of σ_x acting alone on the element, the stress σ_y acted alone giving a stress array F_2, where

$$F_2 = \begin{bmatrix} 0 & 0 & 0 \\ 0 & \sigma_y & 0 \\ 0 & 0 & 0 \end{bmatrix}$$

F_2 would cause a strain array S_2, where

$$S_2 = \frac{1}{E}\begin{bmatrix} -\nu\sigma_y & 0 & 0 \\ 0 & \sigma_y & 0 \\ 0 & 0 & -\nu\sigma_y \end{bmatrix} \tag{3.112}$$

Similarly, a stress array F_3 (σ_z acting alone) would cause a strain array S_3, where

$$F_3 = \begin{bmatrix} 0 & 0 & 0 \\ 0 & 0 & 0 \\ 0 & 0 & \sigma_z \end{bmatrix}$$

and

$$S_3 = \frac{1}{E} \begin{bmatrix} -v\sigma_z & 0 & 0 \\ 0 & -v\sigma_z & 0 \\ 0 & 0 & \sigma_z \end{bmatrix} \qquad (3.113)$$

If all the three direct stresses act simultaneously, thus $(F_1 + F_2 + F_3)$ causes a state of strain described by $(S_1 + S_2 + S_3)$. That is, the state of stress

$$\begin{bmatrix} \sigma_x & 0 & 0 \\ 0 & \sigma_y & 0 \\ 0 & 0 & \sigma_z \end{bmatrix}$$

causes the state of strain

$$\begin{bmatrix} \varepsilon_x & 0 & 0 \\ 0 & \varepsilon_y & 0 \\ 0 & 0 & \varepsilon_z \end{bmatrix} = \frac{1}{E} \begin{bmatrix} \sigma_x - v(\sigma_y + \sigma_z) & 0 & 0 \\ 0 & \sigma_y - v(\sigma_x + \sigma_z) & 0 \\ 0 & 0 & \sigma_z - v(\sigma_x + \sigma_y) \end{bmatrix} \qquad (3.114)$$

Equation (3.114) can be written as

$$\varepsilon_x = \frac{1}{E} \{\sigma_x - v(\sigma_y + \sigma_z)\}$$

$$\varepsilon_y = \frac{1}{E} \{\sigma_y - v(\sigma_x + \sigma_z)\} \qquad (3.115)$$

$$\varepsilon_z = \frac{1}{E} \{\sigma_z - v(\sigma_x + \sigma_y)\}$$

or, in matrix form:

$$\begin{bmatrix} \varepsilon_x \\ \varepsilon_y \\ \varepsilon_z \end{bmatrix} = \frac{1}{E} \begin{bmatrix} 1 & -v & -v \\ -v & 1 & -v \\ -v & -v & 1 \end{bmatrix} \begin{bmatrix} \sigma_x \\ \sigma_y \\ \sigma_z \end{bmatrix} \qquad (3.116)$$

Equations (3.115) and (3.1116) are statements of Hooke's Law, which is true to a first order of accuracy, whether or not σ_x, σ_y and σ_z are principal stresses (see Section 3.5 on the analysis of strain).

Examination of Eq. (3.115) reveals that, even when only one of the stresses σ_x, σ_y or σ_z is non-zero, an unconstrained element will experience strains in all three directions.

EXAMPLE 3.12

A mild steel plate ($E = 200 \text{ kN/mm}^2$, $v = 0.3$) lies in the plane xy (see Fig. 3.55). It has dimensions $250 \text{ mm} \times 250 \text{ mm} \times 10 \text{ mm}$. A compressive stress of 210 N/mm^2 is applied to the plate in the x direction. Find the changes in dimensions of the plate.

$$\sigma_x = -210 \text{ N/mm}^2, \qquad \sigma_y = 0, \qquad \sigma_z = 0$$

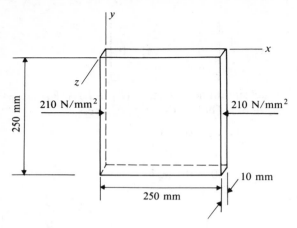

Figure 3.55   Example 3.12

From Eq. (3.115):

$$\varepsilon_x = \frac{\sigma_x}{E} = \frac{-210}{200} \times 10^{-3} = -1.05 \times 10^{-3}$$

$\therefore$ Change in x dimension $= 250 \times (-1.05 \times 10^{-3}) = -0.2625$ mm

$$\varepsilon_y = \frac{-\nu\sigma_x}{E} = -0.3 \times (-1.05 \times 10^{-3})$$

$\therefore$ change in y dimension $= -0.3 \times (-0.2625) = 0.0788$ mm

$$\varepsilon_z = \frac{-\nu\sigma_x}{E}$$

$\therefore$ change in z dimension $= \dfrac{10}{250} \times 0.0788 = 0.0032$ mm

The new dimensions of the plate are 249.7375 × 250.0788 × 10.0032 mm.

EXAMPLE 3.13

The mild steel plate in Example 3.12 is again subjected to $\sigma_x = -210 \text{ N/mm}^2$, but now the top and bottom edges of the plate are prevented from moving in the y-direction by two smooth walls which allow movement to take place in both the x and z directions (see Fig. 3.56, where the thickness of the plate is not shown for simplicity's sake). What are the changes in the plate's dimensions?

In this case, when the compressive stress is applied in the x direction, the plate would like to get thicker (which it is free to do, as before) and the y dimension would like to get bigger, but this change is totally prevented by the two walls. Therefore, $\varepsilon_y = 0$. As we shall see (and it might be intuitively obvious) this will cause a compres-

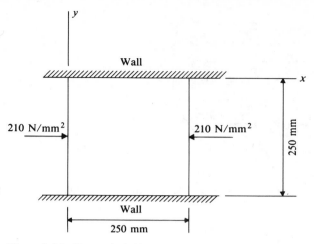

Figure 3.56   Example 3.13

sive stress to be created in the y direction. There will be no stress in the z direction because the plate is not constrained in this direction. So, in this case:

$$\sigma_x = -210 \text{ N/mm}^2, \qquad \varepsilon_y = 0, \qquad \sigma_z = 0$$

and the remaining quantities, namely, σ_y, ε_x and ε_z, have to be found. We can use Hooke's Law to solve the problem.

From Eq. (3.115):

$$\varepsilon_y = \frac{10^{-3}}{200} (\sigma_y - 0.3(-210)) = 0 \qquad \therefore \sigma_y = -63 \text{ N/mm}^2$$

$$\therefore \quad \varepsilon_x = \frac{10^{-3}}{200} (-210 - 0.3(-63)) = -0.9555 \times 10^{-3}$$

$$\therefore \text{ change in } x \text{ dimension} = -0.2389 \text{ mm}$$

and $\qquad \varepsilon_z = \dfrac{10^{-3}}{200} (-0.3(-210 - 63)) = 0.4095 \times 10^{-3}$

$$\therefore \text{ change in } z \text{ dimension} = 0.0041 \text{ mm}$$

Stress–strain relations for shear

The element in Fig. 3.57, when subjected to pure shear stress τ_{xy} $(=\tau_{yx})$ in the xy plane, will deform as shown. The total shear strain is γ_{xy}. The elastic relationship between τ_{xy} and γ_{xy} is given by

$$\frac{\tau_{xy}}{\gamma_{xy}} = G \qquad\qquad (3.117)$$

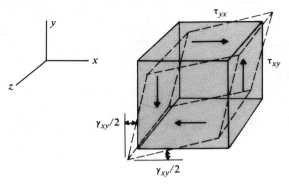

Figure 3.57 Deformation of a 3-dimensional element subjected to shear stress only in the x and y directions

where G is the shear modulus for the material (sometimes referred to as the modulus of rigidity).

Similar, independent, relationships exist in the other two planes, namely

$$\frac{\tau_{xz}}{\gamma_{xz}} = G$$

and (3.118)

$$\frac{\tau_{yz}}{\gamma_{yz}} = G$$

The general statement of Hooke's Law

Combining Eqs (3.115), (3.117) and (3.118), we have the general statement of Hooke's Law for an isotropic, elastic element, namely

$$
\left.
\begin{aligned}
\varepsilon_x &= \frac{1}{E}\{\sigma_x - v(\sigma_y + \sigma_z)\} \\[4pt]
\varepsilon_y &= \frac{1}{E}\{\sigma_y - v(\sigma_x + \sigma_z)\} \\[4pt]
\varepsilon_z &= \frac{1}{E}\{\sigma_z - v(\sigma_x + \sigma_y)\} \\[4pt]
\gamma_{xy} &= \tau_{xy}/G \\[4pt]
\gamma_{xz} &= \tau_{xz}/G \\[4pt]
\gamma_{yz} &= \tau_{yz}/G
\end{aligned}
\right\}
\qquad (3.119)
$$

Inversion of Eq. (3.119) gives stresses in terms of strains as follows:

$$\left.\begin{array}{l}
\sigma_x = \dfrac{E}{(1+v)(1-2v)}\{(1-v)\varepsilon_x + v(\varepsilon_y + \varepsilon_z)\} \\[2mm]
\sigma_y = \dfrac{E}{(1+v)(1-2v)}\{(1-v)\varepsilon_y + v(\varepsilon_x + \varepsilon_z)\} \\[2mm]
\sigma_z = \dfrac{E}{(1+v)(1-2v)}\{(1-v)\varepsilon_z + v(\varepsilon_x + \varepsilon_y)\} \\[2mm]
\tau_{xy} = G\gamma_{xy} \\[2mm]
\tau_{xz} = G\gamma_{xz} \\[2mm]
\tau_{yz} = G\gamma_{yz}
\end{array}\right\} \qquad (3.120)$$

Stress–strain relations for plane stress

A state of plane stress in the xy plane is defined by $\sigma_z = \tau_{xz} = \tau_{yz} = 0$ whilst σ_x, σ_y and τ_{xy} may have non-zero values.

In this case, therefore, Eq. (3.119) reduces to:

$$\left.\begin{array}{l}
\varepsilon_x = \dfrac{1}{E}(\sigma_x - v\sigma_y) \\[3mm]
\varepsilon_y = \dfrac{1}{E}(\sigma_y - v\sigma_x) \\[3mm]
\varepsilon_z = -\dfrac{v}{E}(\sigma_x + \sigma_y) \\[3mm]
\gamma_{xy} = \tau_{xy}/G
\end{array}\right\} \qquad (3.121)$$

Solving for σ_x, σ_y and τ_{xy} from Eq. (3.121) gives:

$$\begin{array}{l}
\sigma_x = \dfrac{E}{(1-v^2)}(\varepsilon_x + v\varepsilon_y) \\[3mm]
\sigma_y = \dfrac{E}{(1-v^2)}(v\varepsilon_x + \varepsilon_y) \qquad (3.122\text{a}) \\[3mm]
\tau_{xy} = Gv_{xy}
\end{array}$$

In matrix notation, these relationships become

$$\begin{bmatrix} \sigma_x \\[3mm] \sigma_y \\[3mm] \tau_{xy} \end{bmatrix} = \begin{bmatrix} \dfrac{E}{(1-v^2)} & \dfrac{vE}{(1-v^2)} & 0 \\[3mm] \dfrac{vE}{(1-v^2)} & \dfrac{E}{(1-v^2)} & 0 \\[3mm] 0 & 0 & G \end{bmatrix} \begin{bmatrix} \varepsilon_x \\[3mm] \varepsilon_y \\[3mm] \gamma_{xy} \end{bmatrix} \qquad (3.122\text{b})$$

Note: The Eqs (3.121) and (3.122) are true for a state of plane stress. They are *not true* for a state of plane strain.

Problem 3.9   Show that, for a state of *plane stress*:

$$\varepsilon_z = \frac{-v}{(1 - v)} (\varepsilon_x + \varepsilon_y)$$

Stress–strain relations for plane strain

A state of plane strain in the xy plane is defined by $\varepsilon_z = \gamma_{xz} = \gamma_{yz} = 0$ whilst ε_x, ε_y and γ_{xy} may all have non-zero values.

In this case, therefore, Eq. (3.119) becomes:

$$\varepsilon_x = \frac{1}{E} \{\sigma_x - v(\sigma_y + \sigma_z)\}$$

$$\varepsilon_y = \frac{1}{E} \{\sigma_y - v(\sigma_x + \sigma_z)\}$$

$$0 = \frac{1}{E} \{\sigma_z - v(\sigma_x + \sigma_y)\}$$ (3.123)

$$\tau_{xy} = \tau_{xy}/G$$

From the third line of Eq. (3.123), it follows that

$$\sigma_z = v(\sigma_x + \sigma_y)$$ (3.124)

Problem 3.10   Show that, for a state of *plane strain*:

(i) $\sigma_x = \dfrac{E}{(1 + v)(1 - 2v)} \{(1 - v)\varepsilon_x + v\varepsilon_y\}$

(ii) $\sigma_y = \dfrac{E}{(1 + v)(1 - 2v)} \{v\varepsilon_x + (1 - v)\varepsilon_y\}$

(iii) $\sigma_z = \dfrac{vE}{(1 + v)(1 - 2v)} (\varepsilon_x + \varepsilon_y)$

Volumetric stress–strain relations and the bulk modulus K

Figure 3.58 shows a cube subjected to stresses σ_x, σ_y and σ_z. As a result, the x, y and z dimensions of the cube, each of which had an original length equal to unity, suffer strains ε_x, ε_y and ε_z respectively. The new lengths of the sides therefore become $(1 + \varepsilon_x)$, $(1 + \varepsilon_y)$ and $(1 + \varepsilon_z)$.

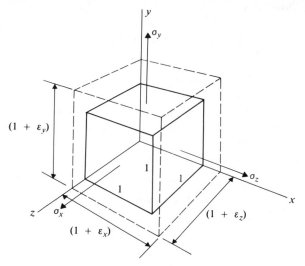

Figure 3.58   A cube subjected to direct stresses only in the x, y and z directions

From Eq. (3.119):

$$\varepsilon_x = \frac{1}{E}\{\sigma_x - v(\sigma_y + \sigma_z)\}$$

$$\varepsilon_y = \frac{1}{E}\{\sigma_y - v(\sigma_x + \sigma_z)\}$$

$$\varepsilon_z = \frac{1}{E}\{\sigma_z - v(\sigma_x + \sigma_y)\}$$

The volume of the cube has changed from unity to $(1 + \varepsilon_x)(1 + \varepsilon_y)(1 + \varepsilon_z)$. The change in volume is therefore:

$$\Delta V = (1 + \varepsilon_x)(1 + \varepsilon_y)(1 + \varepsilon_z) - 1$$

$$= (1 + \varepsilon_x + \varepsilon_y + \varepsilon_z + \varepsilon_x\varepsilon_y + \varepsilon_x\varepsilon_z + \varepsilon_y\varepsilon_z + \varepsilon_x\varepsilon_y\varepsilon_z) - 1$$

i.e. $\Delta V \simeq \varepsilon_x + \varepsilon_y + \varepsilon_z$ (ignoring all products of strains)

Therefore, the volumetric strain ε_v is given by:

$$\varepsilon_v = \text{change in volume/original volume}$$

$$= (\varepsilon_x + \varepsilon_y + \varepsilon_z)/1$$

i.e. $\varepsilon_v = \varepsilon_x + \varepsilon_y + \varepsilon_z$ (3.125)

From Eq. (3.119), it follows that:

$$\varepsilon_v = \frac{1}{E}\{\sigma_x + \sigma_y + \sigma_z - v(2\sigma_x + 2\sigma_y + 2\sigma_z)\}$$

i.e.

$$\varepsilon_v = \frac{(1 - 2v)}{E}(\sigma_x + \sigma_y + \sigma_z) \qquad (3.126)$$

The bulk modulus K is the elastic modulus which relates volumetric stress to volumetric strain (corresponding to E for one-dimensional stress and strain and G for two-dimensional shear stress and strain).

K is defined by putting $\sigma_x = \sigma_y = \sigma_z = \sigma$, then

$$K = \frac{\text{'volumetric' (or hydrostatic) stress}}{\text{volumetric strain}} = \frac{\sigma}{\varepsilon_v} \qquad (3.127)$$

where

$$\varepsilon_v = \frac{(1 - 2v)}{E}(3\sigma) \quad \text{from Eq. (3.126)}$$

$$\therefore \quad K = \frac{\sigma E}{3\sigma(1 - 2v)} = \frac{E}{3(1 - 2v)} \qquad (3.128)$$

Note: From Eq. (3.128), as v approaches a value of 0.5, the denominator approaches zero and K approaches infinity. An infinite value of K would mean that the material is incompressible, which is not possible. It follows that 0.5 is a limiting value of Poisson's ratio for an elastic material.

The relationships between the elastic constants E, G, K and v

The elastic constants are all empirical but they are not independent of one another and knowledge of any two of them will enable the values of the remaining two to be deduced.

Consider the element abcd, Fig. 3.59a, lying in the xy plane and subjected to principal stresses $\sigma_x = \sigma_1$, $\sigma_y = -\sigma_1$ and $\sigma_z = 0$, i.e., a state of plane stress.

The corresponding strains are given by Eq. (3.119) as

$$\varepsilon_x = \varepsilon_1 = \frac{1}{E}(\sigma_1 + v\sigma_1) = \frac{\sigma_1}{E}(1 + v)$$

$$\varepsilon_y = \varepsilon_2 = \frac{1}{E}(-\sigma_1 - v\sigma_1) = -\frac{\sigma_1}{E}(1 + v) \qquad (3.129)$$

$$\varepsilon_z = \varepsilon_3 = \frac{1}{E}(0) = 0$$

(*Note:* This is a special case in which a state of plane stress ($\sigma_z = 0$) creates also a state of plane strain ($\varepsilon_z = 0$).)

The corresponding Mohr's circles of stress and strain in the xy plane are shown in Figs 3.59b and 3.59c respectively.

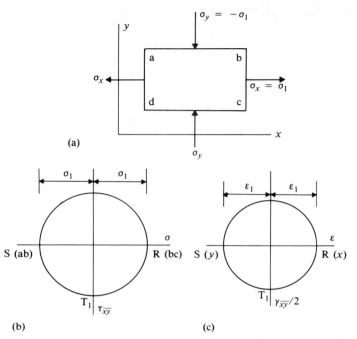

Figure 3.59 An example used to demonstrate the relationships between the elastic constants

At the point T_1,

$$\tau_{\bar{x}\bar{y}\,\text{max}} = \sigma_1$$

$$\frac{\gamma_{\bar{x}\bar{y}\,\text{max}}}{2} = \varepsilon_1$$

and

$$\frac{\tau_{\bar{x}\bar{y}\,\text{max}}}{\gamma_{\bar{x}\bar{y}\,\text{max}}} = G, \text{ the shear modulus}$$

$$\therefore \quad \frac{\sigma_1}{\varepsilon_1} = 2G \tag{3.130}$$

$\therefore$ from Eqs (3.129) and (3.130):

$$2G = \frac{\sigma_1}{\varepsilon_1} = \frac{E}{(1 + v)}$$

$$\therefore \quad G = \frac{E}{2(1 + v)} \tag{3.131}$$

and, repeating Eq. (3.128):

$$K = \frac{E}{3(1 - 2v)} \tag{3.132}$$

The relationships of Eqs (3.131) and (3.132) are always true for an isotropic, elastic material.

EXAMPLE 3.14

The state of stress at a point in a material with $E = 100 \, \text{KN/mm}^2$ and $v = 0.25$ is given by F. Find the corresponding strain array and deduce the maximum and minimum principal strains in the plane xy and their directions relative to the x-axis.

$$F = \begin{bmatrix} -60 & -40 & 0 \\ -40 & 50 & 0 \\ 0 & 0 & 0 \end{bmatrix} \text{N/mm}^2 \text{ with respect to the axes } xyz$$

From F:

$$\sigma_x = -60 \, \text{N/mm}^2, \qquad \sigma_y = 50 \, \text{N/mm}^2, \qquad \tau_{xy} = -40 \, \text{N/mm}^2$$

Also, $\sigma_z = 0$, so F describes a state of plane stress. τ_{xz} and τ_{yz} are also zero.

From Eq. (3.131): $G = \dfrac{E}{2(1 + v)} = \dfrac{100 \times 10^3}{2(1 + 0.25)} = 40 \times 10^3 \, \text{N/mm}^2$

From Eq. (3.121):

$$\varepsilon_x = (-60 - 0.25(50))/100 \times 10^3 = -0.725 \times 10^{-3}$$

$$\varepsilon_y = (50 - 0.25(-60))/100 \times 10^3 = 0.650 \times 10^{-3}$$

$$\varepsilon_z = -0.25(-60 + 50)/100 \times 10^3 = -0.025 \times 10^{-3}$$

$$\gamma_{xy} = -40/40 \times 10^3 = -1.0 \times 10^{-3}; \qquad \frac{\gamma_{xy}}{2} = -0.500 \times 10^{-3}$$

$$\gamma_{xz} = \gamma_{yz} = 0$$

The corresponding strain array is therefore

$$S = \begin{bmatrix} -0.725 & -0.500 & 0 \\ -0.500 & 0.650 & 0 \\ 0 & 0 & -0.025 \end{bmatrix} \times 10^{-3}$$

To find the principal strains, draw the Mohr's circle of strain in xy in accordance with the strain array S. This is shown in Fig. 3.60. From Fig. 3.60:

$$\text{OC} = \frac{1}{2}(\varepsilon_x + \varepsilon_y) = \frac{1}{2}(-0.725 + 0.650) \times 10^{-3} = -0.0375 \times 10^{-3}$$

$$\therefore \ r = (0.6875^2 + 0.5^2)^{1/2} \times 10^{-3} = 0.8501 \times 10^{-3}$$

$$\tan 2\alpha = -0.500/0.6875, \quad \therefore \ 2\alpha = 143.97° \quad \therefore \ \alpha = 71.99°$$

$$\varepsilon_1 = (-0.0375 + 0.8501) \times 10^{-3} = 0.8126 \times 10^{-3}$$

$$\varepsilon_2 = (-0.0375 - 0.8501) \times 10^{-3} = -0.8876 \times 10^{-3}$$

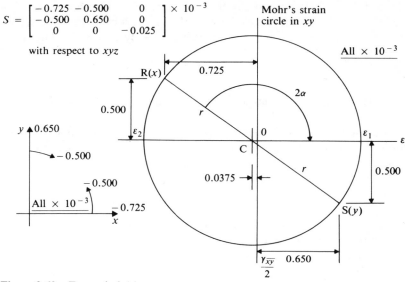

$$S = \begin{bmatrix} -0.725 & -0.500 & 0 \\ -0.500 & 0.650 & 0 \\ 0 & 0 & -0.025 \end{bmatrix} \times 10^{-3}$$

with respect to xyz

Figure 3.60   Example 3.14

Thus the maximum principal strain ($\varepsilon_1 = 0.8126 \times 10^{-3}$) occurs at 71.99° clockwise from the direction x, and the minimum principal strain ($\varepsilon_2 = -0.8876 \times 10^{-3}$) occurs at 90° to ε_1.

Problem 3.11   Repeat Example 3.14 but, instead of drawing the strain circle, draw the stress circle corresponding to the stress array F. From the stress circle, deduce the directions of the maximum and minimum principal stresses (which coincide with the directions of the principal strains), find the values of the principal stresses and, using Hooke's Law, calculate the principal strains. Check the answers against those given in Example 3.14.

3.7 Stresses and strains due to combined forces

Beam web subjected to bending and shear

In a loaded beam, it is usually true that every cross-section of the beam is subjected to both bending moment and shear force, i.e., M and S.

In Section 3.1, we investigated the distribution, across the depth of the cross-section, of the stress due to the bending moment M. We found that it consisted of a stress σ_x which varied linearly with distance y from the principal axis, in accordance with Eq. (3.8), namely

$$\sigma_x = -\frac{M}{I}y \qquad (3.133)$$

where I is the second moment of area of the cross-section about Gz.

In Section 3.2, it was found that the vertical (y-direction) and horizontal (x-direction) shear stress τ, due to the presence of a vertical (y-direction) shear force S, is given at any distance y_1 from the principal axis by Eq. (3.27). That is:

$$\tau = \frac{S}{It}\,\overline{A}\overline{y} \tag{3.134}$$

where $\overline{A}\overline{y}$ is the first moment of area, about the principal axis Gz of that part of the cross-section beyond y_1 (Fig. 3.14a).

Let us now consider the cantilever shown in Fig. 3.61a and, more particularly, the total stress condition at the point A in the web. A is situated distance x_A from the root of the cantilever and at y_A from the principal axis Gz.

At the cross-section containing the point A:

$$M_A = P(L - x_A) \quad \text{and} \quad S_A = P$$

Because P is a positive load (in accordance with our sign convention), M_A and S_A will both be positive at all sections along the beam.

The normal and shear stresses at the point A are given by Eqs (3.133) and (3.134) respectively as

$$\sigma_{xA} = -\frac{M_A}{I}\,y_A = -P\frac{(L - x_A)}{I}\,y_A \tag{3.135}$$

and

$$\tau_{xyA} = \frac{S_A}{It}\,\overline{A}\overline{y} = \frac{P}{It}\,\overline{A}\overline{y} \tag{3.136}$$

In Eq. (3.136), $\overline{A}\overline{y}$ is the first moment of area about Gz of that part of the beam cross-section for which $y > y_A$, and t is the web thickness.

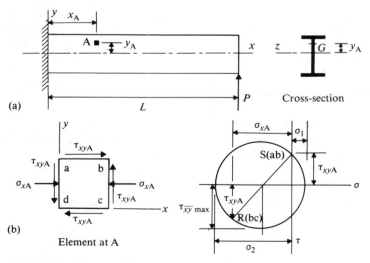

(a)

(b)

Element at A

Figure 3.61   Stresses at a point in the web of a beam

In this case, σ_{xA} will be negative (i.e., compressive) and τ_{xyA} will be positive. These signs have been recognized on the sketch of the stressed element at A shown in Fig. 3.61b. The vertical normal stress, σ_{yA}, is zero.

Having found the values of σ_{xA} and τ_{xyA}, we are now faced with a very familiar problem, namely, to find the principal stresses and the maximum shear stress on the element. This is done by drawing Mohr's circle of stress as shown in Fig. 3.61c. Knowing the values of σ_{xA} and τ_{xyA}, the circle is fully defined and the values of σ_1, σ_2, $\tau_{\bar{x}\bar{y}max}$, θ and α can be found.

The procedure described above is illustrated in Example 3.15.

EXAMPLE 3.15

The simply-supported beam AD in Fig. 3.62a carries loads of 150 kN and 100 kN at B and C as shown. The beam has a uniform cross-section which is drawn in Fig. 3.62e.

Find the principal stresses and the maximum shear stress in the web at the point where it joins the top flange and at the cross-section of the beam just to the left of B.

In order to find the normal and shear stresses at any point on the cross-section, we first need to know the values of M and S at the section. In Figs 3.62b and 3.62c, the shear force diagram and bending moment diagrams are drawn for the beam. This is not strictly necessary because the shear and bending moment at any section can be calculated without drawing these diagrams, but they do illustrate that, just to the left of the section at B, there exists a combination of the maximum shear force and the maximum bending moment.

From Figs 3.62b and 3.62c, we see that, just to the left of B, $M = 405$ kN m and $S = -135$ kN.

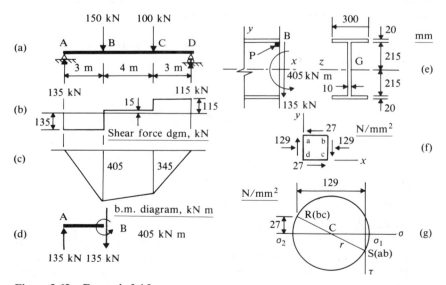

Figure 3.62   Example 3.15

The cross-section shown in Fig. 3.62e is similar to that used in Example 3.3 where it was established that the second moment of area, $I = 674 \times 10^6$ mm^4.

We are asked to investigate the stress conditions at the point p, indicated in Fig. 3.62e.

At the point p, $y = 215$ mm. Therefore, from Eq. (3.133), the x-direction normal stress at p is given by σ_{xp} where

$$\sigma_{xp} = -\frac{M}{I} y = -\frac{405 \times 10^6}{674 \times 10^6} \times 215 = -129 \text{ N/mm}^2$$

The vertical and horizontal values of the shear stress at p, namely τ_{xyp}, are given by Eq. (3.27).

At p:

$$\bar{A} = (300 \times 20) \text{ mm}^2 = \text{area of cross-section above p}$$

$$\bar{y} = (215 + 10) \text{ mm} = \text{distance from G to the centroid of } \bar{A}$$

$$t = 10 \text{ mm} = \text{web thickness at p}$$

Therefore, from Eq. (3.27):

$$\tau_{xyp} = \frac{S}{It} \bar{A}\bar{y} = \frac{-135 \times 10^3}{674 \times 10^6 \times 10} (300 \times 20)(225) = -27 \text{ N/mm}^2$$

The state of stress at p, namely $\sigma_{xyp} = -129$ N/mm^2 and $\tau_{xyp} = -27$ N/mm^2, is illustrated in Fig. 3.62f. The corresponding Mohr's circle of stress is drawn in Fig. 3.62g.

(Note that, because τ_{xyp} is negative, the point R, representing the stresses on face bc of the element, is plotted *above* the horizontal axis.)

Let
$$r = \text{radius of the Mohr's circle}$$

$$r = \sqrt{\left[27^2 + \left(\frac{129}{2}\right)^2 \right]} = 70 \text{ N/mm}^2$$

$$\therefore \quad \sigma_1 = -\frac{129}{2} + 70 = 5.5 \text{ N/mm}^2$$

$$\sigma_2 = -\frac{129}{2} - 70 = -134.5 \text{ N/mm}^2$$

and
$$\tau_{\bar{x}\bar{y}\text{max}} = r = 70 \text{ N/mm}^2$$

Problem 3.12   The beam used in Example 3.15 is subjected to a concentrated load of 250 kN at the middle of the 10 m span. If it can be assumed that the shear force at any cross-section of the beam is distributed as a uniform shear stress in the web (see the end of the solution to Example 3.3), find the maximum tensile stress in the web of the beam.

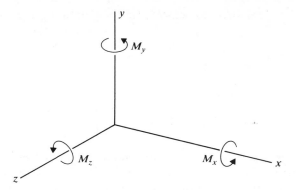

Figure 3.63   Clockwise screw-rule for positive moments applied to positive faces

Member subjected to bending moment and axial force

Before proceeding with this topic, it is necessary to say a word about the sign convention adopted to define positive moments acting upon the positive face of an element. These positive moments are defined in Fig. 3.63. M_x, M_y and M_z act about the x, y and z axes respectively, and each is defined as positive when it turns in a clockwise direction about its own axis, i.e., a clockwise screwrule defines their positive directions.

Figure 3.64 shows positive M_z and positive M_y applied to the positive cross-sectional face of a beam. Thus, as M_z is viewed from a point along the z-axis, it appears as an anticlockwise moment acting in the xy plane. This is in accord with the convention used throughout the previous treatment of the bending of beams and the stresses caused by a bending moment M_z which, until now, we have usually simply called M.

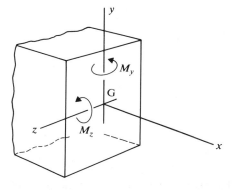

Figure 3.64   Positive M_z and M_y acting on a positive face

Thus, Eq. (3.8) relates the bending moment M and the normal (x-direction) stress σ_x caused by M;

$$\sigma_x = -\frac{M}{I}y$$

This can be written more specifically as

$$\sigma_x = -\frac{M_z}{I_z}y \qquad (3.137)$$

Equation (3.137) indicates that positive M_z causes negative (i.e., compressive) σ_x at positive values of y measured from the principal axis Gz. This is clear from Fig. 3.64, which also shows the application of a positive M_y, and it will be noted that this will cause *positive* (i.e., tensile) σ_x at positive values of z measured from the principal axis Gy. It follows that

$$\sigma_x = \frac{M_y}{I_y}z \qquad (3.138)$$

If positive M_z and positive M_y act together, then the value of σ_x at any point on the cross-section is given by

$$\sigma_x = -\frac{M_z}{I_z}y + \frac{M_y}{I_y}z \qquad (3.139)$$

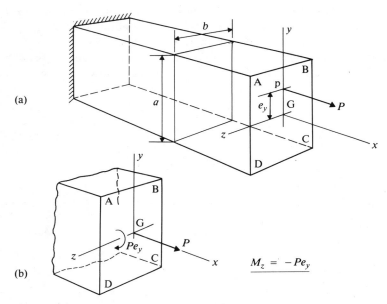

Figure 3.65 Stresses in a member subjected to a force parallel to the longitudinal axis but eccentric to the z-axis

We can now turn to Fig. 3.65a which shows a prismatic member of rectangular cross-section with dimensions $a \times b$, built-in at one end and subjected to a positive load P acting parallel to the x-axis.

P acts through the point p, which is located on the principal axis Gy at a distance e_y from the centroid G of the cross-section. What is the distribution of the stress σ_x on any cross-section of the member due to P?

To answer this question, it is first noted that P creates a clockwise moment equal to Pe_y about the principal axis Gz. If, therefore, the line of action of P is transferred to pass through G (Fig. 3.65b), and this transfer is accompanied by the application of a moment Pe_y about Gz, then we have, in Fig. 3.65b, a combination of loading which is precisely equivalent to that in Fig. 3.65a.

The load P acting through the centroid of the cross-section will cause a uniform x-direction tensile stress equal to P/A over the whole cross-section, where A is the area of the cross-section.

The moment Pe_y will cause a stress distribution in accordance with Eq. (3.137) where $M_z = -Pe_y$. Therefore the total stress condition is given by

$$\sigma_x = \frac{P}{A} - \frac{M_z}{I_z} y$$

i.e.

$$\sigma_x = \frac{P}{A} + \frac{Pe_y}{I_z} y \qquad (3.140)$$

Equation 3.140 is illustrated in Fig. 3.66. At the top fibres (at $y = a/2$) of the cross-section, the maximum value of σ_x will occur. At the bottom fibres (at $y = -a/2$) the minimum value of σ_x will occur. That is, along AB:

$$\sigma_x = \frac{P}{A} + \frac{Pe_y}{I_z} \frac{a}{2}$$

and, along CD:

$$\sigma_x = \frac{P}{A} - \frac{Pe_y}{I_z} \frac{a}{2}$$

It has been assumed in the illustration in Fig. 3.66 that σ_x along CD is negative, that is

$$\frac{Pe_y}{I_z} \frac{a}{2} > \frac{P}{A}$$

By definition, σ_x is zero along the neutral axis whose position is therefore defined, at $y = y_N$, from Eq. (3.140), by

$$0 = \frac{P}{A} + \frac{Pe_y}{I_z} y_N$$

i.e.

$$y_N = -\frac{I_z}{Ae_y} \qquad (3.141)$$

If $y_N < -a/2$, there will be no neutral axis on the cross-section and σ_x will be positive (tensile) at every point.

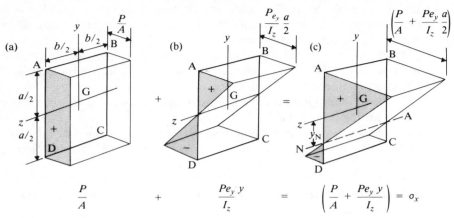

$$\frac{P}{A} \quad + \quad \frac{Pe_y\,y}{I_z} \quad = \quad \left(\frac{P}{A} + \frac{Pe_y\,y}{I_z}\right) = \sigma_x$$

Figure 3.66   Stress distribution due to the load P in Fig. 3.65

We now investigate the stress distribution on the cross-section when P is applied eccentrically to both Gz and Gy as shown in Fig. 3.67a. Its point of application, p, is now defined by the coordinates e_y, e_z. This is equivalent to applying P at G together with a moment Pe_y about Gz and a moment Pe_z about Gy, as shown in Fig. 3.67b.

In this case, the x-direction stress σ_x at any point on the cross-section is given by Eq. (3.139), in which, by reference to Figs 3.64 and 3.67b, it will be seen that

$$M_z = -Pe_y \quad \text{and} \quad M_y = Pe_z$$

Substituting these values into Eq. (3.139), it follows that

$$\sigma_x = \frac{P}{A} + \frac{Pe_y}{I_z}\,y + \frac{Pe_z}{I_y}\,z \tag{3.142}$$

Thus, σ_x varies, not only with y, but also with z.

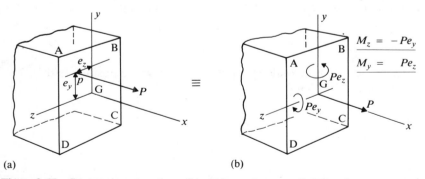

Figure 3.67   Stresses in a member subjected to a force applied parallel to the longitudinal axis but eccentric to the two transverse axes

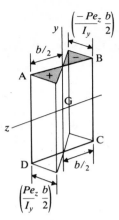

$$\left(\frac{-Pe_z}{I_y}\frac{b}{2}\right)$$

$$\left(\frac{Pe_z}{I_y}\frac{b}{2}\right)$$

Figure 3.68 Stress distribution due only to the moment effect of the force in Fig. 3.67 about the y-axis

Figure 3.68 shows the σ_x-distribution due to $M_y = Pe_z$ acting alone. The addition of this distribution to that of Fig. 3.65c therefore represents Eq. (3.142) and is shown on Fig. 3.69.

In Fig. 3.69, σ_x at $A = \sigma_{xA} = Aa$ is given by putting $y = a/2$ and $z = b/2$ in Eq. (3.142). Similarly, using Eq. (3.142):

$$\sigma_{xB} = Bb \text{ at } y = a/2 \quad \text{and} \quad z = -b/2$$

$$\sigma_{xC} = Cc \text{ at } y = -a/2 \quad \text{and} \quad z = -b/2$$

$$\sigma_{xD} = Dd \text{ at } y = -a/2 \quad \text{and} \quad z = b/2$$

Because σ_x varies linearly with both y and z, the points a, b, c and d are joined by straight lines to form the plane abcd. The value of σ_x at any point on the cross-section ABCD is now represented by the x-direction distance to the plane abcd. If the plane

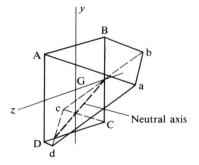

Figure 3.69 Stress distribution due to the combined moment and axial force effects for the case in Fig. 3.67

abcd intersects ABCD, the line of intersection represents $\sigma_x = 0$ and is therefore the neutral axis. This is shown in Fig. 3.69.

The location of the neutral axis is found by putting Eq. (3.142) to zero, i.e.,

$$0 = \frac{1}{A} + \frac{e_y}{I_z} y + \frac{e_z}{I_y} z \qquad (3.143)$$

defines the neutral axis. It is obviously a straight line.

If a member is subjected to an axial load P and bending moments M_y and M_z (which might be quite independent of the load P), then

$$\sigma_x = \frac{P}{A} - \frac{M_z}{I_z} y + \frac{M_y}{I_y} z \qquad (3.144)$$

The non-axially-loaded column

An important problem which frequently arises in practice is that of a column subjected to axial compressive load accompanied by bending moments acting about one or both principal axes. The bending moments might be transmitted from the rest of the structure, of which the column forms a part, or they might arise because the compressive load is applied eccentrically (i.e., not along the centroid) as in our discussion above.

Frequently, both of these circumstances exist, but we will concentrate attention on

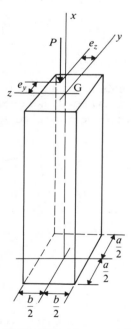

Figure 3.70 An eccentrically-loaded column

the non-axially loaded column. In fact, we have already solved this problem in Eq. (3.142). All we have to do is to reverse the sign of P and to rotate the member of Fig. 3.65 from the horizontal to the vertical, and we have a column subjected to non-axial compressive loading. This is shown in Fig. 3.70 and, in order that Eq. (3.142) remains applicable, the x, y and z axes have also been rotated for this special case.

Recognizing that P is now a negative load, the σ_x-distribution at any cross-section along the length of the column in Fig. 3.70 is given from Eq. (3.142) as

$$\sigma_x = -\frac{P}{A} - \frac{Pe_y}{I_z} y - \frac{Pe_z}{I_y} z \qquad (3.145)$$

The use of this equation will be illustrated in Example 3.16. (If P is a tensile load, the signs in Eq. (3.145) will all be positive.)

EXAMPLE 3.16

A column of rectangular cross-section of dimensions 200 mm × 100 mm in the y and z directions respectively is subjected to a compressive load of 800 kN which acts at the point $e_y = -50$ mm, $e_z = 20$ mm. Find the distribution of σ_x on the cross-section of the column and locate the neutral axis.

The situation is illustrated in Fig. 3.71a.

$$\text{Area of cross-section} = A = 200 \times 100 \text{ mm}^2 = 2 \times 10^4 \text{ mm}^2$$

$$\text{Second moment of area about } Gy = I_y = (200 \times 100^3)/12 = 16.7 \times 10^6 \text{ mm}^4$$

$$\text{Second moment of area about } Gz = I_z = (100 \times 200^3)/12 = 66.7 \times 10^6 \text{ mm}^4$$

$$e_y = -50 \text{ mm}$$

$$e_z = 20 \text{ mm}$$

We recognize that the applied load of 800 kN is negative and use Eq. (3.145) to find the value of σ_x at the corners A, B, C and D of the cross-section. The calculation will be carried out in N and mm units.

At A: $y = 100$ mm, $z = 50$ mm

$$\therefore \sigma_{xA} = -\frac{800 \times 10^3}{2 \times 10^4} - \frac{800 \times 10^3}{66.7 \times 10^6}(-50)(100) - \frac{800 \times 10^3}{16.7 \times 10^6}(20)(50) \text{ N/mm}^2$$

i.e., $\sigma_{xA} = -40 + 0.600(100) - 0.958(50) = -27.9 \text{ N/mm}^2$

At B: $y = 100$ mm, $z = -50$ mm

$$\therefore \sigma_{xB} = -40 + 0.600(100) - 0.958(-50) = 67.9 \text{ N/mm}^2$$

At C: $y = -100$ mm, $z = -50$ mm

$$\therefore \sigma_{xC} = -40 + 0.600(-100) - 0.958(-50) = -52.1 \text{ N/mm}^2$$

At D: $y = -100$ mm, $z = 50$ mm

$$\therefore \sigma_{xD} = -40 + 0.600(-100) - 0.958(50) = -147.9 \text{ N/mm}^2$$

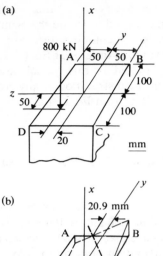

(a)

(b)

Figure 3.71 Example 3.16

The stress distribution is shown in Fig. 3.71b. The precise value of σ_x at any point (y, z) on the cross-section can be found by inserting the values of these coordinates into Eq. (3.145).

The neutral axis is now found by putting $\sigma_x = 0$ in Eq. (3.145). Thus, along the neutral axis:

$$0 = -40 + 0.600y - 0.958z$$

Using this equation, when $y = 100$ mm, $z =$ 20.9 mm (along AB)

when $x = -50$ mm, $y = -13.2$ mm (along BC)

The neutral axis is drawn in Fig. 3.71b.

It should be clear that if a compressive load P kN of any value is applied at the same location as the 800 kN used here, all the σ_x stresses calculated above can be factored by the ratio $(P/800)$ and the position of the neutral axis will not change.

The core or kernel of a cross-section is defined as the area of a cross-section within which a compressive load must act if it is not to cause tensile stress at any point on the cross-section.

It can be illustrated by finding the core of a *rectangular cross-section*. Consider the cross-section ABCD of dimensions $a \times b$ shown in Fig. 3.72. Suppose a compressive,

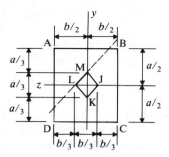

Figure 3.72 The core of a rectangular cross-section

x-direction load P is applied at some point in the north-west quadrant of the cross-section such that e_y and e_z are both positive. If the load is confined to this quadrant, the first point to experience tensile stress, as P moves away from G, is C. If, therefore, the stress at C must not experience tensile (i.e., positive) stress, then, at the limit, $\sigma_{xc} = 0$, i.e., from Eq. (3.142):

$$\sigma_{xc} = \frac{P}{A} + \frac{Pe_y}{I_z}\left(\frac{-a}{2}\right) + \frac{Pe_z}{I_y}\left(-\frac{b}{2}\right) = 0 \qquad (3.146)$$

For the rectangular cross-section:

$$A = ab; \qquad I_z = \frac{ba^3}{12}; \qquad I_y = \frac{ab^3}{12}$$

Substituting these values into Eq. (3.146) will give an equation in e_y and e_z which will define the limit of the allowable movement of P away from G in the north-west quadrant to ensure that the stress at C does not become tensile. Thus

$$\frac{1}{ab} - \frac{a}{2}\frac{12}{ba^3}e_y - \frac{b}{2}\frac{12}{ab^3}e_z = 0$$

that is

$$\frac{e_y}{a} + \frac{e_z}{b} = \frac{1}{6} \qquad (3.147)$$

Equation (3.147) represents the line LM in Fig. 3.72. By symmetry, there are lines MJ, JK and KL which define the limit of the movement of P within the north-east, south-east and south-west quadrants such that tension does not occur at D, A and B respectively. The area LMJK defines the core of the cross-section. It is easy to show how this leads to the so-called *middle-third rule*, because a compressive load applied outside LJ along the z-axis will cause tension along either BC or AD and, similarly, a compressive load applied outside MK on the y-axis will cause tension along either AB or CD.

Problem 3.13 (i) show that if a column has a solid, circular cross-section of diameter D, the point of application of a compressive load P must not be greater than $D/8$ from the centre of the cross-section if there is to be no tensile stress at any point on the cross-section.

(ii) If the column has a hollow circular cross-section of outside diameter D and inside diameter d, show that the radius of the core is equal to $(D^2 + d^2)/8D$.

Problem 3.14 Figure 3.73 shows a box-section column with a bracket welded to it. The bracket carries a load of 50 kN at a point on the Gz axis and 150 mm from the centroid G of the column cross-section. A second load $\bar{P}$ acts through G. Find the maximum allowable value of $\bar{P}$ if the compressive stress must not exceed 150 N/mm² (i.e., $\sigma_x \nleq -150$ N/mm²) at any point on the column cross-section.

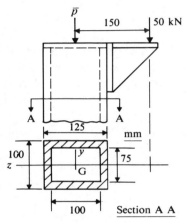

Figure 3.73 Problem 3.14

Problem 3.15 The beam ABCD in Fig. 3.74a carries a vertical load of 20 kN which can move to any point between A and D. The beam is simply-supported at B and C and has a hollow box cross-section with relative dimensions as shown in Fig. 3.74b. Deduce the actual dimensions of the cross-section if the bending stress in the beam must not exceed 250 N/mm² in magnitude.

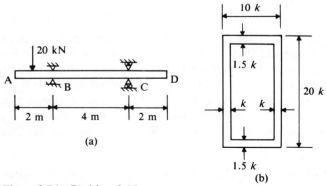

Figure 3.74 Problem 3.15

Problem 3.16   (i) Show that, for a thin-walled circular tube of mean diameter d and wall thickness t, the second moment of area of the cross-section about the diameter is approximately $\pi d^3 t/8$.

(ii) Figure 3.75 shows the elevation of a mast made of two thin-walled tubes A and B. Tube A has a mean diameter of 100 mm and a wall thickness of 3 mm. Tube B has a mean diameter of 150 mm. The mast has a total height of 3.5 m and, applied through the centre-line at the top, carries a load of 6 kN inclined at 60° to the horizontal. If the maximum *vertical* compressive stress is not to exceed 200 N/mm² in either tube, find the maximum length of tube A and the minimum wall thickness of tube B.

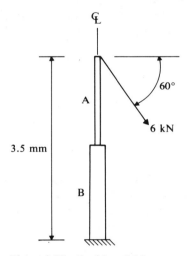

Figure 3.75   Problem 3.16

Problem 3.17   A shaft consists of two concentric, thin-walled circular tubes fixed to each other at their ends through stiff circular discs. The outer tube is copper and has a mean diameter of 100 mm and a wall thickness of 3 mm. The inner tube is of steel. Find the mean diameter and the wall thickness of the steel tube such that, when a torque is applied to the assembly, the consequent shear stresses in the steel and copper are in the ratio 2:1 and the torque is shared equally between the two tubes. For steel, $G = 83 \times 10^3$ N/mm². For copper, $G = 26 \times 10^3$ N/mm².

Problem 3.18   At a point in the web of a plate girder, a given set of loads produces a vertical shear stress of 90 N/mm² together with a horizontal compressive stress of 120 N/mm². Assuming that an increase in the loads causes the stresses to increase in the same proportion, by what factor can the loads be increased before the maximum shear stress at the point in the web reaches 135 N/mm²?

Problem 3.19   A uniform beam is of symmetrical I-section with flanges 125 mm × 15 mm and a web 300 mm × 10 mm. It is supported as a cantilever in a horizontal position and carries a vertical load of 200 kN at its tip, acting in the plane of the web.

Estimate the magnitudes and directions of the principal stresses and the maximum shear stress in the web at a section distance 0.5 m from the applied load:

(a) at a point on the central longitudinal axis,
(b) at a point near the junction of the web and the top flange.

Problem 3.20   A rectangular plate (Fig. 3.76) of dimensions $3l \times l$ lies in the xy plane and is subjected to uniform stress in its own plane. As a result, the area of the plate increases by 0.1 per cent and the length of its perimeter also increases by 0.1 per cent. Also, the angle bad increases by 1.5×10^{-3} radians.

Deduce the principal strain array for the plate and sketch an element subjected to principal strains in the xy plane. (Ignore products of strain as negligibly small.)

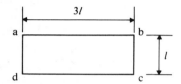

Figure 3.76   Problem 3.20

Problem 3.21   The plate shown in Fig. 3.77 is 3 mm thick and, when unloaded, just fits between edge constraints which allow movement in the y direction but not in the x direction. A compressive load P is applied in the y direction, causing a strain of -200×10^{-6} in the direction m at 45° to the axes.

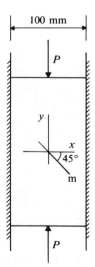

Figure 3.77   Problem 3.21

Draw the Mohr's circles of stress and strain, find the value of P, and write the principle strain array. Sketch an element of the plate subjected to the maximum shear strain and show its orientation to the x-axis.
For the plate, $E = 210 \times 10^3$ MN/m^2 and $v = 0.3$.

Problem 3.22 A tension specimen of aluminium alloy has a gauge length of 200 mm and a diameter of 12 mm. A load of 37.5 kN develops a tensile stress in the bar equal to the proportional limit of the material. At this load, the gauge length has increased to 200.914 mm and the diameter has decreased to 11.983 mm.

Determine the following properties of the material: the modulus of elasticity, the proportional limit, Poisson's Ratio, the shear modulus and the bulk modulus.

Problem 3.23 A column of cruciform cross-section (Fig. 3.78) is made from two I-sections of size 467 mm × 193 mm. The column is fabricated by cutting one of the I-sections down the centre of the web and welding the resulting T-sections A and B to either side of the other I-section as shown.

Obtain the equation of the line QQ defining the limit of the zone within which a compressive load of 1300 kN, applied parallel to the x-axis, must act if the compressive stress at the point e is not to exceed 250 N/mm^2, i.e. $\sigma_x \not< -250$ N/mm^2. (The increase in depth due to cutting and welding of the reconstituted section AB, and the area of the welds, may be neglected.)

The properties of each I-section about its own principal axes are $I_{max} = 457 \times 10^6$ mm^4, $I_{min} = 22 \times 10^6$ mm^4, and the cross-sectional area of each is $A = 12.52 \times 10^3$ mm^2.

Indicate the entire area within which the compressive load must lie if the specified stress is not to be exceeded at any point on the cross-section.

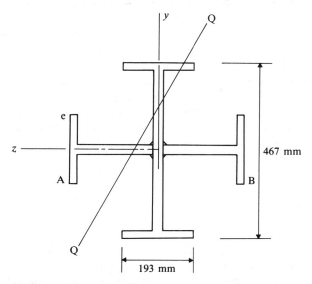

Figure 3.78 Problem 3.23

4. Structural materials: steel and concrete

The structural engineer has a wide range of materials for use in construction and a thorough knowledge of their properties is essential in order to understand the behaviour of structures and structural members. The two materials foremost in importance are steel and concrete, and attention in this book will be confined to these.

It is necessary for the student of structural engineering to have a knowledge of the following properties of a material:

- tensile and compressive strengths;
- elasticity;
- ductility;
- resistance to corrosion and fire;
- durability.

It will be interesting to note how so vastly different steel and concrete are in relation to these properties. In particular, one should note that when one material has a property which appears disadvantageous for use in structures, the corresponding property in the other material is often advantageous. It is, therefore, not too surprising to learn that concrete and steel are frequently used in combination, and reinforced and prestressed concrete are two such combinations which will be introduced here.

Again it is emphasized that it is only the intention to present an introduction to these materials with the main aim of providing the student with a sufficient basic understanding to allow him to follow some of the arguments and discussions within the professional periodicals. More advanced topics relating to these and other important structural materials are included in the book relating to second year studies.

4.1 Introduction to steelwork

Properties of steel

Although the materials in a structure are often subjected to complex two-dimensional or three-dimensional states of stress, the basic structural properties relating to strength and deformation are determined from simple tensile or compressive tests. In

304

a   b

Figure 4.1 Typical steel specimens used for tensile testing: (a) a thin flat specimen with enlarged ends for gripping; (b) a specimen of circular cross-section (*Source*: Manchester University Engineering Department)

the case of steel, tensile test specimens of the shape shown in Fig. 4.1 are used in conjunction with the type of testing machine shown in Fig. 4.2.

A variety of different steels having a wide range of strength and ductility is available for use in structures. Their properties depend on their chemical composition and on the heat and mechanical treatments to which they have been subjected. The extreme of this range of steels, and the one which finds most frequent use in civil engineering and building structures, is mild steel. Its somewhat curious stress–strain curve is shown in Fig. 4.3.

Referring to the curve in Fig. 4.3, the part OA represents the linear elastic range. In this region, the strains are very small and are proportional to the stress. The modulus

Figure 4.2   A typical laboratory testing machine (*Source*: Manchester University Engineering Department)

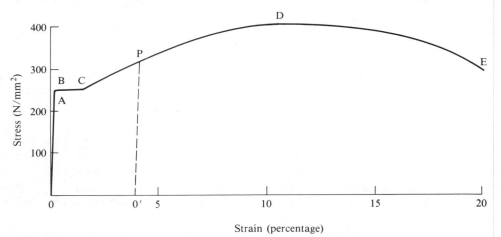

Figure 4.3   Typical stress–strain curve for mild steel

of elasticity is determined by the ratio of stress to strain in this region and is approximately 205 kN/mm². Up to this elastic limit A, the strains disappear when the load is removed and so the test specimen returns to its original length.

If, however, the stress is increased further, the point B is soon reached at which the strains increase substantially without any increase in stress. The part BC is known as the plastic range and the stress causing this behaviour is known as the yield stress. The material is now no longer totally elastic, and unloading from a point within BC would leave the specimen permanently strained.

From the point C, the mild steel specimen now requires additional stress to cause further strain. The region CD is referred to as the strain hardening range of the steel. Fracture of the specimen occurs at E following the gradual 'necking' of the specimen (Fig. 4.4). It is, in fact, this reduction of the area of the specimen caused by the necking which results in the somewhat artificial drooping part of the curve, as the stress during that stage is still calculated on the original cross-sectional area.

Although the ultimate strength of the steel is clearly substantially higher than the yield stress of the steel, the correspondingly high permanent deformation means that such high stresses cannot be used in mild steel structures. It is essential that the deformations of actual structures remain both elastic and small, and so the stresses up to the service load must be kept within the linear elastic range. Thus, for all intents and purposes, the yield stress is effectively the maximum value that can be adopted in practical design.

It is of interest to note that if the specimen were unloaded at some point P between C and D, the stress–strain graph would drop along a line O′P parallel to OA. The intercept on the horizontal axis indicates the permanent strain in the specimen. A subsequent test on the same specimen would give a new stress–strain curve which would correspond approximately to the curve O′PDE. In other words, there is now no plastic range and there is an extended elastic range O′P. This has important application in the production of cold-worked steels which then clearly have a higher effective strength than does mild steel.

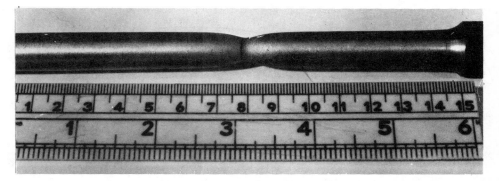

Figure 4.4 A steel specimen after testing showing the typical necking of mild steel (*Source*: Manchester University Engineering Department)

Although the ultimate strength of the mild steel has, as stated earlier, little application in design, the strain at failure is of importance. It is a measure of the ductility of the steel, and mild steel is clearly highly ductile. Its importance comes in the reassurance one has with steel structures that sudden fracture could not occur and that there would generally be ample obvious warning of potential structural failure. Moreover, this property of ductility allows a redistribution of stress to occur at positions where there would tend to be a concentration of stress, around holes for example. There are exceptions to this, though, and these are described later.

The stress–strain curve for mild steel is shown also in Fig. 4.5 together with those of

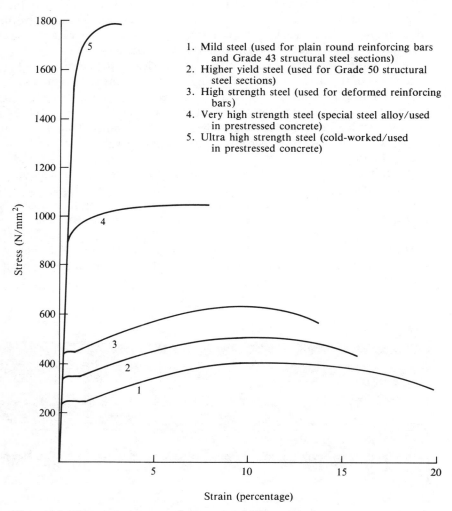

1. Mild steel (used for plain round reinforcing bars and Grade 43 structural steel sections)
2. Higher yield steel (used for Grade 50 structural steel sections)
3. High strength steel (used for deformed reinforcing bars)
4. Very high strength steel (special steel alloy/used in prestressed concrete)
5. Ultra high strength steel (cold-worked/used in prestressed concrete)

Figure 4.5   Stress–strain curves for a range of different steels

a range of different steels. It is of particular interest to note how the ductilities of the steels decrease with increase in strength. Only the lower strength steels are available in the form of structural sections for use in construction. In Britain, those most frequently used are designated Grade 43 and Grade 50, the number following the continental practice of expressing the ultimate strength of the steel in kgf/mm^2.

The higher strength steels are available in bar or wire form for use in concrete construction. The highest strengths are the result of cold-working. For these steels there is no distinct yield plateau, their stress–strain diagram being characterized by a gradual curvature from the initial straight region. A nominal yield point is defined for such steels as that stress at which the test specimen would have a permanent strain of 0.2 per cent on removal of the load. This is termed the 0.2 per cent proof stress.

The compressive stress–strain relationships of steels are also of importance. Tests indicate that for the steels used in structural steelwork, the yield stress and the modulus of elasticity are identical in both tension and compression. Whilst this simplifies some design in that it permits the use of geometrically symmetrical sections in many applications, the possibility of the buckling of members in compression often places severe restrictions on the compressive stresses permitted in steel members. Such behaviour is discussed later.

Steelwork sections

Structural steel is supplied in the form of hot-rolled steel sections. The various shapes available are illustrated in Fig. 4.6. They are manufactured by passing white-hot steel ingots through a succession of rolls in a rolling mill. All are available in a wide range of sizes—the British I-section, for example, is available in depths ranging from 76 mm to 914 mm.

The most appropriate section for use as a beam is the I-section, either in the form of universal beams (depths from 203 to 914 mm) in which the surfaces of the flanges are parallel, or in the form of the smaller joists (depths from 76 to 254 mm) which have tapered flanges. Needless to say, the I-section is only efficient as a beam if arranged to

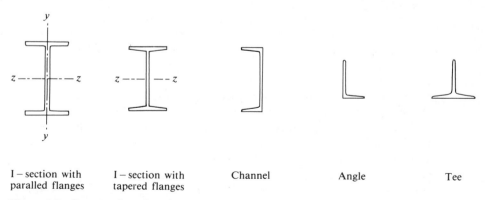

I – section with   I – section with   Channel   Angle   Tee
paralled flanges   tapered flanges

Figure 4.6   Structural steel sections

bend about its z-axis. (Designation of the z-axis here follows from the earlier orientation of the axes used in analysis in the earlier chapters, but it should be noted that tables issued by the steelwork industry usually depicts this axis as the x–x axis.) The large proportion of the material contained in the flanges ensures a high second moment of area about the horizontal axis passing through the centroid of the section (the major principal axis—see Section 3.1 in Chapter 3). Reference to the beam expression:

$$f = \frac{My}{I}$$

clearly indicates that the I-section would be the most efficient at resisting bending.

A word of warning is necessary here. In Chapter 1, attention was drawn to the buckling tendency of members stressed in compression. A beam is stressed partly in compression and partly in tension, and there is therefore a tendency for the compression region of a beam to move sideways (Fig. 4.7). The second moment of area of the compression zone about the y-axis is a measure of the ability of the beam to resist such movement. However, the rectangular flange is not so efficient in this respect. This means that it is sometimes necessary to reduce the maximum permissible compressive bending stress in order to avoid an instability type failure.

An alternative solution is to enhance the buckling strength of the compression flange. The welding of a channel to the top flange (as in Fig. 4.8) is one method of improving the lateral stability and hence of maintaining a high permissible compressive stress. However, many cases arise in practice when the steel beam is restrained from buckling sideways by other parts of the structure. For example, the concrete floor of a building (Fig. 4.9), though supported by the beam, may well provide restraint to the sideways movement of the compression flange by virtue of the friction at the interface and its own rigidity in that direction.

For such cases, the I-section is a very efficient form of beam. Its efficiency can, however, be further improved by the process shown in Fig. 4.10. Using automatic cutting methods, the universal beam section is cut along the zig-zag line shown in Fig. 4.11a. The two parts are then rewelded to form the beam shown in Fig. 4.11b whereby a universal beam of, say, 406 mm becomes a castellated beam of 609 mm depth. Not only is the efficiency of the beam thereby improved, but the resulting holes are often advantageous for accommodating and supporting services such as conduits carrying electric cables (Fig. 4.11c), an important consideration in the design of buildings.

The I-section is also frequently used for the columns of a structure. However, to reduce the buckling problem, which of course is much more severe than in beams, the universal column section is proportioned somewhat differently. In comparison with the universal beam, the flanges of the universal column are much thicker and much wider (Fig. 4.12). This gives an enhanced second moment of area about the y-axis, and so an enhanced resistance to buckling about this weaker axis. Even so the I-shape is not ideal for use in columns and it may sometimes be necessary to adopt a combination of sections connected together to act as a single column. Figure 4.13

Figure 4.7   The tendency for members in compression to deflect sideways applies also to the compression flange of I-beams, resulting in a tendency to twist the beam

Figure 4.8   The resistance to sideways deflection of a compression flange of an I-beam is enhanced by welding a channel to the flange

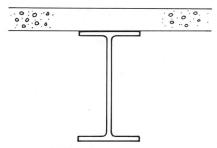

Figure 4.9   Friction at the interface of a concrete floor slab and steel beam will provide restraint to the buckling tendency of the compression flange of the beam

Figure 4.10   Automatic cutting of an I-beam in the manufacture of a castellated beam (*Source*: Steel Construction Institute)

shows a column formed from pairs of channels joined together by suitably spaced batten plates and this arrangement has an improved resistance to buckling about the weaker axis.

An alternative to such compound members are the steel hollow sections manufactured in circular and square or rectangular forms (Fig. 4.14). Seamless sections can be produced by a special hot-rolling process, but the larger sections usually require a welded seam. Having good second moment of areas about both principal axes, the sections are of particular use for columns and struts. Moreover, they have a particularly pleasing appearance, and their smaller surface areas compared with the I-section reduce painting and fire-protection costs.

The angle section shown previously in Fig. 4.6 is sometimes used for the roof-truss type of structure, but it finds its greatest use in steelwork connections.

Steelwork connections

There are basically two methods of joining members together to form a completed structure—bolting and welding. Formerly, riveting was also an important method but has now been largely superseded for civil engineering and building structures.

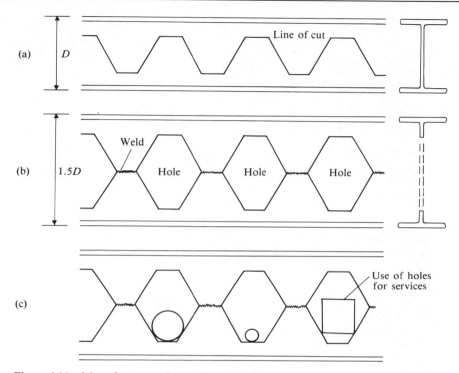

Figure 4.11   Manufacture and use of castellated beams

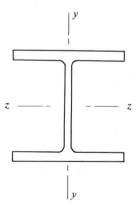

Figure 4.12   Typical proportions of a steel I-section used for columns. The flanges are thicker and wider than those used for beams to give the section a greater resistance to buckling about the y–y axis

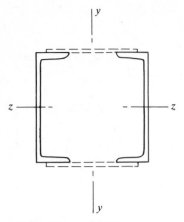

Figure 4.13 An efficient column having good resistance to buckling can be formed from a pair of channels by welding suitably spaced plates to their flanges

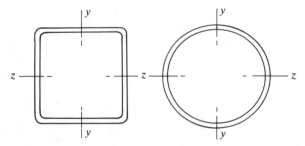

Figure 4.14 Hollow steel sections have good second moment of area about both orthogonal axes and are therefore very efficient as compression members

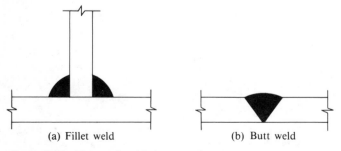

(a) Fillet weld   (b) Butt weld

Figure 4.15 Types of welded connections

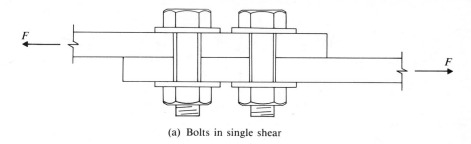

(a) Bolts in single shear

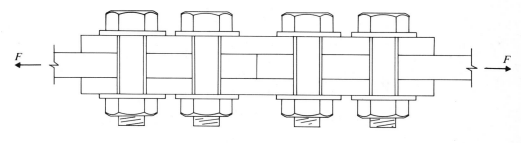

(b) Butt splice—bolts in double shear

Figure 4.16   Types of bolted connections

Welding is a process of joining by fusion. The most common method of providing the necessary heat is an electric arc. A pool of molten metal is formed adjacent to the arc with additional metal deposited by the melting of the metallic electrode. Welds are categorized as either fillet welds or butt welds. The former are deposited in the angle between adjoining plates (Fig. 4.15a) whilst with butt welds the deposited material lies mainly within the gap between adjacent aligning plates (Fig. 4.15b). Although fillet welds require no edge preparation, butt welds do, in general, require some initial preparation of the surfaces to be joined.

Bolts transfer forces between members by their action in shear and tension. Where the connected plates tend to separate, the bolts will be stressed in tension; where the connected plates tend to slide relative to one another, the bolts are stressed in shear. The bolts are said to be in single shear (Fig. 4.16a) if failure across one section of each bolt would cause failure of the connection, and in double shear (Fig. 4.16b) if failure across two sections of each bolt would be necessary to cause failure.

The greater care that is required with welding makes it a difficult and expensive site operation. Although that does not entirely preclude its use on site, the usual practice is to restrict welding to connections that can be made in the workshops and to effect site connections by bolting. Examples of the way these procedures are used for steelwork connections are given in Fig. 4.17. Particular note should be made of the positions of the workshop welding and the site bolting.

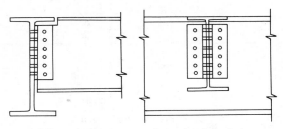

(a) Beam to beam connection using bolted cleats

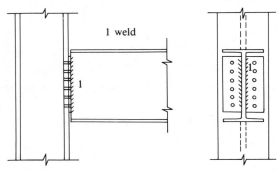

(b) Beam to column connection using end plate welded to web of beam

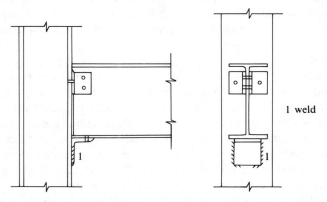

(c) Beam to column connection using angle seat welded to column

Figure 4.17   Some typical steelwork connections

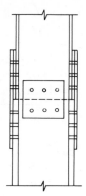

(d) Column to column connection using bolted splices

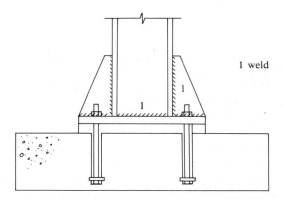

1 weld

(e) Column to concrete column base connection

Fig. 4.17   (*continued*)

4.2 Problems with steel structures

Corrosion

Steel has a grave disadvantage—in many environments it will rust. As a result it usually has to be given some form of protection.

Corrosion, or rusting, is a chemical reaction involving both oxygen and water, although environmental pollutants such as sulphur compounds and chlorides also play an important part. However, water is an essential part of the process and it is found that when the relative humidity of the atmosphere is less than about 70 per cent, the rate of corrosion is negligible. It follows that in a heated building there is little need for protection, but for use in other situations in temperate climates protection will be essential.

Protective systems can be categorized as 'active' or 'passive'. The latter is provided by a coating (e.g., paint) which simply prevents the moist air penetrating to the surface of the steel. Unfortunately, the smallest of cracks in this coating provides entry for the air which will then attack the steel. The volumetric expansion as a result of the formation of the rust further disrupts the protection and the corrosion can proceed rapidly.

An 'active' form of protection can be provided by a coating containing metals which themselves combine with the oxygen and render it incapable of attacking the steel. Such functions are provided by certain kinds of paints (known as primers, and, for example, the lead in red lead paint and the zinc in zinc chromate paint have this capability) and by the metallic coats obtained by galvanizing. In order that these active systems remain fully effective over a period of many years, it is desirable that their coatings are themselves protected from the continuing action of the oxygen. A combination of both the passive and the active systems therefore forms the basis of many treatments.

Fire protection

The possibility of fire plays an important part in the design of buildings, particularly of multi-storey buildings. Whilst steel is itself incombustible, its mechanical properties, such as the yield point and the modulus of elasticity, are a function of temperature. For example, Grade 43 steel has a yield stress of approximately 275 N/mm^2 at normal room temperature, but this drops to 165 N/mm^2 at 550°C and is virtually zero at a temperature of 800°C.

Tests indicate that a steel member subjected to its full design load would collapse in approximately 15 minutes after the commencement of a fire of 'standard' severity. Such a period would clearly not be sufficient to allow complete evacuation of a multi-storey building or to allow fire fighting to be effective. It therefore becomes necessary to improve the fire resistance of the steel structure by protecting the steel members from the fire. The degree of protection required is specified in terms of hours—1, 2 or 4 hours are typical of most buildings. These are the nominal periods a structure must be able to withstand a fire without collapsing, the particular period required depending on the type and size of the building.

Whilst there is a wide variety of materials and methods available for protecting steel, the majority fall into two main categories:

– boarded systems;
– spray applied systems.

A typical boarded system is shown in Fig. 4.18. Most are based on either a mineral fibre or naturally occurring vermiculite in combination with a cement or silicate binder. The boards are fixed in place by gluing, stapling or screwing. Their thickness depends on the degree of fire resistance required. To overcome the relatively slow rate

Figure 4.18   Boarded type of fire protection applied to a steel structure (*Source*: Cape Boards Limited)

of fixing on site, some systems are preformed to the correct size and shape prior to delivery on site. They can be supplied in a variety of finishes which may not require further treatment.

The spray systems have the advantage that they are fast to apply and therefore the labour cost is relatively low. However, they have the disadvantage of being wet and messy, and it is difficult to control the thickness of the cover on the steelwork. For the vermiculite or fibre materials with cement binders, the finish is rough (Fig. 4.19) and generally can only be used where they are hidden from view as, for example, by suspended ceilings.

Alternative forms of coating which can be applied by spraying are the intumescents. These are applied quite thinly (and so brushing or trowelling are alternative methods of application) but have the characteristic of swelling under the influence of heat to produce an insulating layer, sometimes 50 times thicker than the original coating. Although proving an economical form of fire protection, intumescents are at present limited to providing fire ratings of up to 2 hours.

Figure 4.19 Spray type of fire protection applied to a steel structure (*Source*: British Steel Corporation)

At one time, an important method of protecting steel columns from fire was by encasing in concrete (Fig. 4.20). The fact that such encasement also increased the load-carrying capacity was initially ignored, although it can now be allowed for in design. However, the need to provide timber formwork to contain the concrete whilst it hardens, and also the need for a cage of reinforcement within the concrete, necessitate time-consuming operations which, in these modern times, detract from the main advantage of steelwork, namely quick erection.

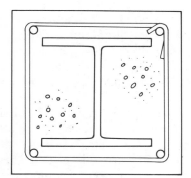

Figure 4.20 Steelwork can be encased with concrete as a means of fire protection

A modern method of fire protection that has been applied to hollow steel sections is by filling the sections with water. Circulation of the water transfers the heat away from the members subject to the fire and prevents the steel from becoming over-heated.

Local buckling

Steel is a material which, because of its relatively high cost, must be used sparingly. Its high strength, together with the ease with which it can be shaped into sections having thin flanges and thin webs, facilitates economy in the use of the material. Unfortunately, problems arise because of this. As a result of high compressive stresses in the thin flanges and webs, local buckling can occur (Fig. 4.21). Whilst the dimensions of the various hot-rolled sections are proportioned to give good stability from this standpoint, some care must be taken in design as, for example, not all sections are sufficiently 'compact' to be able to sustain flexural compressive yield stresses throughout the compression flange. Moreover, of those sections which could attain such stresses, because of local buckling not all could accommodate the high plastic strains implied by the stress–strain diagram of Fig. 4.3. The problem is exacerbated in steel members specially fabricated from thin plates for purposes of economy.

The buckling of webs in regions of high shear is a comparable problem. It can best be explained by reference to the state of stress on an element of a universal beam at the position of the neutral axis. The longitudinal stresses at this position are zero and therefore the element will be subjected to a state of pure shear stress (Fig. 4.22). However, pairs of such stresses acting on the horizontal and vertical faces of the element are equivalent to direct stresses acting in directions at 45° to the axis of the

Figure 4.21   Local buckling of a flange of steel I-beam (*Source*: Manchester University Engineering Department)

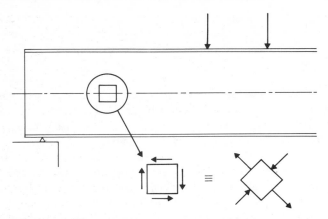

Figure 4.22   Illustrating the development of inclined tensile and compressive stresses as a result of shear. In a thin-webbed steel beam it is the compressive stresses which cause the problem

Figure 4.23   Buckling of a panel of a steel box-girder. The buckles are at the positions where the white lines appear curved (*Source*: Manchester University Engineering Department)

Figure 4.24   The main girders of this steel bridge have been strengthened in shear by welding vertical stiffeners to the web (*Source*: Centre Belgo-Luxembourgeois d'Information de l'Acier)

beam (see Section 3.4, Chapter 3). Whilst it is the inclined tensile stresses which cause the problem in concrete structures (see page 335 for further discussion), it is the inclined compressive stresses which are the problem in steel structures as their action in such thin webs may cause buckling (Fig. 4.23). To avoid this happening, it is necessary in design to limit the maximum shear stress to a value known to be safe, or alternatively to strengthen by incorporating web stiffeners (Fig. 4.24).

Brittle fracture

As the high ductility of steel has been described as one of its advantages, a word of caution is necessary. Circumstances do arise when the ductility is not available and sudden failure has been known to occur by complete fracture of the steel at stresses well below the yield stress.

Fortunately, such brittle fracture is rare. Investigations of actual failures indicate that there had been a defect of some kind (for example, a small crack within the material produced during manufacture) which had happened to occur adjacent to a stress concentration (for example, a weld or hole as a result of fabrication). Other factors influencing the problem are temperature and the rate of straining—the lower the temperature and the higher the strain rate, the greater the propensity for brittle fracture. Thus, although brittle fracture is rare in practical structures, where low temperatures or impact loading or very high stresses are a possibility, it is necessary for the designer to pay particular attention to the type of steel and to the fabrication details.

4.3 Materials for concrete

Concrete can be likened to an artificial stone. It is a composite material obtained by mixing (Fig. 4.25) cement, water and aggregates in specific proportions. It is the aggregates which form the bulk of the concrete and so their properties have a considerable bearing on the overall characteristics of the concrete. Nevetheless, it is the cement which must be considered the most important constituent since it is its chemical reaction with the water which gradually translates the fluid mass into a solid. The hardened cement paste so formed by the chemical reaction of water and cement acts as a matrix binding together the aggregate particles and transmitting stress from particle to particle. Concrete is a very complex material and the fact that it is often produced on site requires the structural engineer to have some knowledge of the constituent materials.

Figure 4.25   Concrete being mixed (*Source*: British Cement Association)

Cement

Portland cements, so called originally because of the resemblance of the set cement to a high quality natural stone quarried in Portland, Dorset, are the most widely used types of hydraulic cements. They are produced by burning either limestone or chalk (which provides CaO) with either clay or shale (which provide SiO_2 and Al_2O_3). The resulting clinker is ground to a very fine powder. Whilst the complete chemistry is too complex to receive attention here, it is worth while noting the main compounds (see Table 4.1) together with the shorthand notation used in cement chemistry. Moreover, it is necessary to mention that, whilst Table 4.1 covers the main compounds, there are numerous other compounds. Some of these, for example, the alkaline oxides of sodium and potassium, can have important deleterious effects as will be discussed later.

Table 4.1 Main compounds in Portland cements

Name	Empirical formula	Oxide formula	Short formula	Heat evolution and hardening
Tricalcium silicate	Ca_3SiO_5	$3CaO.SiO_2$	C_3S	Fast
Dicalcium silicate	Ca_2SiO_4	$2CaO.SiO_2$	C_2S	Slow
Tricalcium aluminate	$Ca_3Al_2O_6$	$3CaO.Al_2O_3$	C_3A	Fast
Calcium aluminoferrite	$2Ca_2AlFeO_5$	$4CaO.Al_2O_3.Fe_2O_3$	C_4AF	Very slow

It will be readily understood that variations in the basic materials and in the production processes will produce variations in the amounts of the various compounds in different cements. This has some influence on the rate of heat evolution during the hydration of the cement and on the rate of hardening of the concrete. The following gives a few examples of the types of Portland cements available, and their use will become more apparent in the subsequent discussions:

- *Ordinary Portland cement*   This is the most widely used of all hydraulic cements. It has a medium rate of hardening (and hence of heat evolution) and can be used for a wide range of structures.
- *Rapid-hardening Portland cement*   Of similar composition to ordinary Portland cement, its greater rapidity of hardening is due to the fact that the particles of cement have been more finely ground. This facilitates the hydration of the cement and is, therefore, used in situations where it is useful to develop enhanced strengths at early ages.
- *Low-heat Portland cement*   As the name implies, this cement has a low rate of heat evolution and hence a low rate of strength development. This is a result of it containing less tricalcium silicate and rather more dicalcium silicate than ordinary Portland cement.

Aggregates

Although a mixture of water and cement would harden to form a solid, it would not be useful in construction for the following reasons:

- it would be too costly to provide the necessary bulk;
- excessive shrinkage would occur, giving rise to severe cracking;
- the modulus of elasticity of the material would be too low, giving rise to excessive deflection of the members when subjected to load.

By incorporating inert fillers capable of transmitting stress, these objections can be overcome. Whilst there are different types of fillers depending on the particular use of the concrete, the great majority of concrete is formed from natural aggregates such as sands, gravels and crushed rock.

In order that the aggregates occupy as large a part of the concrete as possible, it will be clear that there must be a variation of size of particles in order to achieve close packing. To facilitate this, the aggregates are normally provided in two categories:

- coarse aggregate (defined as particles retained on a 5 mm sieve and including gravel and crushed rock);
- fine aggregate (defined as particles passing a 5 mm sieve such as natural sand and the fine fraction resulting from crushing).

These are used in such proportions and gradings that, together with the cement paste formed from the cement and water, a dense concrete is obtained.

Water

The water added to a concrete mix must be clean and free from harmful impurities. For example, water containing organic acids, sugars, sulphates and chlorides should not be used.

The amount of water required to complete the hydration of the cement is approximately 25 per cent of the cement by weight. However, it is always necessary to add extra water to the mix in order to maintain sufficient fluidity, or workability as it is called, during the placing and compaction of the concrete in the formwork (Fig. 4.26). This additional water will not, of course, contribute to the strength of the hardened concrete. Indeed, much of it will gradually be drawn to the surface of the concrete where it will evaporate, leaving behind a system of fine capillary pores in the hardened cement paste. It follows that from the strength standpoint, the additional water necessary to obtain workability should be kept as low as possible.

4.4 Properties of concrete

Setting of concrete

It will be understood that there is a gradual change from the stage at which the concrete can be regarded as a fluid to the stage at which it is regarded as a solid. The end of the first stage is termed the *initial set* and is approximately 45 minutes from the onset of mixing; the beginning of the final stage is termed the *final set*. Although these are somewhat arbitrary in definition (and, of course, depend on the type of cement), they have considerable importance in practice. For example, it is important that the

Figure 4.26 The fluid concrete needs to be contained by formwork until it has hardened (*Source*: British Cement Association)

concrete has been transported to, and placed in, the formwork prior to the concrete stiffening to the extent defined as initial set. It is, on the other hand, important that the strength of the concrete develops sufficiently fast to allow further construction to proceed.

Strength of concrete

Concrete continues to harden, and so to develop its compressive strength, over a long period of time as demonstrated in Fig. 4.27. The continuing rise in the strength of the concrete over a period of years assumes that the concrete has not been allowed to dry out and that there is sufficient water retained internally to allow hydration to proceed. For practical purposes, it is usual to adopt the 28 day strength as the basis of strength for design.

For an understanding of its breakdown under compressive stress, concrete (Fig. 4.28a) can be likened to a continuous three-dimensional micro-structure. The

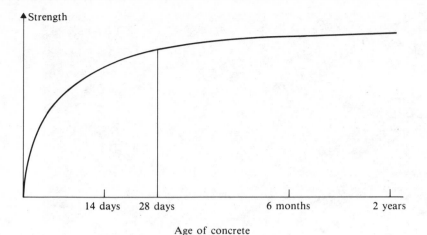

Figure 4.27 Concrete continues to develop strength over a long period of time. For most practical purposes, the 28-day strength is used as the basis of design

hardened cement paste conveys the forces through the concrete from aggregate particle to aggregate particle, the latter tending to attract the forces as a result of their higher modulus of elasticity. The structure behaves in a similar way to a framework when loaded in compression, in that members of the framework orientated transversely to the direction of the compression force will be stressed in tension (Fig. 4.28c). It is the members of the framework stressed in tension which are the weaker, with failure being initiated at the junction of aggregate and cement paste (Fig. 4.28b).

This micro-cracking develops gradually during the application of the compressive forces and commences long before it becomes sufficiently widespread to permit visible tensile cracking to be observed at the surface of the concrete. Such visible cracking for concrete stressed in compression will normally signify the imminent rupture of the concrete.

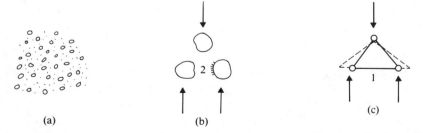

Figure 4.28 Concrete (a) can be likened to a framework (c) and members (1) orientated transversely to the direction of the applied compressive stress are stressed in tension. It is the tension members within the concrete which are the weak link, with failure being initiated at the junction of the aggregate and cement paste (2)

It is important to realize that the failure of concrete in compression requires a tensile strain in the direction orthogonal to the direction of compression. Where such tensile strain is inhibited, so the compressive stress causing failure is affected. Although this has great importance in practice, further discussion of it is delayed until the second year studies.

Deformation of concrete

A typical relationship between compressive stress and compressive strain for the case where there is no restraint on the tensile strain in the transverse direction is shown in Fig. 4.29. There are several important points to observe:

- it is not a straight line;
- unloading from any point on it leads to a permanent strain;
- rupture of concrete occurs at approximately a strain of 0.0035.

The first two points are related to the gradual development of the micro-cracking under the increasing stress. For example, as the micro-cracking occurs, so the stiffness of the micro-structure of the concrete reduces, giving greater deformation for each further increment of load.

It is of interest to compare the compressive stress–strain characteristic for concrete with that for mild steel. Figure 4.30 shows that, in comparison with mild steel, concrete has:

- a much lower strength;
- a much reduced modulus of elasticity;
- a much lower strain capacity.

However, one should not jump to too many conclusions. For example, as concrete can be regarded as a cheap material, the relatively low compressive strength can often be overcome simply by using larger cross-sectional areas. For compression members, such as columns, this can be regarded as advantageous as it reduces the buckling tendency.

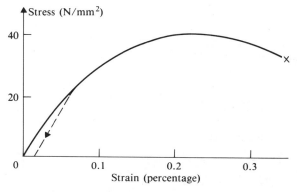

Figure 4.29   Typical stress–strain relationship for concrete

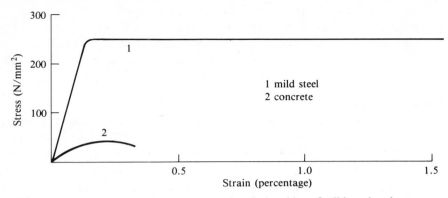

Figure 4.30 Comparison of the stress–strain relationships of mild steel and concrete

A further difference between steel and concrete concerns the effect of time on deformation. Whilst the duration of the load normally has virtually no effect on the strain caused in steel, it has considerable effect on the compressive strain in concrete. This phenomenon is illustrated in Fig. 4.31. The eventual strain may be 2 to 3 times the instantaneous strain caused at the time of application of the load.

The major part of the increase in strain is caused by *creep* and is due to the effect of the applied stress on the multi-molecular forces by which the uncombined water is held within the cement paste. In effect, the water is being 'squeezed' into the empty pores.

The other part of the increase in strain with time is caused by *shrinkage* and is due to the gradual drying out of the concrete as the free water in the pores is drawn to the surface. This in turn upsets the multi-molecular equilibrium and causes part of the physically-held water to be drawn into the empty pores, allowing the cement paste particles to become more compact. Such shrinkage strain will occur even in the

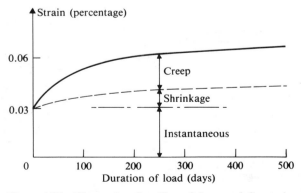

Figure 4.31 Illustrating the effect of time on deformation of concrete

absence of applied stress. The effects of creep and shrinkage have far-reaching consequences in design.

4.5 Reinforced concrete

Development of flexural cracking

The previous section on strength and deformation of concrete concentrated entirely on concrete subjected to compressive stress. This is because the tensile strength of concrete is so low, of the order of one-tenth of its compressive strength, that it is largely ignored in the design of structures. In this respect it is similar to natural stone, and would have similar limitations were it not for the fact that it is possible to reinforce the regions of a concrete member subjected to tensile stress. This is achieved by fixing round steel bars in the appropriate position within the mould prior to placing the fluid concrete (see Fig. 4.36). After hardening, the concrete and steel will be bonded together and will act together as composite material.

The behaviour of reinforced concrete is introduced by reference firstly to a simply supported unreinforced concrete beam (Fig. 4.32a). When subjected to a load, the

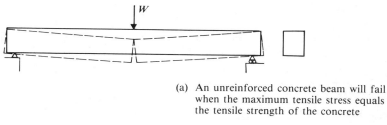

(a) An unreinforced concrete beam will fail when the maximum tensile stress equals the tensile strength of the concrete

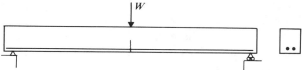

(b) In a beam reinforced with steel bars, a crack will be initiated at approximately the same load but does not now cause failure

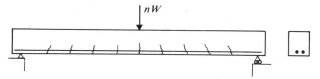

(c) Further load will cause additional cracks in the tensile region of the beam

Figure 4.32   Illustrating the behaviour of reinforced concrete beams under load

bending of such beams causes longitudinal tensile stresses in the lower region and longitudinal compressive stresses in the upper region. If the applied load is gradually increased, the plain concrete beam will soon collapse. As soon as the maximum tensile stress reaches a value equal to the tensile strength of the concrete, a crack will be initiated. This will instantly pass completely through the section, causing total collapse of the beam.

For a similar beam with reinforcing bars positioned in the concrete close to the bottom face, cracking would be initiated at approximately the same load. At the position of the crack, however, the reinforcing bars take over from the concrete in resisting the internal tensile force of the bending moment. In this connection, it will be recalled that the tensile strength of the steel is more than 100 times the tensile strength of the concrete. The crack can now only pass through the lower region of the concrete beam, and the upper region remains intact (Fig. 4.32b) and capable of resisting the compressive stresses caused by the bending moment.

At the position of this initial crack there will be a tendency for the stretched fibres of the concrete to return to their original unstressed length. This would necessitate the concrete sliding along the bars, but such sliding is resisted by the bond between the bars and the concrete. Nevertheless, there is some relief of tensile stress in the concrete in the immediate neighbourhood of this initial crack (Fig. 4.33), so that now the maximum tensile stress in the concrete is a short distance away from the position of maximum bending moment.

As the load is increased, so this maximum tensile stress increases in response to the higher bending moment, and when it attains a value equal to the tensile strength of the concrete an additional crack is formed. Further load will cause the pattern of cracks shown in Fig. 4.32c. The positions of flexural cracks and the main reinforcing bars in other types of beams are summarized in Fig. 4.34.

It may come as a surprise to learn that the structural engineer is willing to accept a structural material which is expected to crack in service. However, it is important to realize that such flexural cracks are so fine as to be hardly visible. Their width is of the order of 0.1 mm. Such cracks would not normally be visible from distances of 1 m or beyond, and it is necessary to examine the surface of a beam very closely to ascertain the pattern of cracking. It is most desirable that students undertake such tasks whenever the opportunity to examine a reinforced concrete structure presents itself. It will be found that multi-storey car parks present the best opportunity for studying

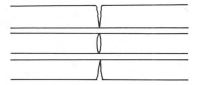

Figure 4.33 Sectional plan at the level of the reinforcing bars. Although the concrete is restrained at the interface of bars from returning to its unstressed length, the concrete further from the bars does lose its tensile strain and a crack on the surface becomes visible

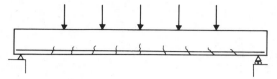

A simply supported beam is subjected to
sagging bending and hence must incorporate
reinforcing bars near the soffit

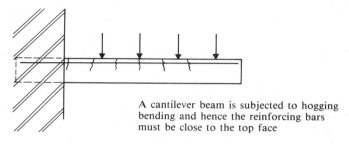

A cantilever beam is subjected to hogging
bending and hence the reinforcing bars
must be close to the top face

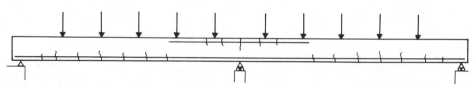

A continuous beam will be subjected to sagging bending for most of the
span and hogging bending in the region of internal supports, so both top and
bottom reinforcing bars are essential.

Figure 4.34   Illustrating the positions of main flexural reinforcing bars in reinforced concrete
beams

flexural cracking, as their low floor height facilitates close inspection of the beams. (If,
on close examination of such concrete structures, no cracks can be detected, it is likely
that the material is not reinforced concrete but prestressed concrete—see page 343
for further discussion.)

Notwithstanding the previous description of cracking, all students will have seen
reinforced concrete structures in which the cracking has been all too obvious.
However, such cracking is unlikely to be the simple flexural cracking described.
Unfortunately, there are many causes of cracking of concrete structures and some of
these are discussed later.

Bond between the concrete and the reinforcing bars

At this stage it is desirable to look more closely at the problem of bond. From a
consideration of the forces acting on the part of the beam to the left of section XX in
Fig. 4.35, it will be seen that there is a tendency for the bars to be pulled through the

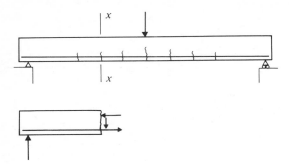

Figure 4.35   Illustrating the need for bond between the reinforcing bars and the concrete

concrete. It is the *bond* between the concrete and the steel bars which resists this, the bond forces being partly frictional in character and partly due to adherence or chemical bonding between the concrete and steel.

The most common type of reinforcing bars are of high strength (nominal yield strength 460 N/mm^2), and to ensure that such high stresses can be achieved the bars are normally provided with surface deformations as illustrated in Fig. 4.36. Such projections provide a high degree of interlocking between the steel and the concrete,

Figure 4.36   Deformations on the surface of reinforcing bars enhance the bond between the steel and the concrete and allow higher steel stresses to be adopted. Note that in this instance the concrete is being placed by pumping it through a pipe; see also Fig. 5.57 (*Source*: Pozzolanic Lytag Limited)

Figure 4.37 When plain round bars are used as main reinforcement, the bars would normally be bent at the ends to form hooks to eliminate the possibility of a bond failure

and so give additional restraint to relative movement. When the lower strength mild steel (nominal yield stress 250 N/mm^2) is used it is permissible to adopt plain round bars, although if such bars are used as the main reinforcement in beams, the bars would normally be provided with hooks at their ends (Fig. 4.37) to eliminate the possibility of a complete bond failure. The cost of providing such hooks is not insignificant and, as hooks are not normally required when using deformed bars, the high strength deformed bars are currently the most common type of reinforcing steel.

Problem of shear

An alternative form of cracking in reinforced concrete beams is that due to shear. It will be recalled from the discussions of problems in steelwork that, as a result of shear, normal compressive and tensile stresses occur in directions inclined at 45° to the axis of the beam. It is possible for the tensile stresses to promote diagonal cracking near the neutral axis of a beam at positions of high shear force and low bending moment (Fig. 4.38a).

More usually though, the diagonal cracks develop from the interaction of shear and bending whereby the flexural cracks extend up the beam in an inclined direction (Fig. 4.38b). Under further increments of load, these inclined cracks can develop alarmingly quickly to cause the type of diagonal cracking collapse shown in Fig. 4.39. When calculations indicate that such cracking is a possibility, it is necessary to incorporate vertical reinforcement in the beam in addition to the longitudinal flexural reinforcement. This reinforcement, termed shear reinforcement, takes the form of stirrups (alternatively termed links) which envelop the longitudinal reinforcement (Fig. 4.40).

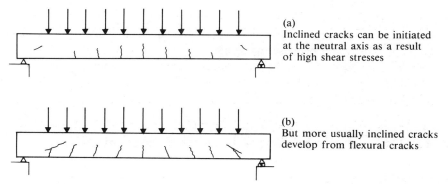

(a)
Inclined cracks can be initiated at the neutral axis as a result of high shear stresses

(b)
But more usually inclined cracks develop from flexural cracks

Figure 4.38 Illustrating the effect of high shear stresses on the cracking of reinforced concrete beams

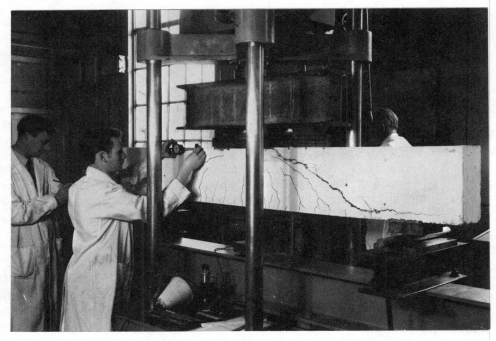

Figure 4.39　A reinforced concrete beam which has failed as a result of the diagonal cracking caused by shear forces (*Source*: Manchester University Engineering Department)

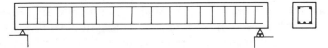

Figure 4.40　Reinforcement against shear usually takes the form of vertical stirrups enveloping the longitudinal reinforcing bars

In the past, a number of structural failures have occurred in practice when stirrups were not provided because the design calculations indicated that diagonal cracking would not occur. It is, therefore, now the rule that all main beams must incorporate some stirrups, irrespective of whether the calculations indicate that diagonal cracking is likely or not.

4.6　Elastic analysis of a reinforced concrete section

Basis of the theory

In the design of structural members it is essential to be able to estimate the strength of the members. To achieve this for reinforced concrete members, it is necessary to allow

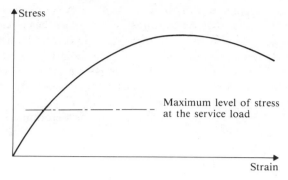

Figure 4.41   Up to the level of stress at the service load, the stress–strain relationship for concrete can be considered linear

for the non-linear stress–strain behaviour illustrated previously for concrete in compression in Fig. 4.29. Such analysis will form part of the second year studies.

At this stage, the analysis will be limited to a linear elastic method which has been used for reinforced concrete for many years and can be considered applicable to the service load conditions (Fig. 4.41). The safety factors adopted in design to establish the permissible service load on a member ensure that the compressive stresses in the concrete are within the range where, for all practical purposes, a linear stress–strain relationship can be assumed. The method is applicable to the calculation of stresses in beams and remains an important part of reinforced concrete studies.

Conditions applicable to the analysis of a cracked section of a reinforced concrete beam are summarized in Fig. 4.42. An important assumption is that there is no relative movement of the reinforcing bars and the concrete. This may be difficult to accept as flexural cracking seems to imply a distinct relative movement. It should be recalled, however (see Fig. 4.33), that the crack observable on the face of a beam does not correspond to the conditions at the position of the reinforcing bars where the bond between the concrete and the bar resists the opening of the crack.

1. Region of concrete in compression
2. Reinforcing bars (cross-sectional area A_s)

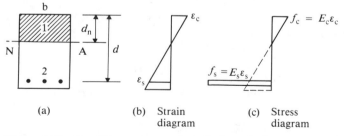

Figure 4.42   Conditions at a cracked section of a beam

The assumption that, at the position of the cracks, plane sections remain plane under the action of the load implies that the compressive strains in the concrete above the neutral axis and the tensile strains in the reinforcing bars below the neutral axis are proportional to the distances from the neutral axis (Fig. 4.42b). The linear stress–strain relationships for both concrete and steel determine the stress diagram (Fig. 4.42c). It will be readily understood that at a crack there can be no tensile stress in the concrete.

The intensities of stress depicted in Fig. 4.42c can now be determined using the standard beam equation

$$f = -\frac{My}{I}$$

where f = longitudinal stress at a distance y from the neutral axis;
  M = bending moment acting on the section;
  I = second moment of area of the cracked cross-section.

It will be recalled that this equation was developed (Chapter 3) for linear elastic conditions in a beam and will, therefore, apply to this analysis of the reinforced concrete beam. However, before it can be used, it will be necessary to explain how to determine the position of the neutral axis and how to calculate the second moment of area.

Position of the neutral axis

Earlier problems related to stress in beams were concerned with beams of a single homogeneous material. For such cases it was proved that, for sections symmetrical about the y-axis and subjected to moments in the x–y plane, the neutral axis always passed through the centroid of the cross-section and this was a useful starting point in analysis. However, reinforced concrete incorporates two very different materials, so this proposition on the position of the neutral axis does not apply. Nevertheless, the fundamental principle from which the proposition was developed, namely, that the forces of compression and tension on the section must be equal, does still apply and so this is now the starting point for determining an expression for the position of the neutral axis (i.e., the depth of the compression zone) of the reinforced concrete section.

Referring to Fig. 4.42:

Force in concrete in compression F_c = average stress × area of compression zone

$$= 0.5 f_c b d_n$$

As

$$f_c = \varepsilon_c E_c$$

then

$$F_c = 0.5 \varepsilon_c E_c b d_n$$

Force in steel in tension F_s = stress in bars × area of bars

$$= f_s A_s$$

As

$$f_s = \varepsilon_s E_s$$

and by direct proportion
$$\varepsilon_s = \frac{d - d_n}{d_n} \varepsilon_c$$

then
$$F_s = \frac{d - d_n}{d_n} \varepsilon_c E_s A_s$$

For longitudinal equilibrium
$$F_c = F_s$$

thus
$$0.5\varepsilon_c E_c b d_n = \frac{d - d_n}{d_n} \varepsilon_c E_s A_s$$

Writing $A_s = rbd$ (where r represents the ratio of the cross-sectional area bd occupied by the reinforcing bars) and $E_s/E_c = m$ (where m is termed the modular ratio), and rearranging, leads to a quadratic equation in d_n/d which gives

$$\frac{d_n}{d} = \sqrt{(m^2 r^2 + 2mr)} - mr \tag{4.1}$$

This expression is the starting point for the elastic analysis of reinforced concrete sections. In the past, to avoid the somewhat awkward calculation, Eq. (4.1) was expressed in a graphical form from which the neutral axis depth ratio (d_n/d) could be read directly for any value of the ratio of the area of the tension reinforcement r. In modern times, the use of electronic calculators makes such a graphical form unnecessary.

Transformed section

An alternative method of analysis makes use of the 'transformed' section. In this, the reinforced concrete section is changed to an imaginary equivalent section of single material only.

In the case of reinforced concrete, it is usual to transform the section to one totally of concrete. The method is illustrated in Fig. 4.43. Compared with the actual section, the area of the reinforcing bars has been replaced by the equivalent area of concrete, namely, mA_s. This area of concrete is 'equivalent' to the reinforcing bars in terms of the force that would develop for a particular strain.

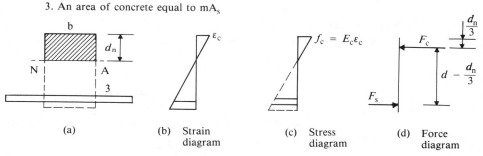

Figure 4.43 Use of the transformed section

The strain and stress diagrams corresponding to such a section would now be as shown in Figs 4.43b and 4.43c. The corresponding forces are

in compression $$F_c = 0.5 E_c \varepsilon_c b d_n$$

in tension $$F_s = E_c \varepsilon_c \frac{(d - d_n)}{d_n} m A_s$$

Examination will show that these are identical to the expressions determined previously and so, clearly, the same equation for the neutral axis would result. However, in using the transformed section one can now use the relationship for single homogeneous materials that the position of the neutral axis passes through the centroid of the section, and this allows a slightly easier derivation. Thus, referring to Fig. 4.43a and taking moments of areas about the neutral axis:

$$bd_n \times \frac{d_n}{2} = m A_s (d - d_n)$$

Substituting $A_s = rbd$ leads directly to the quadratic in d_n/d, and so to Eq. (4.1).

Second moment of area

In order to calculate the stresses in the transformed section caused by a bending moment M, it is necessary to determine its second moment of area about the neutral axis. This is obtained from Fig. 4.43a by summing the separate contributions due to the compression zone and due to the area of concrete representing the reinforcing bars. Thus:

$$I_t = \frac{bd_n^3}{12} + bd_n \left(\frac{dn}{2} \right)^2 + m A_s (d - d_n)^2$$

$$= \frac{bd_n^3}{3} + m A_s (d - d_n)^2 \tag{4.2}$$

Calculation of stress

Using the value of I_t obtained from the above expression, it is now possible to calculate stresses using the standard beam expression. For example, the maximum compressive stress in the concrete compressive zone is given by

$$f_c = \frac{M d_n}{I_t} \tag{4.3}$$

The tensile stress in the concrete of the transformed beam at the level of the reinforcing bars would then be calculated using the similar triangles in Fig. 4.43c to give

$$\frac{f_c (d - d_n)}{d_n}$$

which in turn leads to the tensile stress in the reinforcement of the actual beam as

$$f_s = \frac{mf_c(d - d_n)}{d_n}$$

Although the above procedure has utilized the beam expression developed for beams of homogeneous materials in Chapter 3, it is desirable to draw attention to an alternative procedure based more directly on fundamental principles. As the lever arm between the internal tensile and compressive forces is $d - \dfrac{d_n}{3}$ (see Fig. 4.43d), then since:

external moment M = compressive force × lever arm

$$= 0.5f_c b d_n \left(d - \frac{d_n}{3} \right)$$

then

$$f_c = \frac{2M}{b d_n \left(d - \dfrac{d_n}{3} \right)} \tag{4.4}$$

This particular approach to the calculation of stresses is somewhat easier and is generally preferred.

Problem 4.1   Prove that the value of f_c given by Eqs (4.1), (4.2) and (4.3) is identical to that given by Eq. (4.4).

Modular ratio

A comment is necessary on the value to be adopted for the modular ratio m in the above expressions. Whilst the modulus of elasticity of the steel is accurately known, the particular value to be adopted for the concrete is less straightforward. Although, as stated earlier, the stress–strain characteristic can be regarded as linear in the range of stress up to service load conditions, the effect of creep as a result of sustained stress affects the degree of strain. As a result, it is possible for the value of m to vary from perhaps 8 for short-term loading to 24 for long-term loading. In practice, structures are subjected to loading of which part is long term (e.g., dead load) and part is of variable duration (imposed load).

Fortunately, the effect of the change of m over quite a large range has only a modest effect on the value of the calculated stresses. The choice of an intermediate value in the range 10 to 20 (depending on the ratio of the long-term/short-term loading) will be sufficiently accurate for most purposes.

EXAMPLE 4.1
The reinforced concrete beam of rectangular cross-section shown in Fig. 4.44 is

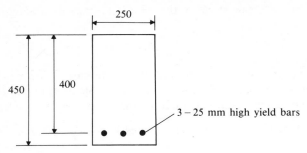

Figure 4.44   Section of a reinforced concrete beam used to illustrate the calculation of stresses in Example 4.1

subjected to a maximum bending moment of 100 kN m. Calculate the maximum stresses in the reinforcement and the concrete assuming a modular ratio of 10.

Area of reinforcement A_s (from tables) $= 1470$ mm^2

Ratio of reinforcement $r = \dfrac{A_s}{bd} = \dfrac{1470}{250 \times 400} = 1.47 \times 10^{-2}$

$$mr = 0.147$$

Using Eq. (4.1) $\dfrac{d_n}{d} = 0.415$

$$d_n = 0.415 \times 400 = 166 \text{ mm}$$

Depth of centroid of compression $= \dfrac{d_n}{3} = 55$ mm

Lever arm between tension and compressive forces $= 400 - 55 = 345$ mm

Force in tension $F_s = \dfrac{\text{bending moment}}{\text{lever arm}} = \dfrac{100 \times 10^3}{345} = 290$ kN

Stress in reinforcement $= \dfrac{F_s}{A_s} = \dfrac{290 \times 10^3}{1470} = \underline{197 \text{ N/mm}^2}$

Maximum stress in concrete $= 2 \times$ average stress $= \dfrac{2 \times 290 \times 10^3}{250 \times 166}$

$$= \underline{14 \text{ N/mm}^2}$$

The calculation is now repeated using the transformed section method:

$$I_t = \dfrac{250 \times 166^3}{3} + [10 \times 1470 \times (400 - 166)^2] = 1186 \times 10^6 \text{ mm}^4$$

$$\text{Maximum stress in concrete} = \frac{Md_n}{I_t} = \frac{100 \times 10^6 \times 166}{1186 \times 10^6} = \underline{14 \text{ N/mm}^2}$$

$$\text{Stress in reinforcement} \quad = \frac{mM(d - d_n)}{I_t} = \frac{10 \times 100 \times 10^6 \times (400 - 166)}{1186 \times 10^6}$$

$$= \underline{197 \text{ N/mm}^2}$$

4.7 Prestressed concrete

There are two main characteristics of reinforced concrete which detract from its usefulness. They are:

- the formation of flexural cracks;
- the limitation of the reinforcement to the relatively low strength steels.

Although it has been argued that flexural cracks are normally so fine as to be largely unnoticeable, there are circumstances in which even such cracks are undesirable. This occurs, for example, with water-retaining structures, and also for external construction where subsequent weathering and discolouration at the cracks may detract from the appearance.

Figure 4.45   A prestressed concrete bridge (*Source*: British Cement Association)

The reason for the limitation of the reinforcement to the relatively low strength steels (Fig. 4.5) is also connected with serviceability. It should be noted that the elastic modulus of steel is independent of the strength of the steel, with the result that a concrete beam reinforced with ultra high strength steels would have substantially wider cracks and higher deflections in comparison with a similar beam reinforced with equivalent amounts of lower strength steels. As the latter are the more expensive on a cost/strength basis, the restriction in reinforced concrete to the use of the lower strength steels is a distinct economic drawback.

Prestressed concrete overcomes both these disadvantages. As a result, it becomes possible to use longer and more slender spans (Fig. 4.45).

Principles

The basic idea behind prestressed concrete can be illustrated with a row of books (Fig. 4.46). Clearly, there is zero tensile strength between one book and the next, and yet they can be transformed into a 'beam' capable of supporting its own weight if sufficient pressure is applied at the ends. It will, moreover, be found to be advantageous to apply the pressure below the mid-height of the books in preference to, say, applying the pressure near the top, as 'cracks' would then tend to open

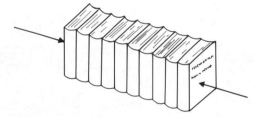

Figure 4.46 The principles of prestressing can be illustrated with a row of books

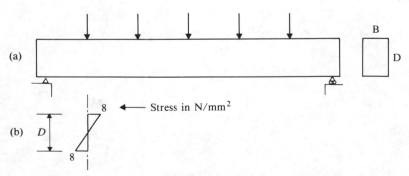

Figure 4.47 An assumed distribution of flexural stresses across the mid-span section of the beam due to the bending moment caused by the loads

between the books near the soffit. These points can be clarified from a consideration of the effect of the forces on the stress distribution across a section of a beam.

Consider a simply supported beam of rectangular cross-section (Fig. 4.47a). Let us assume that the maximum flexural stress caused by the load is 8 N/mm². The stress distribution across the section subjected to the maximum bending moment is, therefore, as shown in Fig. 4.47b with compressive stresses in the upper half of the beam and tensile stresses in the lower half. Such maximum stresses would not, of course, be possible in a plain concrete beam as failure would have occurred when the tensile stress attained a value of approximately 3 N/mm².

Let us now assume that, prior to applying the load to the beam, compressive forces are applied along the axis of the beam (Fig. 4.48a), resulting in a uniform stress distribution of 8 N/mm² throughout the beam (Fig. 4.48b).

If the same load as in the previous case is now applied (Fig. 4.49a), the stress distributions of the two cases are superimposed. Thus, at the mid-span section the resulting stress distribution will be as shown in Fig. 4.49b. The main point to note is that the longitudinal stresses on the section are now entirely compressive. This applies to all the sections of the beam and no tensile stresses have occurred as a result of applying the load. Clearly then, it would have been possible to use plain concrete

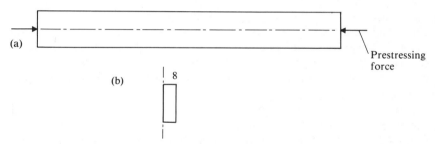

Figure 4.48 If, prior to the application of the load, end forces are applied to act along the axis, a uniform prestress acts on all sections

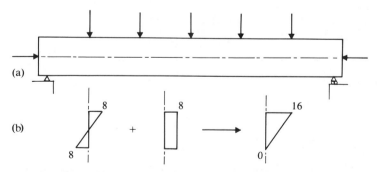

Figure 4.49 For the combination of the load and the axial force, the stresses at mid-span must be superimposed. No tensile stresses now occur

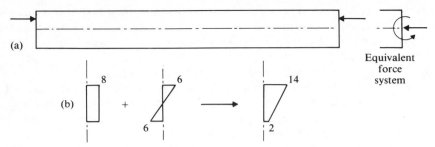

Figure 4.50   If the pretressing force is applied above the axis of the beam, the moment effect of the eccentric force will increase the compressive stresses above the axis and decrease them below the axis

for such a beam as there is no possibility of tensile cracking occurring. Prestressing the concrete has, therefore, transformed it into a useful structural material without the need for reinforcing bars.

It is now desirable to look at the situation comparable to applying the pressure to the books above mid-depth. If, therefore, instead of applying the prestressing forces along the axis of the beam, they are applied above the axis (Fig. 4.50a), a different distribution of prestress results. As discussed in Chapter 3, an eccentric force is equivalent to an axial force acting in conjunction with a moment. In this case, the stresses caused by the moment will be compressive above the axis and tensile below the axis. Let the extent of the stresses be as indicated in Fig. 4.50b. The resulting distribution of prestress is still entirely compressive but is no longer uniform over the section.

The application of the load (Fig. 4.51a) will now result in the stress distribution shown in Fig. 4.51b. Clearly, such a distribution would not be acceptable in a concrete beam. Not only has the load now caused high tensile stresses in the lower part of the

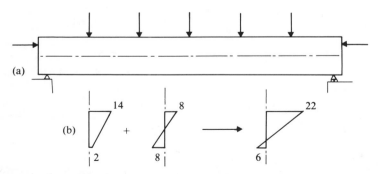

Figure 4.51   This is now unacceptable. The effect of the load superimposed on the prestress due to the force applied above the axis causes very high compressive stresses in the top fibres and tensile stresses in the bottom fibres at mid-span

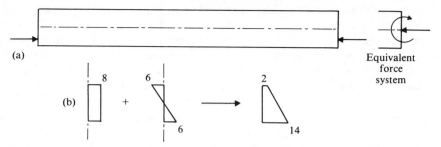

Figure 4.52 When the pretressing force is applied below the axis, the moment effect of the eccentric force will decrease the compressive stresses above the axis and increase them below the axis

beam, but in addition the compressive stresses in the top fibres have become undesirably large. Obviously, this form of prestress is not so advantageous as the previous case.

On the other hand, a more advantageous prestress in the beam is obtained by positioning the prestressing force below the axis of the beam (Fig. 4.52a). It will be noted that the stresses caused by the moment effect of the eccentric load are now opposite in kind to those due to the applied load (Fig. 4.52b).

The superposition of the stresses due to such a prestressing force and the stresses due to the same applied load now result in an overall distribution of compressive stresses (Fig. 4.53b) with a stress of 6 N/mm² in the bottom fibres. This is clearly a better arrangement in comparison with applying the prestressing forces along the axis since it means that a substantially higher load could be applied to the beam without any tensile stresses, and hence tensile cracking, occurring.

The previous cases have considered the effect of the prestress on the stress distribution across the section subjected to the maximum bending moment. Whilst

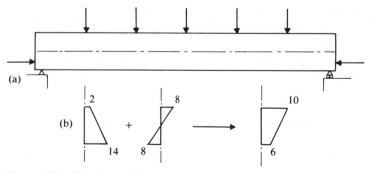

Figure 4.53 This is an advantageous arrangement. The combination of the load and prestress when the prestressing force is below the axis results in an acceptable distribution of compressive stresses across the mid-span section. Indeed a much higher load could now be applied without causing tensile stresses

that is an important consideration in design, it is equally important to study other cross-sections along the length of the beam. In beams of constant section and straight prestressing tendons, the stresses due to prestress will be the same at all sections. At sections where the bending moment caused by the load is zero, the stress distribution will not be modified by the effect of the load. In the previous example where the tendons were advantageously placed below the axis, the stresses at the ends of the beam will remain as shown in Fig. 4.52b. It therefore becomes apparent that the designer must take care to position the tendons so that in the end regions tensile stresses do not occur in the *top* fibres of the beam and, moreover, that the compressive stresses in the *bottom* fibres do not become dangerously high. In other words, the eccentricity of the centroid of the prestressing force must not be excessive. The balancing of the conditions near mid-span, where a high eccentricity is of advantage, with the conditions near the ends, where a low eccentricity is desirable, present the designer with a difficult problem, one that is often overcome by varying the eccentricity of the prestressing force along the length of the beam.

Prestressing techniques

The question now arises: how best can the prestressing force be applied? In theory, jacks pressing against strong abutments (Fig. 4.54) could be adopted, but this is rarely a practical or economical proposition. All current methods utilize tensioned steel *tendons*, their tensile force being resisted by the concrete which, as a result, develops an equal compressive force. The steel tendons can take the form of:

– single cold-drawn wires (usually not greater than 8 mm in diameter and of characteristic strength of approximately 1600 N/mm^2);
– strands (essentially cables made up by twisting wires together, usually in groups of 7 or 19 wires);
– cold-worked bars (of characteristic strength approximately 1000 N/mm^2).

It is important to note the strength of the steel in relation to that used for reinforced concrete. Much of the economy of prestressed concrete derives from this use of very high strength steels.

There are two separate techniques for applying the prestress—pretensioning and post-tensioning.

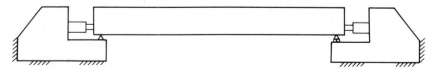

Figure 4.54   In theory, jacks reacting against strong abutments could be used to apply the prestressing force to concrete beams

Pretensioning

In pretensioning, the tendons pass through the mould and are tensioned prior to the concrete being placed, being anchored against abutments positioned outside the mould (Fig. 4.55a). The concrete is now placed within the mould and around the tendons (Fig. 4.55b). After the concrete has hardened sufficiently, the tendons are slowly released from the restraint of the abutments (Fig. 4.55c). As the tensioned tendons try to return to their original unstressed length, they are restrained by the bond between the concrete and the steel. There will be some resultant shortening of the concrete as it becomes stressed in compression. An equilibrium position will be reached when the compressive force in the concrete is in equilibrium with the new tensile force in the tendons, now slightly lower than its previous value.

It is of interest to note how the Poisson ratio effect plays a part in assisting the bond between the concrete and steel. During the tensioning of the tendons, not only do they increase in length but there is a corresponding decrease in diameter. The opposite will occur following the release of the tendons and they will tend to increase in diameter as the longitudinal stress tends to decrease. Indeed, at the ends of the tendons outside the prestressed concrete unit the tendons will now have zero stress and so will return completely to their original diameter. Thus, at the ends of the units there will also be a wedge type action slightly assisting the bond in transmitting the tendon force to the

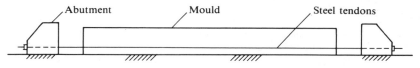

(a) Steel tendons are tensioned prior to placing the concrete and anchored against abutments

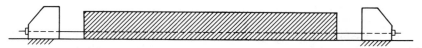

(b) The concrete is now placed in the mould

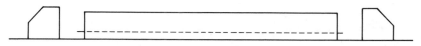

(c) After the concrete has hardened, the tendons are released from the abutments and cut to length

Figure 4.55   Illustration of the principles of the pretensioning system of prestressing concrete

concrete. Notwithstanding, a deformed type of wire (either crimped or indented) is often used to further assist bond.

Post-tensioning

In post-tensioning the tendons are tensioned after the concrete has been placed and the concrete has hardened. The tendons may be positioned in the mould prior to the concrete being placed, but it is then essential to prevent bond occurring by enclosing the tendons in a protective metal sheath. Alternatively, ducts are formed in the concrete (Fig. 4.56a) using removable rubber formers through which the tendons are subsequently passed. When the concrete has attained sufficient compressive strength, the steel is stressed using jacks reacting against the concrete member itself (Fig. 4.56b) and the tendons are fixed in positions using special anchorages concreted into the ends of the unit. Once all the tendons have been stressed and anchored, the ducts are filled with a cement grout (Fig. 4.56c). This is partly to prevent corrosion of the steel but also to ensure some bond between the tendons and the concrete, which in turn ensures better structural interaction in the event of flexural cracking caused by overload.

The use of the pretensioning system is particularly useful for producing precast units in the factory where permanent long-line stressing beds can be used for the manufacture of several similar units at the same time (Fig. 4.57). The post-tensioning

(a) A duct is formed in the concrete beam during manufacture

(b) Steel tendons are threaded through the ducts and stressed using jacks reacting against the beam

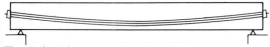

(c) The steel tendons are anchored at the ends and the ducts filled with cement grout

Figure 4.56   Illustration of the principles of post-tensioning system of prestressing concrete

Figure 4.57   A long-line prestressing bed used for the manufacture of precast concrete units. (*Source*: Dow-Mac Concrete Limited)

system can also be used for precast factory-made units and becomes particularly appropriate when the size of the unit makes it desirable to utilize curved tendons to vary the degree of eccentricity. For prestressing carried out on site, the use of the post-tensioning system is preferred in order to avoid the need for providing temporary abutments. A particularly useful constructional operation incorporating post-tensioning, and previously illustrated with the use of the books, is the prestressing together of precast segments to form a large single unit (Fig. 4.58).

4.8 Elastic analysis of a prestressed concrete section

Introduction to the design problems

In selecting a suitable section in the design of a prestressed concrete beam, there are several conflicting criteria. On the one hand, it is desirable (from the standpoint of ultimate load considerations which are discussed in the book related to the second

Figure 4.58 The use of precast concrete segments prestressed together to form a bridge (*Source*: British Cement Association)

year studies) to place the steel tendons as low in the section as possible. On the other hand (and from the same standpoint), one needs a substantial area of concrete in the upper region of the beam in order to resist the compressive force in regions of sagging bending moments. The latter tends to raise the centroid of the section, so giving high eccentricities of the prestressing tendons with the probability, when using straight tendons, of causing tensile stresses at sections of low bending moment.

One method of overcoming this problem is to construct the beam in different stages, and this is frequently used in practice. Part of the beam is precast and prestressed, and so shaped as to have a low centroid. This enables the majority of the prestressing tendons to be placed near the soffit of the beam without causing tensile stresses. The mass of concrete required to resist the compressive stresses caused by the applied load is added later on site in what is termed *in-situ* construction.

An example of this type of construction often used for bridges is shown in Figs 4.59 and 4.60. The factory-made precast prestressed beams are transported to site and placed side by side in their permanent position. *In-situ* concrete is then placed between and over the precast units to form a solid slab-type structure. This method of construction obviates the need for temporary timber formwork, often an important

Figure 4.59   The use of precast prestressed concrete inverted-T beams in the construction of a bridge. The space between the webs of the beams will be infilled with concrete (*Source*: British Cement Association)

consideration in the construction of bridges where access may be difficult. The subsequent composite action of the precast concrete and the hardened *in-situ* concrete is ensured by extending the stirrups from the precast units into the *in-situ* concrete.

During the construction, and prior to the hardening of the *in-situ* concrete, the precast units will have to support not only its own dead load but also the additional

1 Precast beams
2 *In-situ* concrete
3 Transverse reinforcing bars

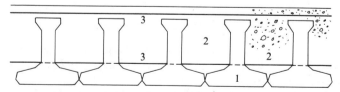

Figure 4.60   Illustrating the method of constructing bridges using a combination of precast and *in-situ* concrete

dead load of the fluid *in-situ* concrete. It is this part of the analysis which will now be illustrated.

Example of calculation

It is desired to calculate the stress distribution across the mid-span section of the precast unit shown in Fig. 4.61 caused by the placing of the *in-situ* concrete in the construction of a bridge of 12 m span.

If the dimensions of the precast units were decided entirely by the designer, the first steps of the calculation would be to determine:

– the cross-sectional area of the section;
– the position of the centroid;
– the second moment of area about the horizontal axis through the centroid;
– the self-weight of the beam.

However, the frequent use of such beams has enabled a degree of standardization whereby the designer selects a beam from a range of beams available from the precast concrete manufacturers. The sectional properties of such beams are then already available to the designer, and for the beam selected for this example, the properties are given in Fig. 4.61a. The positions of the prestressing tendons are shown in Fig. 4.61b.

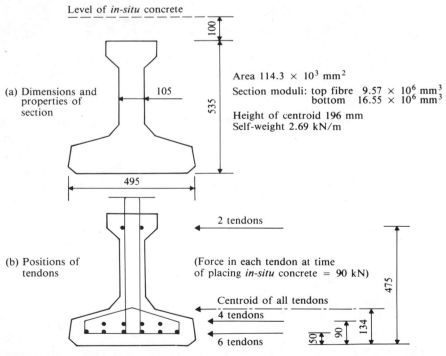

Figure 4.61   Properties of the prestressed concrete beam used to illustrate the calculation

It has become the usual practice in prestressed concrete design to designate compressive stresses as positive and tensile stresses as negative, and this will be adopted here.

Eccentricity of centroid of prestressing tendons $= 196 - 134 = 62$ mm

$$\frac{\text{Stress in top fibres}}{\text{due to prestress only}} = \frac{12 \times 90 \times 10^3}{114.3 \times 10^3} - \frac{12 \times 90 \times 10^3 \times 62}{9.57 \times 10^6} = 2.5 \text{ N/mm}^2$$

$$\frac{\text{Stress in bottom fibres}}{\text{due to prestress only}} = \frac{12 \times 90 \times 10^3}{114.3 \times 10^3} + \frac{12 \times 90 \times 10^3 \times 62}{16.55 \times 10^6} = 13.5 \text{ N/mm}^2$$

$$\frac{\text{Bending moment at midspan}}{\text{due to self-weight of units}} = \frac{2.7 \times 12^2}{8} = 48.6 \text{ kN m}$$

$$\frac{\text{Stress in top fibres due}}{\text{to self-weight}} = \frac{48.6 \times 10^6}{9.57 \times 10^6} = 5.1 \text{ N/mm}^2$$

$$\text{Stress in bottom fibres} = -\frac{48.6 \times 10^6}{16.55 \times 10^6} = -2.9 \text{ N/mm}^2$$

The resulting stress distribution across the mid-span section of the precast units prior to placing the *in-situ* concrete is summarized in Fig. 4.62c.

$$\frac{\text{Cross-sectional area of } in\text{-}situ}{\text{concrete supported by one unit}} = (635 \times 500) - 114\,300 = 203\,200 \text{ mm}^2$$

$$\frac{\text{Weight of } in\text{-}situ \text{ concrete}}{\text{supported by one unit}} = 23.5 \times 0.203 \qquad = 4.77 \text{ kN/m}$$

(a) Stress distribution due to prestress only
(b) Stress distribution due to weight of precast unit only
(c) Stress distribution prior to placing the *in-situ* concrete
(d) Stress distribution due to weight of *in-situ* concrete only
(e) Resulting stress distribution after placing the *in-situ* concrete

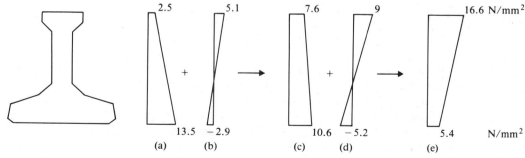

Figure 4.62 Stress distribution across the mid-span section at different stages in the construction

$$\text{Bending moment at mid-span due to weight of } \textit{in-situ} \text{ concrete} = \frac{4.77 \times 12^2}{8} = 86 \text{ kNm}$$

$$\text{Stress in top fibres due to weight of } \textit{in-situ} \text{ concrete} = \frac{86 \times 10^6}{9.57 \times 10^6} = 9 \text{ N/mm}^2$$

$$\text{Stress in bottom fibres due to weight of } \textit{in-situ} \text{ concrete} = -\frac{86 \times 10^6}{16.55 \times 10^6} = -5.2 \text{ N/mm}^2$$

The resulting final distribution of stress at mid-span is summarized in Fig. 4.62e. After the *in-situ* concrete has hardened, all subsequent imposed load is resisted by the *in-situ* and precast concrete acting compositely (i.e., together) as a slab.

4.9 Problems in concrete structures

Corrosion of reinforcement

In view of the earlier description of one of the main problems in steelwork structures, it might be thought that the corrosion of the reinforcement in concrete structures would also be a major problem, since the concrete cover cannot be regarded as providing a complete barrier to the ingress of moisture and oxygen. In fact it is not a major problem, and this is due to the alkaline nature of the hydrated cement compounds in concrete which inhibits the corrosion of the embedded steel.

Unfortunately, corrosion problems can occur when insufficient attention is paid to the positioning of the bars within the concrete. This arises as a result of the attack on the surface of concrete members from the acidic solutions of carbon dioxide and sulphur dioxide present in moist atmospheric conditions. The resultant reactions, termed carbonation, within this outer region of the concrete reduces its alkalinity and consequently its effectiveness in preventing corrosion. Fortunately, in good dense concrete the depth of the carbonation of the concrete remains small, and as long as the reinforcing bars have a sufficient cover to ensure that they are embedded in concrete which remains alkaline, there will be no corrosion. The depth of cover required varies with the grade of concrete and the conditions of exposure of the structure, but would be of the order of 40 mm for many external structures.

If a bar is positioned in a region of concrete which becomes carbonated, corrosion of the bar will soon be manifest. The corrosion products occupy a greater volume than the original steel bar and this volumetric expansion imparts tensile forces to the concrete cover, often causing severe cracking along the line of the reinforcing bar. There are, unfortunately, many examples in practice (Fig. 4.63).

An alternative cause of corrosion of reinforcement is the presence of chloride ions in the concrete, and such corrosion can then occur even in an alkaline environment. At one time, calcium chloride was added deliberately to a concrete mix as a means of accelerating the hardening of the concrete, but with the subsequent recognition of its corrosive effect this is not now permitted. Nevertheless, the consequences of such constructional procedures will still be felt for many more years. Indeed, the current use

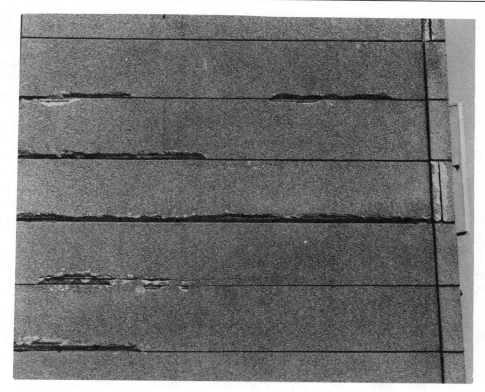

Figure 4.63 Cracking typical of that caused by the corrosion of reinforcing bars (*Source*: Building Research Association)

of de-icing salts on roads in winter provides a source of chloride contamination in concrete bridges which will continue to cause problems of corrosion. To overcome this particular difficulty, it may eventually prove advisable to adopt special measures, for example, the use of stainless reinforcing steels or, alternatively, the use of bars with protective surface coatings.

Special mention should perhaps be made that natural chlorides exist to some extent in most aggregates. For the most part, however, the aggregates available in the British Isles contain insufficient chloride to cause any significant ill effect. However, this does not apply to aggregates available in some areas of the Middle East, where very high chloride contents occur and have been the cause of severe deterioration of many concrete structures.

Cracking due to shrinkage and temperature effects

Although flexural cracks in concrete structures normally remain quite fine and therefore acceptable, wide unsightly cracking due to other causes is a recurring problem. Two of the causes are shrinkage and temperature effects. It is essential that designers obtain a clear understanding of the basic mechanism of such cracking in order to avoid these problems.

Thermal cracking of concrete is most likely to occur at a very early stage in the life of a concrete structure. Heat is generated during the hydration of the cement. In thick sections of concrete, most of this heat is temporarily retained by the concrete and so the temperature rises. In the first few hours after casting, and whilst the concrete remains in a fluid or viscous state, its resulting expansion can be accommodated without causing compressive stresses.

After a further period of time, say 24 hours, the rate of hydration reduces and the rate of heat loss to the surroundings exceeds the rate of heat generation. As a result, the concrete will cool and tend to contract. However, by this stage, the concrete has set and hence, in situations where there is restraint to contraction, tensile stresses will be created. As after such a short period of time the tensile strength of the concrete will be very low, cracking will occur. If there is insufficient reinforcement passing through the cracks, as may be the case if the engineer has designated that particular direction as requiring only a nominal amount of reinforcement, the widths of the cracks are likely to be substantial.

This type of cracking can occur, for example, in the thick retaining walls of the type illustrated in Fig. 4.64. The wall is normally cast on to the previously cast base and it is this sequence of construction which has an important bearing on this problem as the hardened base provides restraint to the horizontal thermal contraction of the wall. As a result of this restraint, wide vertical cracks can occur. It should be noted that the main reinforcing bars resisting the flexural tensile stresses of the cantilever wall run vertically and so do not contribute to the restriction of thermal cracking. As the amount of horizontal reinforcement passing through the thermal cracks plays an important part in their resultant width, it is clear that this reinforcement must not be stinted, *even though it is not required from bending moment considerations*.

1 cracks resulting from restraint to thermal contraction.

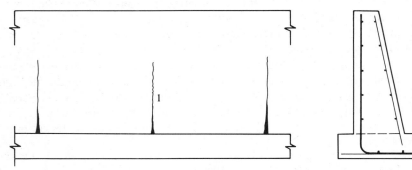

1

Figure 4.64   Elevation and section of a reinforced concrete retaining wall, illustrating the type of cracking that can occur as a result of high temperatures caused in thick members by the heat of hydration

The cracks initiated in this way in the early life of the structure gradually become wider with age because of the further phenomenon of shrinkage as a result of the concrete drying out. It is, therefore, important to avoid their formation. This can be achieved by reducing the initial rise in temperature of the concrete. There are a number of measures which can be taken to accomplish this, and these are discussed in the book relating to second year studies.

Deterioration of concrete

It is possible for concrete to deteriorate and gradually disintegrate under certain circumstances. There are several sources of deterioration, but a worrying one that has come to the fore in recent years is that known as alkali–aggregate reaction. It results from the chemical interaction between alkaline pore fluid originating from the Portland cement and certain particles of silica present in the aggregate. This reaction forms a gel which absorbs water and produces a volume expansion. The resulting internal disruptive bursting forces cause widespread random cracking throughout the concrete which becomes visible on the surface of the concrete (Fig. 4.65).

The problem of alkali–aggregate reaction was largely unknown in the British Isles prior to 1977, but since then several cases have been confirmed. Whilst this may have occurred because new sources of aggregate have contained the alkali reactive

Figure 4.65   An example of the deterioration of concrete as a result of alkali–aggregate reactions (*Source*: Building Research Association)

minerals, there is increasing concern that slight changes have occurred in the chemical compounds of modern cements. It is becoming increasingly important for structural engineers to have a greater understanding of the complex relationship between cement compounds and concrete durability, and also of the possible benefits to be gained by the inclusion of certain additions to the mix such as pozzolanas and microsilica.

Problem 4.2 The steelwork beam–column connection illustrated in Fig. 4.17b is assumed to transmit shear force only. For the case where the thickness of the flange of the column is 12 mm, the thickness of the end-plate of the beam is 10 mm, and the bolts are 19 mm diameter, calculate the maximum design shear force if the maximum permissible shear stress in the bolts is 160 N/mm^2 and the maximum permissible bearing stress on the bolts is 430 N/mm^2.

I – beam		Channel	
Depth	457 mm	Depth	229 mm
Thickness of flanges	15 mm	Thickness of web	7.6 mm
Thickness of web	9 mm	Centroid p	20 mm
Area	85.3 cm^2	Area	33.2 cm^2
I_{zz}	28 522 cm^4	I_{zz}	2610 cm^4
I_{yy}	829 cm^4	I_{yy}	159 cm^4

Figure 4.66 Problem 4.3

Problem 4.3 The crane gantry girder shown in Fig. 4.66 is made up from a universal beam and a channel, these having the sectional properties indicated. Calculate the elastic moduli of the girder about its horizontal axis. A horizontal force applied by the crane is considered to be resisted by the composite top flange. Calculate the relevant elastic modulus of this top flange.

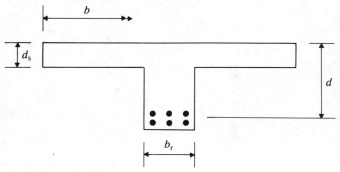

Figure 4.67 Problem 4.4

Problem 4.4 For beam and slab structures in reinforced concrete, a beam resisting sagging bending moments is considered to be of the form shown in Fig. 4.67. Assuming linear elastic conditions, obtain the quadratic equation for the depth of the neutral axis when the neutral axis falls below the slab.

5. Structural design and construction

Structural engineering is the science and art of designing and constructing economic and elegant structures to safely resist the forces to which they may be subjected.

Within this concise definition of structural engineering there are several important phrases, and these will be discussed in this chapter. The first phrase to be noted by the student is 'science and art'. Undoubtedly structural engineering involves science, since much of it is based on the principles of observation, theories, mathematical deduction and laboratory experimentation. Structural analysis has clearly developed from such foundations.

However, structural design, whilst it incorporates analysis, is also much concerned with art. In the first place, it involves making decisions on structural arrangements and materials, and it is this skill in the art of conceiving a structure which must be acquired largely on the basis of experience and practice. A first step towards this is the development of knowledge of some practical structural arrangements and of construction. This will introduce the student to an intuitive understanding of structural engineering which will subsequently allow him to develop it further by direct observation of actual structures.

Although some of the structures discussed in the following sections (for example, continuous beams and portal frames) are statically indeterminate (and so their analysis is not covered in detail in this book), their description here is desirable in order that the student can begin to have some understanding of their behaviour and their use in practice. Nonetheless, it is important to realize that not all the seemingly complex arrangements discussed are statically indeterminate. For purposes of design, many of the structures can be considered as assemblages of structural elements which function independently and without mutual interaction.

5.1 Structural arrangements for resisting vertical loading

Floors with one-way spanning beams

The floors in multi-storey buildings are invariably of reinforced or prestressed concrete, and the manner of their construction is discussed later. For certain special types of industrial structures, floors can be constructed using steel plate but these are

362

not suitable for buildings, partly because of the problem of sound transmission (for acceptable sound insulation, the floor needs to have a sufficient mass) and partly because of the problem of fire resistance. The following illustrations will therefore assume concrete floors, although beams and columns may be either concrete or steel.

A possible arrangement of slabs, beams and columns for transmitting the loads to the foundations is shown in Fig. 5.1. The floor slab (1) spans between the main beams (2) which in turn span between the columns (3). In the case of the steelwork, secondary beams (4) will be required to connect the columns in the longitudinal direction. For the one-way spanning floor slab illustrated here, these secondary beams are not required as supports for the slab but are essential during the constructional stage to provide a degree of stability to the bare framework (Fig. 5.2). Moreover, the secondary beams may have an important part to play in restraining the steelwork columns from bending about their weak axis under the action of the vertical load. In the case of an *in-situ* reinforced concrete structure, such longitudinal beams will not be required for purposes of stability during construction, but may be required along the edge of the slab (5) for supporting the external walls.

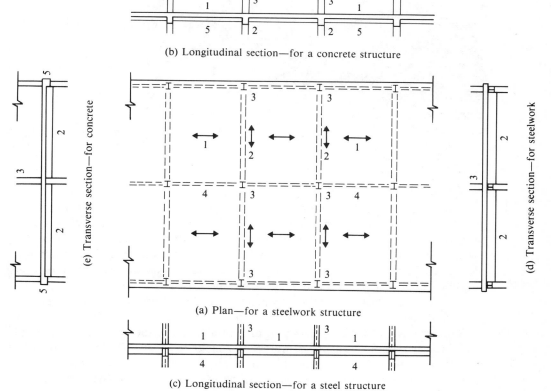

(b) Longitudinal section—for a concrete structure

(e) Transverse section—for concrete

(d) Transverse section—for steelwork

(a) Plan—for a steelwork structure

(c) Longitudinal section—for a steel structure

Figure 5.1 A structural arrangement for floors using longitudinally spanning concrete slabs with transversely spanning beams

Figure 5.2   A steel framework for a multi-storey structure (*Source*: British Steel Corporation)

It is pertinent to draw attention here to an essential difference between the steelwork and reinforced concrete frames (see Figs. 5.3a and 5.3b). For the structure based on the steelwork frame, the steel beams and the concrete slab are considered to behave quite separately. For example, the top fibres of the steel beam are assumed to be able to shorten under their bending compressive stresses without interference or restraint from the concrete slab. This is not the case with the concrete structure. Indeed, the concrete beam and the concrete slab are deliberately made integral by concreting them at the same time and, to further ensure their interaction, the reinforcement from the beam extends into the slab. As a result, the region of the slab adjacent to the beam acts as the compression zone of the beam, and the overall depth of the beam now includes the depth of the slab. The concrete slab in this region is therefore acting structurally in two separate ways—it is acting as part of the flexural component spanning to the beam and is also acting as part of the beam spanning to the columns.

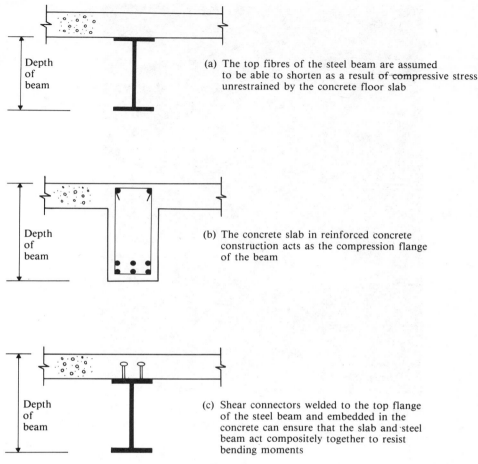

(a) The top fibres of the steel beam are assumed to be able to shorten as a result of compressive stress unrestrained by the concrete floor slab

(b) The concrete slab in reinforced concrete construction acts as the compression flange of the beam

(c) Shear connectors welded to the top flange of the steel beam and embedded in the concrete can ensure that the slab and steel beam act compositely together to resist bending moments

Figure 5.3   Composite action between the slab and its supporting beam is inherent in *in-situ* concrete but not in steelwork

This is advantageous compared with the steelwork arrangement as a smaller overall depth of construction of the floor is possible, and this has important consequences for the cost of the vertical elements of the building. Indeed, it is of historical interest to mention that it was this 'invention' of the T-beam which provided the greatest single economic advantage that reinforced concrete held over steel. It is, however, now possible to tie a steel beam and concrete slab together by welding shear connectors to the top face of the steel beam (Fig. 5.4) prior to the placing of the concrete. These shear connectors will ensure that as the top flange of the steel beam is stressed in compression under the influence of sagging bending moments, so also is the concrete in the region of the beam. In other words, the steel and concrete act together as a

Figure 5.4 Shear connectors welded to the top flange of steelwork ensure the composite action between the steel beams and the concrete floor (*Source*: Centre Belgo-Luxembourgeois d'Information de l'Acier)

composite beam with advantages comparable to the reinforced concrete beam (Fig. 5.3c).

In buildings where internal columns are undesirable, as for example in multi-storey car parks, a possible arrangement would be for the beams to span between the external columns. In order to accommodate the longer span, it will usually be necessary to reduce the spacing of the beams (Fig. 5.5) and therefore also of the columns. This not only reduces the area of the floor to be supported by a particular beam, but the reduction of the span of the slab allows a reduction in its thickness, with a consequent reduction in the intensity of the dead load. Even so, it will normally be necessary to use deeper beams than is possible for the case where internal columns are acceptable.

An interesting structural innovation involving floors with one-way spanning beams was used for a multi-storey hotel in Minneapolis, USA. It is illustrated in Fig. 5.6.

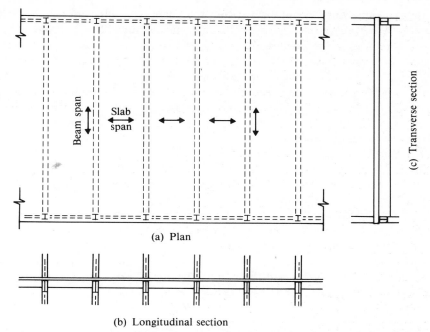

(a) Plan

(b) Longitudinal section

Figure 5.5   For the structural arrangement with no internal columns, the spacing of the frames will normally be reduced. Whilst this reduces the span of the slab, so permitting a thinner slab, deeper beams will be necessary

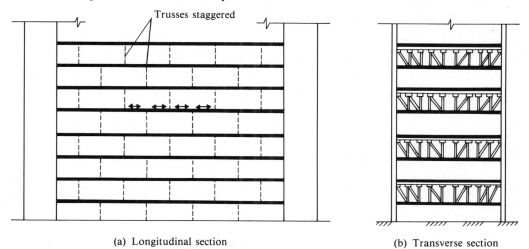

(a) Longitudinal section

(b) Transverse section

Figure 5.6   Illustrating the principles behind an interesting innovation used for the construction of the Radisson Hotel, Minneapolis, USA. The transverse-spanning storey height trusses are staggered, enabling the floor slab to span longitudinally from the top of one truss to the bottom of the neighbouring truss. The span of the slab is therefore one-half of the spacing of the trusses

Storey-height trusses contained within the partition walls between the hotel rooms span 16 m transversely between the external columns. The unusual aspect of the structure is that the trusses are staggered longitudinally, with the concrete floors spanning from the *top* of one truss to the *bottom* of the neighbouring truss on the next floor. Thus the span of the concrete floors is half the spacing of the trusses. As a result of this arrangement, the effective structural depth of the floors is extremely small, resulting in considerable economy in the cost of the vertical elements of the building.

Floors with two-way beam system

A drawback of the arrangement illustrated in Fig. 5.5 is that the facades of the building have to incorporate closely spaced columns. If it is desirable to adopt more widely spaced columns whilst retaining closely spaced beams, an arrangement using a two-way spanning beam system is often used (Fig. 5.7). As in the previous case, the floor slab spans between the transverse spanning beams, but whereas the beams 1 span directly to the columns, the beams 2 span to beams 3 running longitudinally between the columns. The beams 3 are designed to carry the reactions from beams 2 to the columns. It is of interest to note that whilst the beams 1 apply loads to the columns eccentrically, the loads from the pairs of beams 3 are symmetrically applied to the columns, this latter being less stringent.

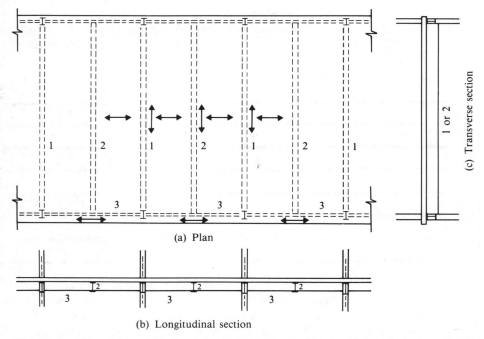

(a) Plan

(b) Longitudinal section

(c) Transverse section

Figure 5.7   When it is architecturally undesirable for the facade columns to be too close, longitudinal floor beams may be used to support alternate transverse beams and carry their load to the widely-spaced columns

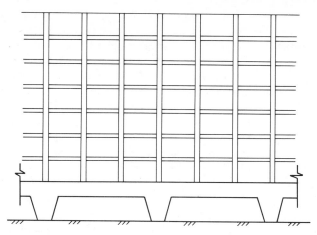

Figure 5.8   In order to retain closely-spaced columns in the facade but widely-spaced columns at the ground floor, deep intercepting beams can be used at first floor level

It may be that the objection to the closely spaced columns in Fig. 5.5 applied only to conditions at the ground floor level where, for example, more widely spaced columns may be required for purposes of access. In such circumstances, rather than the solution in Fig. 5.7, it may be preferable to retain the closely spaced facade columns for all the upper floors but with the majority of these terminating on very deep longitudinal intercepting beams at first floor level. These beams then carry the loads to the more widely spaced ground floor columns (Fig. 5.8).

Such longitudinal beams need not be at first floor level and Fig. 5.9 illustrates an

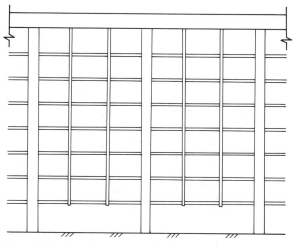

Figure 5.9   An alternative solution is to carry the floor loads upwards, using hangers supported by deep longitudinal beams at roof level

arrangement with the main longitudinal beam at the top of the building. In this case, the loads from the transverse beams at each floor are carried *upwards*, with the load-carrying members of the facades acting in tension. This has the advantage that the members can be very narrow (acting in tension, there is no tendency for buckling), but has the disadvantage that the loads from all the floors must be transmitted to the foundations by the main columns throughout the full height of the building.

Floors with a triple beam system

When the width of the building is large and yet it is still desirable to omit internal columns, the previous arrangements may not be suitable because of the long span of the beams supporting the floor. Rather than attempt to reduce the depth of such beams to a minimum, it may be preferable to adopt widely spaced, deep main beams with both secondary and tertiary beams carrying the floor loads to these beams. Such an arrangement is shown in Fig. 5.10. In such circumstances it would usually be preferable to adopt lattice girders for the main beams (1). These support the secondary beams (2) which in turn support the tertiary beams (3). The latter support the slab and are closely spaced to allow a shallow depth of concrete slab. Although in Fig. 5.10 the

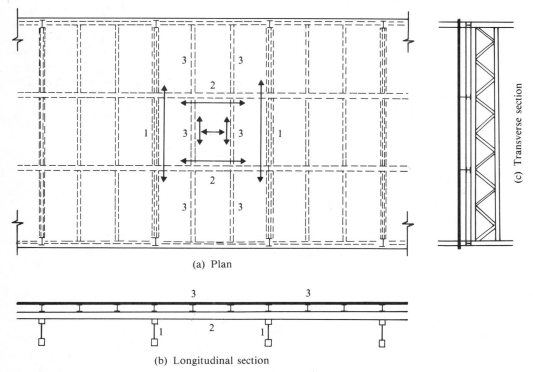

(a) Plan

(c) Transverse section

(b) Longitudinal section

Figure 5.10   For very wide buildings with internal columns omitted, it may be necessary to adopt a triple beam system

various beams are shown at different levels (for example, the secondary beams under the tertiary beams), this is largely for ease of illustration and it would be more usual to include all within the depth of the main girder. It would then be possible to omit those tertiary beams between a pair of columns by allowing the top chord of the lattice girder to support the slab directly.

Arrangements with cantilever beams

A greater freedom in the architectural treatment of the cladding of a building is obtained by avoiding columns in the outer walls. This can be achieved by extending the beams of the frames beyond the columns to support the floors in this region by cantilever action (Fig. 5.11). However, a feature of the cantilever beam is that the advantages of beam–slab interaction are lost. In reinforced concrete, for example, the advantageous T-beam action (illustrated in Fig. 5.3b) is lost, the compression forces resulting from the hogging bending moments in the cantilever having to be resisted by the relatively small area of concrete in the lower part of the web (Fig. 5.12a). In the case of steelwork, the frictional restraint between the upper steel flange and the concrete slab, advantageous in regions of sagging bending, plays no part in preventing lateral buckling of the bottom compression flange in regions of hogging bending (Fig. 5.12b).

When the architect's requirement for the type of structure previously illustrated in Fig. 5.8 is for the ground floor columns to be set back from the external walls, the arrangement would be as shown in Fig. 5.13. The columns above the first floor are retained at the position of the external walls and are supported at first floor level by the deep reinforced concrete slab cantilevering transversely beyond the columns. This arrangement, it will be observed, retains the advantages of beam–slab interaction for the floors above the first floor. An example is shown in Fig. 5.14.

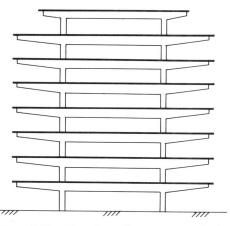

Figure 5.11 Use of cantilevers to transfer load to internal columns allows a greater architectural freedom in the treatment of the facade

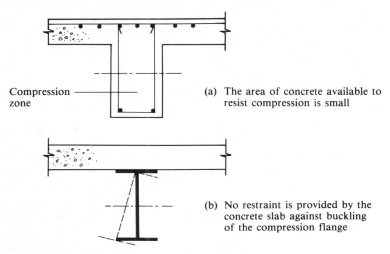

Compression zone   (a) The area of concrete available to resist compression is small

(b) No restraint is provided by the concrete slab against buckling of the compression flange

Figure 5.12   However, there are disadvantages with floor arrangements which require resistance to hogging bending moments

A corresponding variation for the type of structure previously illustrated in Fig. 5.9 is shown in Fig. 5.15. The loads from the various floor beams are now carried up through the hangers at the external walls to cantilever girders at roof level which transmit the loads to the supports within the building. This structural system is most appropriate to buildings approximately square in plan with the girders cantilevering in the two orthogonal directions from a reinforced concrete central core. An example of such a building is shown in Fig. 5.16.

Figure 5.13   An arrangement similar to Fig 5.8, but with ground floor columns set back, can be obtained by using a deep reinforced concrete slab at first floor level. This avoids the disadvantages of the arrangement shown in Fig. 5.11 (see also Fig. 5.14)

Figure 5.14   An example of the type of building illustrated in Fig. 5.13 (*Source*: British Cement Association)

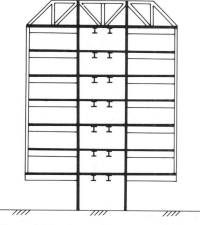

Figure 5.15   An arrangement corresponding to Fig. 5.9 in which the floor loads are carried upwards to deep cantilevers and thence to the concrete core (see also Fig. 5.16)

Figure 5.16   An example of the type of building illustrated in Fig. 5.15 (see also Fig. 5.71) (*Source*: Centre Belgo-Luxembourgeois d'Information de l'Acier)

5.2 Structural arrangements for resisting wind loading

The principal forces from wind on multi-storey buildings are in the horizontal direction and are therefore perpendicular to the forces of gravity. Nevertheless, forces due to wind flow can act in the vertical direction and the considerable uplift forces from aerodynamical effects (compare the aeroplane) have important design implications for roof and light-weight structures. However, this discussion will be related to the horizontal forces. These horizontal forces due to the wind pressure act on the facade of a building, but it is part of the function of the space-enclosing system to transfer these forces to the supporting frames.

Frames with rigid joints

It has been previously explained (Chapter 1) that one method of achieving stable rectangular frames is to incorporate some rigid beam–column connections in the frame (see Fig. 1.42). The term 'rigid' simply implies that the joint has sufficient

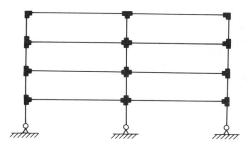

Figure 5.17   A rigidly-jointed framework is stable and is able to resist horizontal forces

strength and stiffness to prevent any relative rotation of the connected members at that position, and, for example, the angle between two orthogonal members remains at 90° at the junction under the action of the load. In theory, one such rigid joint in a particular frame would provide stability, but in practice additional rigid joints would be incorporated in order to reduce the moment capacity required from a particular joint.

Stable multi-storey buildings are formed from suitable superpositioning of rigidly jointed, multi-bay portal frames as shown in Fig. 5.17. However, this particular structural system for resisting horizontal loads in multi-storey buildings in steelwork has severe practical limitations. Firstly, the cost of beam–column moment connections is considerably higher than that for the simpler shear connections. Whilst the latter will have some resistance to the relative rotation of the connected members, their moment of resistance will be very small and for practical purposes can be considered as pin-joints. The equivalent multi-storey frame constructed using such beam–column connections is represented in Fig. 5.18, and this would be much cheaper than the one illustrated in Fig. 5.17. Needless to say, the frame shown in Fig. 5.18 is unstable and unable to resist wind loading, but it is nevertheless an acceptable structural form when protected by the structural arrangements discussed below.

Although these points do not preclude the use of rigidly jointed steelwork frames, indeed they are frequently used for single-storey buildings, it would be unusual to use

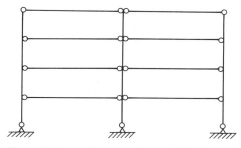

Figure 5.18   A framework in which the beam-column connections are designed to transmit shear only is equivalent to a pin-jointed frame and is therefore inherently unstable

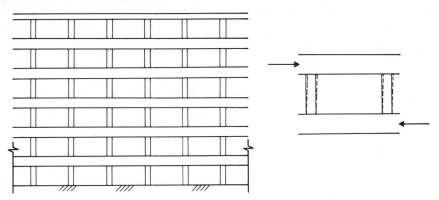

Figure 5.19   The combination of deep spandrel beams and the columns in the facade can act as rigidly-jointed frames to resist the wind forces in the longitudinal direction. Deformation is virtually limited to lengths of the columns between the beams

such frames for obtaining transverse stability in multi-storey buildings higher than three storeys.

However, one method of providing stability in the *longitudinal* direction of a building which comes within the category of 'frames with rigid joints' is the use of deep spandrel beams in the external walls. These deep beams (Fig. 5.19) are connected to the external columns to form a series of interconnected rigid-jointed frames which are able to resist the wind forces acting parallel to these walls. The large depth of the beams means that they have little deformation under the bending moments induced; moreover, their depth ensures that the length of the column over which bending can occur is considerably reduced. The resulting deformations in the direction of the wind are therefore very small.

Frames with diagonal bracing

Bracing in the vertical plane using triangulated trusses cantilevering from the foundations is frequently used as a means of resisting wind forces in steel-framed buildings. It is not used for concrete-framed buildings.

Although the triangulated pin-jointed frames shown in Fig. 5.20 cannot be adopted for the majority of frames of a building for obvious architectural reasons, it is usually possible to adopt such an arrangement in some of the frames, at the ends of a building for example. Moreover, it is possible for the remaining rectangular frames to be designed as pin-jointed (as in Fig. 5.18) provided that the braced frames are designed to withstand the wind forces for the whole of the building and that the wind loading exerted on the rectangular frames can be transferred to the braced frames. For multi-storey buildings, it is possible to utilize the immense strength and stiffness of the floor slabs in the horizontal plane to transfer the wind loading from the rectangular frames to the braced frames.

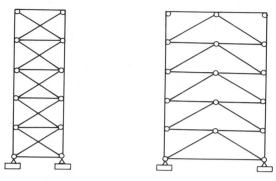

Figure 5.20   Triangulated frames (termed braced frames) are inherently stable but for obvious architectural reasons cannot be incorporated in all transverse frames

The principles are demonstrated in Fig. 5.21 which represents a possible arrangement for a particular storey of a multi-storey structure. The vertical loads are carried by the floor slab (1) spanning between adjacent frames. For the horizontal loads, the floor slab acts as a very deep beam spanning in its own plane between the two end braced frames (2) and transfers the wind loads acting on the internal rectangular pin-jointed frames (3) to these end frames. The stiffness of the floors in this horizontal plane is normally such that there will be virtually no horizontal displacement of the internal frames *relative* to the braced frames.

Possible arrangements for the braced frames of multi-storey buildings are shown in Fig. 5.22. As the open rectangular parts of these do not contribute to resisting the horizontal wind forces, it is only the triangulated parts of these frames (i.e., identical to those in Fig. 5.20) which would feature in the analysis of the effects of the wind load. It

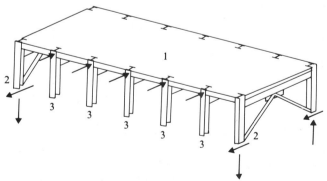

Figure 5.21   Illustrating the combination of braced frames (2) and unbraced frames (3) to produce a stable structure. The floor slab (1) of a particular storey of a multi-storey building is capable of transferring by in-plane diaphragm action the wind forces from the non-braced internal frames to the braced end frames

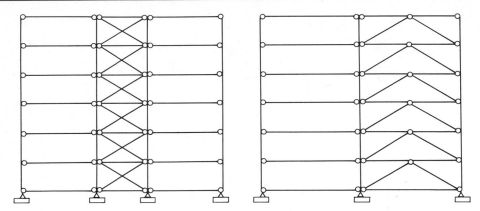

Figure 5.22 Illustration of two possible practical arrangements of transverse frames incorporating bracing. The open rectangular parts of the frames do not contribute to the resistance to horizontal forces. The parts of the frames which do resist horizontal forces are identical to those shown in Fig. 5.20

will be clear that the narrower the width of these triangulated parts, the higher the forces in the members caused by the wind loads and hence the higher the lateral deflection of the frame. Thus one adopts as large a width as circumstances allow. The K-type bracing is used when the adopted width of the frame would otherwise necessitate the diagonal members being too long and insufficiently inclined.

It must also be borne in mind that any tensile forces in the members caused by the wind will tend to counter the compressive forces due to the vertical loads, and indeed

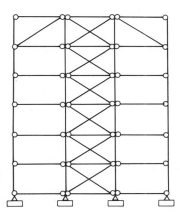

Figure 5.23 A narrow braced frame may allow the horizontal deflections under the wind loads to be excessive. Such frames can be stiffened by bringing the external columns into action by incorporating additional diagonal members across the full width of the building in the top storey

may overcome them completely to give overall tension. This has particular relevance in the design of foundations, as it is then necessary to increase the size of the concrete bases simply to counter the overturning effect. In design calculations it is always necessary to consider the combination of loading which has the worst effect—in this case the combination of *maximum* wind loading and *minimum* vertical loading would be the most adverse.

To reduce the lateral deflections caused by the wind, it is possible for the braced frames to be stiffened by bringing the adjacent external columns into play in resisting the forces. This can be accomplished by incorporating extra diagonal members across the full width at the top of the building (Fig. 5.23), and additionally at other levels if necessary. The tendency for this top storey to rotate as the braced vertical cantilever deflects under the wind load will now cause forces in the outer columns, and this effect reduces the overall lateral deflection of the frames.

Needless to say, reversal of the direction of the wind must be allowed for in design. In the triangulated frames in Fig. 5.20, the inclined members which are in tension for one direction of the wind will be in compression for the reversal of the wind. The compressive forces are the more critical because of the buckling problem and these determine the size of the members. As tension members can be much more slender, it is sometimes more economical to provide duplicate diagonal members which are assumed to be effective only when acting in tension, their contribution when acting in compression being ignored—a system known as counter-bracing. Thus the pairs of diagonals in Fig. 5.24a could be slender tie bars. Their behaviour under the wind loads is, for purposes of design, assumed to be that illustrated in Figs 5.24b and 5.24c.

However, the use of such crossed slender tension members has limited application in tall multi-storey structures because of the need to restrict lateral deflections of the braced frames—slender tension members will undergo greater change in length than the corresponding heavier compression members. In tall multi-storey structures it is

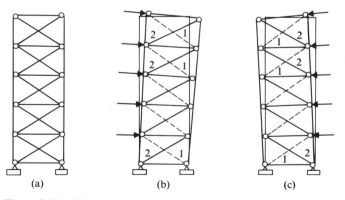

Figure 5.24 Illustrating the principle of counter-bracing. The diagonal members in compression (1) are ignored in the analysis, the frame relying on the diagonal members stressed in tension (2). Very slender diagonal members can therefore be adopted and may be unconnected at their intersection.

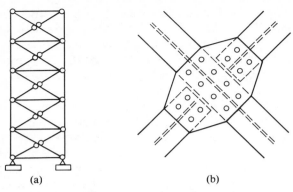

(a)   (b)

Figure 5.25   When heavier diagonal crossed members are required, it becomes necessary to connect them at their intersection—one member generally continues uninterrupted and the other is spliced (see b). Their behaviour in both tension and compression is now allowed for. Note that the interconnection ensures that the tension member of a pair of diagonals provides restraint to the compression member against buckling only in the plane of the frame

therefore necessary to accept diagonal members in compression. As a result, it is sometimes necessary to provide some restraint to long inclined members in order to develop their compressive strength. The use of connected crossed diagonals (Fig. 5.25), whereby the one acting in tension restrains the diagonal in compression against in-plane buckling, is one way of achieving this. On the other hand, special bracing members may be necessary as for the K-type bracing in Fig. 5.26. The additional inclined members are there simply to provide restraint to the main members against in-plane buckling when stressed in compression—the effective length of the inclined compression member (an important parameter relevant to the strength and lateral stiffness of members in compression) has been reduced by the additional member for

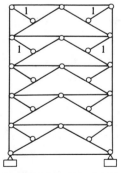

Figure 5.26   Special bracing members (1) may be necessary where long compression members occur. These reduce the effective length of the compression members for buckling in the plane of the frame which then determines the juxtaposition of their major and minor axes (see Fig. 5.27)

1—Universal column section   2—Channel section

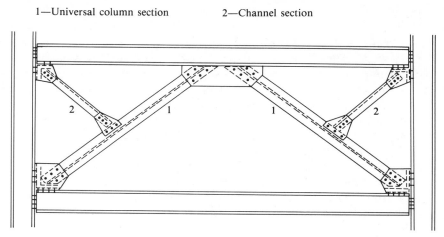

Figure 5.27   Typical steelwork details of bracing. Note the orientation of the major and minor axes of the main bracing members

the case of in-plane bending of the compression member. Note that the effective length of the compression member has not been altered for out-of-plane bending, and this therefore determines the juxtaposition of the major and minor axes, the member being positioned so that the out-of-plane bending would occur about the major axis. (By considering equilibrium at the junction of the inclined members, it will be seen that the additional members play no part in resisting forces due to normal pin-jointed frame action.) Typical steelwork detailing of such an arrangement is illustrated in Fig. 5.27.

When the braced frame needs to be positioned within the building, it is not always possible to adopt straight inclined members. Figure 5.28 shows a typical kinked arrangement which permits door openings. The additional inclined members are now an integral part of the triangulated frame.

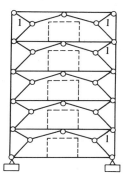

Figure 5.28   A kinked arrangement of bracing may be necessary where door openings are required. Members 1 are now essential in order to effect the change of direction of forces in the adjoining inclined members

Shear walls

In the same way that the concrete floor slab acts as a very strong and stiff beam in its own plane, so the wall is similarly effective in resisting forces in its own plane (Fig. 1.22). As a result, the shear wall, particularly that constructed in reinforced concrete, has an important place in providing stability to structures. Figure 5.29 shows a typical arrangement for a multi-storey structure with reinforced concrete shear walls at the ends of the building. These now act in the same way as the braced frames previously discussed, with the floor slabs transferring the horizontal wind forces to the end walls.

Brickwork and masonry walls, whilst of great importance in low-rise structures, cannot be used to the same extent in high multi-storey structures because of their low tensile strength. Figure 5.30 shows the type of diagonal cracking that occurs when the horizontal forces due to wind are sufficient to overcome the precompression in the wall caused by the vertical loading.

Nevertheless, brickwork and masonry walls can play an important part in

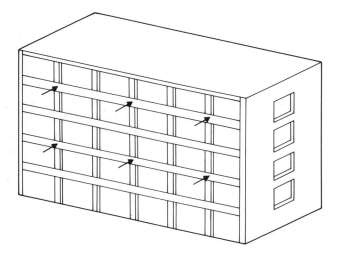

Figure 5.29 Reinforced concrete end walls are an important method of providing overall stability to a building. The floor slabs transfer the wind forces to the end walls which eliminate the need for the type of bracing in Fig. 5.27

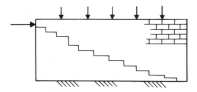

Figure 5.30 Brickwork has a low tensile strength and high transverse forces tend to cause failure by diagonal cracking

stabilizing multi-storey structures if they are used in conjunction with surrounding frames. A typical in-filled frame is shown in Fig. 5.31. The containment of the brickwork by the frame now prevents the diagonal cracking type of failure, and the brickwork in turn strengthens and stiffens the frame by acting as an inclined strut (Fig. 5.32). The behaviour is in effect comparable to the braced frame discussed earlier, and the analysis would be based on similar triangulated pin-jointed frames (Fig. 5.33). Simple rules established from tests enable the strength and stiffness of the diagonal brickwork strut to be estimated.

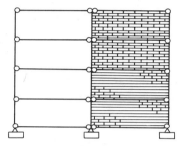

Figure 5.31   Brickwork infilling of rectangular pin-jointed frames is an important form of bracing. The containment of the brickwork by the frame prevents the diagonal cracking type of failure

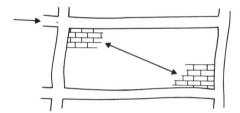

Figure 5.32   The infilled brickwork acts as an inclined strut in resisting the sway deformation of the frame

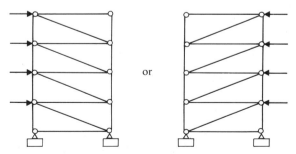

Figure 5.33   Analysis of the infilled frame in Fig. 5.31 would be based on one of these triangulated frameworks, depending on the direction of the wind

Core structures

There are two requirements in the design of multi-storey buildings which can influence the structural system adopted for resisting the wind loading:

– the lifts require a rigidly-constructed surrounding tube;
– the staircases must be provided with fire resisting compartment walls.

These needs can be conveniently met by enclosing both the lifts and stairs within a reinforced concrete core (Fig. 5.34). This core can then also serve as the structure resisting the horizontal forces by acting as a stiff hollow-box cantilevering from the foundations. The floors are again important in transferring the wind forces from the façade to the core. In these structures, the columns and beams (in either steel or concrete) outside the core are required to support only the vertical loads.

Tube system

In order to understand the structural system used to resist the wind loads in very tall buildings, it is necessary to have an understanding of the limitations and disadvantages of the systems previously described.

In the braced frame, for example, it is only those columns that are part of the braced frame which are stressed in tension or compression as a result of resisting the wind forces. For the system illustrated in Fig. 5.23 this also involves those external columns in the same plane as the bracing, but the remainder of the columns of the building, although undergoing the same lateral deflection as the braced frames (as a result of the floor slabs moving with the braced frames), play no part in resisting the wind forces. The floor slab is not sufficiently stiff in the *vertical direction* to ensure that these columns undergo an extension or shortening comparable to those of the braced frames. As a result the braced frame system, whilst of great importance in the design of

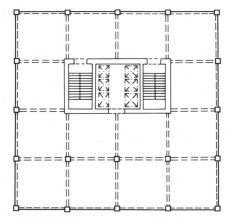

Figure 5.34 Typical floor plan for a tall multi-storey building using the core system of resisting the horizontal wind forces. The reinforced concrete walls providing the requisite fire protection to the stairs and lifts form a stiff hollow box cantilevering from the foundations

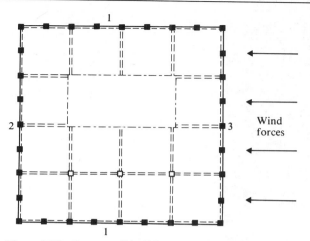

Figure 5.35   In very tall buildings it is essential to utilize the external columns to resist the horizontal wind forces. The problem is how to develop the forces in the columns and make the building behave as a tube

multi-storey buildings, cannot be used for very tall buildings. Similar drawbacks apply to the shear wall and core systems.

For very tall buildings it is necessary to ensure that all the external columns play a part in resisting the wind loads. For example, in Fig. 5.35 not only must the braced outer walls parallel to the direction of the wind provide resistance to the wind but they must be assisted by the outer walls perpendicular to the wind. Thus, as the corner columns at the junctions of walls 1 and 2 are stressed in compression as a result of walls 1 resisting the wind forces, so all the columns along the wall 2 must also become stressed in compression. Similarly, the columns along the wall 3 would become stressed in tension. In effect, the walls 2 and 3 have become the outer chords of the bracing walls 1, or alternatively the building is behaving as a *tube*.

There are several means of establishing such behaviour. That used for the 100-storey John Hancock Center in Chicago (Fig. 5.36), clearly depicted as the diagonal bracing on the four faces of the building, is positioned externally and forms a distinctive, but not necessarily attractive, architectural feature. It is easy to appreciate that as the corner columns become stressed under the effect of the wind forces, so the diagonal bracing in the walls corresponding to 2 and 3 of Fig. 5.35 will ensure that the columns within these walls become similarly stressed.

The method adopted for the 110-storey World Trade Center buildings in New York (Fig. 5.37) incorporates a rather different approach. The four external walls are formed using closely spaced columns connected by very deep spandrel beams. The walls corresponding to 1 in Fig. 5.35 therefore resist the wind loads by the type of rigidly-jointed frame action described earlier, but now with the assistance of walls 2 and 3 as their columns are forced by their deep girders to conform to the corner columns.

Figure 5.36   The John Hancock Center, Chicago
(*Source*: British Constructional Steelwork Association)

It is, of course, possible to combine structural systems to resist wind forces. The combination of the tube and the core to give a tube-in-tube system is one example. Moreover, different systems can be adopted for the two orthogonal directions—for example, the shear wall system may be adopted for lateral wind loading in conjunction with rigidly-jointed frames for the longitudinal direction.

5.3 Aspects of some specialist structures

Single-storey buildings

Steel is the predominant material used for single-storey buildings, principally because in the creation of wide-span structures the resulting dead load is considerably less

Figure 5.37   World Trade Center, New York
(*Source*: British Constructional Steelwork Association)

than with concrete structures. Moreover, the fire-resistance requirements for single-storey buildings are less onerous and fire protection of the steelwork is often not required.

The majority of single-storey buildings for industrial purposes are constructed using the portal frame system. A typical arrangement for medium spans is illustrated in Fig. 5.38 and incorporates solid-web I-shaped members. For roofs of greater span, lattice girders (Fig. 5.39) will normally prove more economical—the greater depth of member required now favours the lattice girder with its more efficient use of steel despite the greater fabricating costs.

The use of rigid joints between the main beams or lattice girders and their supporting columns is adopted in order to obtain stable frames which can resist the wind forces acting parallel to the plane of the frames. However, just as with multi-storey structures, such rigid joints are not essential for stability and indeed in some circumstances, for example, where differential settlement of the foundations may

Figure 5.38   Steel portal frames using solid-web rafters (*Source*: British Steel Corporation)

Figure 5.39   Steel portal frames incorporating lattice girders (*Source*: British Steel Corporation)

Figure 5.40   A case where the roof trusses are simply supported at their ends. In this structure the wind loads are resisted by the near-side columns acting as cantilevers fixed at the base (*Source*: British Constructional Steelwork Association)

occur, rigid joints are undesirable. An example of roof trusses simply supported at their ends is shown in Fig. 5.40. Needless to say, alternative paths for transferring the wind forces to the foundations are now required, although this is not so easily achieved as with the multi-storey structure. In the latter, the concrete floors have ample strength and stiffness to transfer wind forces to the braced end-frames, but the thin light-weight profiled steel sheeting frequently used for the roofs of single-storey buildings is much less efficient in this regard. Whilst undoubtedly the roof cladding is being used increasingly for this purpose, the use of diagonal bracing members in the plane of the roof is still frequently an important feature of such design.

A typical arrangement of the bracing for resisting the forces caused by the wind acting transversely to the building is shown in Fig. 5.41. The forces carried to the roof level by the vertically spanning columns are transferred by the diagonal bracing to the end gables. These end gables incorporate bracing in the vertical plane which, in turn, transfers the wind forces to the foundations. In the design of roof bracing one would normally adopt the counter-bracing principle previously illustrated in Fig. 5.24 in which diagonal members tending to be stressed in compression are ignored in the analysis. Thus for a uniform distribution of forces along the length of the building, as

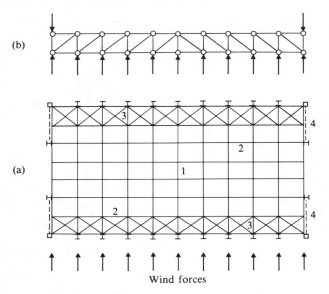

Figure 5.41   Plan of the roof of a single-storey building showing the type of bracing sometimes used to transfer the forces resulting from transverse wind to the vertically-braced gable frames. (a) General arrangement: 1—main girders, 2—purlins, 3—bracing, 4—vertical bracing in the gables. (b) Equivalent truss assumed in the analysis

illustrated in Fig. 5.41a, the truss assumed in the analysis would be the Pratt truss in Fig. 5.41b in which the diagonals act in tension whilst the 'verticals' act in compression. It will be seen that two lines of purlins form the two chords of the truss, whilst parts of the upper chords of the main girders form the 'verticals' of the truss. The forces caused in these members by this additional action must be taken into account in their design. It will be noted that the structural system is statically determinate and so the forces in the members are easily calculated.

A similar arrangement can be used to transfer horizontal forces at roof level as a result of wind acting on the gable ends, forces which tend to make the frames collapse as a pack of cards. Figure 5.42 shows bracing in the roof across the width of the building in the end bays. The forces are thereby transferred to bracing in the vertical plane between pairs of columns in the longitudinal walls. Such vertical bracing would normally be positioned in the same bay as the roof bracing, but that is not essential as long as care is taken to ensure that the forces can be transferred along the building (by the eaves beams, for example) to the position of the vertical bracing.

Two functions of the purlins have now been described—that of supporting the roof sheeting and carrying the vertical loads to the main girders and that of acting, where required, with the diagonal bracing to resist the wind loading. An additional important function is that of bracing (i.e., providing lateral restraint to) the compression side of the main girders. The tendency of members in compression to

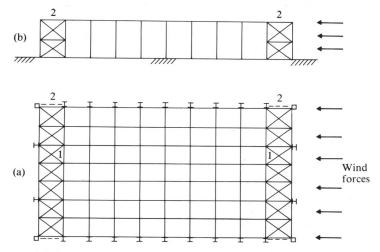

Figure 5.42 Arrangement for transferring wind load acting on gable ends to vertically-braced frames in the longitudinal walls. (a) Plan: 1—roof bracing. (b) Elevation: 2—vertical bracing

buckle sideways must always be kept in mind and suitable provision made. In this case, by connecting the purlins to the main girders and to the transverse roof bracing, sideways buckling of the upper chord of the main girders is prevented. It follows that all purlins are required to act in part as struts.

It is desirable to draw attention to the freedom a designer has in the positioning of transverse and vertical bracing. Needless to say, the architect's requirements regarding doors and glazing must be met, but within such limitations there is often considerable choice. Nevertheless, the various choices will have certain structural implications and it is necessary for these to be understood. For example, the two sets of transverse and vertical bracing as in Fig. 5.42 will tend to resist the effects of temperature change on that part of the building between the sets of bracing. The greater the distance between the vertical bracing, the greater the required deformation of the bracing and therefore the greater the forces induced. The end bays are therefore the worst position for this bracing from this particular standpoint.

In theory, this particular problem could be overcome by adopting a single bay of bracing as, for example, in Fig. 5.43. However, that solution is generally too extreme as it requires too much movement at the gable ends and the transmission of compressive forces from the gables along a considerable length of the purlins. An intermediate solution as in Fig. 5.44 is to be preferred.

With regard to the problem of roof bracing, it is of interest to note the advantages of a space truss. Although the trusses in Fig. 5.45 are one way spanning only, they are three-dimensional in form. Because of the inclination of the girders forming the V-shaped cross-sections, they are able to resist both vertical and horizontal forces and as a result additional bracing in the plane of the roof is not always required.

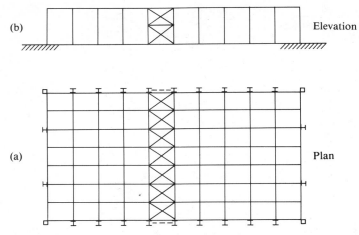

Figure 5.43   Alternative positioning of bracing. This avoids the problem of temperature stresses but requires the purlins to act as struts along the length of the building to carry the wind forces from the gables to the braced bay

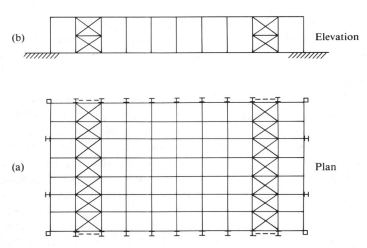

Figure 5.44   A further alternative positioning of bracing intermediate between the previous cases

Bridges

Unlike buildings where the requirements of space, heating, light, ventilation and drainage place the architect and building technologist at the forefront of decision making, in the case of bridge design the structural engineer is dominant. The structural form is often (but not always) apparent, so that bridges form a rich source of information for the student.

Figure 5.45 View from beneath 3-dimensional, but 1-way spanning, roof trusses. Unusually, these particular trusses are in precast prestressed (post-tensioned) concrete (*Source*: British Cement Association)

The type of 'through' bridge shown in Fig. 5.46 is often used to carry railways. Figure 5.47 is used to illustrate diagrammatically the different structural functions, and it is important to note that many of the members have more than one function.

A pair of Warren trusses in the vertical plane (Fig. 5.47a) form the major load-carrying structural system for the vertical loads. Loads on the deck of the bridge are carried first through stringer beams to the main transverse beams and thence to the trusses at the position of a joint.

Lateral forces on these trusses from the wind need to be transferred to the abutments via a truss in the horizontal plane joining the bottom chords of the main trusses (Fig. 5.47b) and a truss joining the top chords (Fig. 5.47c). It should be noted that the top and bottom chords of the main trusses act also as the front and back chords of the bracing trusses, and therefore the forces in these members calculated from the two separate actions need to be superimposed. Similarly, the main transverse beams supporting the deck can also be used as part of the lower bracing truss and the stresses from these two actions need to be superimposed.

Figure 5.46 'Through' bridge using Warren girders being lifted into position by floating cranes (*Source*: Centre Belgo-Luxembourgeois d'Information de l'Acier)

There are several points to note regarding the upper bracing truss. Firstly, it is joining the compression chords of the two main trusses so that its additional important function is to prevent overall buckling of these compression chords. Secondly, when the upper chord is curved, the bracing truss is not planar and the inclined members of the main vertical trusses must, therefore, play a part in altering the line of action of the forces of the bracing. A third point to note is that in 'through' bridges it is necessary to terminate the upper bracing truss at the end joint of the upper chord. From here the accumulated transverse forces are usually transferred to the abutments by utilizing a form of portal frame bracing (Fig. 5.47d).

In most modern highway bridges, the large width of the roadway precludes the use of 'through' bridges and it is usually necessary to position the supporting structural system below road deck level. The use of a concrete slab as the road deck then avoids the need for a separate upper bracing system as the slab itself is able to act in this way. The extensive nature of the subject of highway bridge design makes it necessary to restrict attention here to only a small part of the subject. Some aspects of bridges incorporating members subjected to bending have been chosen.

That simplicity of form does not detract from attractiveness is illustrated by the simply supported beam bridge in Fig. 5.48. In such bridges, the bending moments are

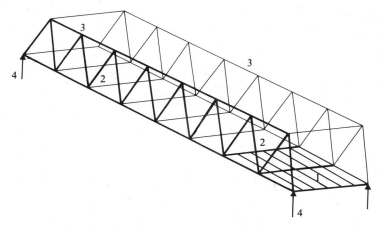

(a) Loads from the bridge deck are transferred by the stringer beams (1) to the transverse main deck beams (2) which transfer the loads to the Warren trusses (3) which transfer the loads to the abutments (4)

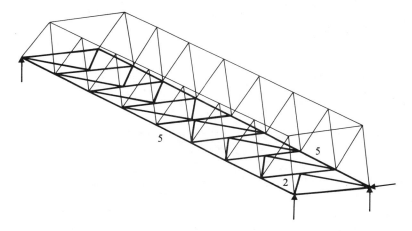

(b) Lower horizontal bracing truss (incorporating the bottom chords (5) of the Warren trusses and the transverse deck beams (2)) transfers wind loads to the abutments

Figure 5.47 Illustrating the 'structure' of a 'through' bridge

continued

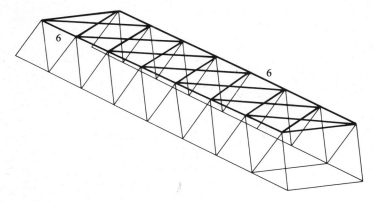

(c) Upper horizontal bracing truss (incorporating the top chords (6) of
the Warren trusses) transfers the wind forces to the end portal
frames and braces the compression chords of the Warren trusses

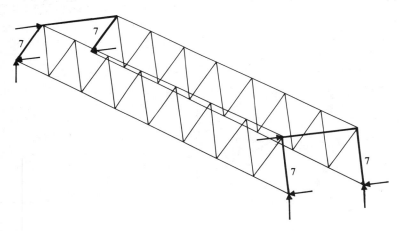

(d) End portal frames (incorporating the end posts (7) of the Warren
trusses) transfer the forces from the upper bracing truss
to the foundations

Figure 5.47   (*continued*)

Figure 5.48   A single-span simply supported bridge. The depth of the beams increases towards midspan where the (sagging) bending moments are greatest (*Source*: British Cement Association)

clearly a maximum at mid-span and in this case the depth of the beams has been increased accordingly. However, such solutions have clear limitations on the span of bridges, as a consideration of the bending moments in simply supported beams clearly demonstrates—for example, for the same intensity of uniformly distributed loading, as the span doubles so the bending moment increases four times; moreover, the dead load of the structure will need to increase as the span increases. It therefore becomes necessary as the spans increase to modify the sagging bending moments by arranging for hogging moments to occur at the ends of the beams (Fig. 5.49). Such hogging

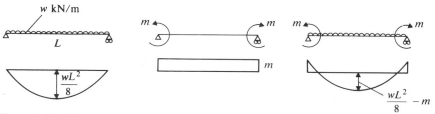

Figure 5.49   Illustrating the reduction of sagging moments in beams as a result of hogging moments at the supports

Figure 5.50   A portal frame bridge (see also Fig. 5.51). The frame action results in high hogging moments at the ends of the beams and low sagging bending moments at mid-span, the logical outcome being the attractive arch-shaped soffit (*Source*: British Cement Association)

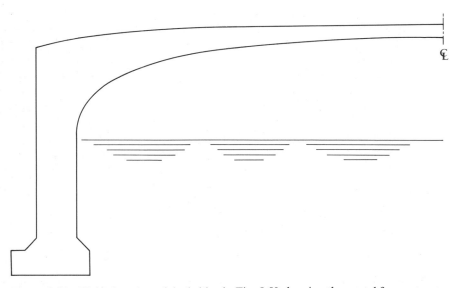

Figure 5.51   Half-elevation of the bridge in Fig. 5.50 showing the portal frame structure

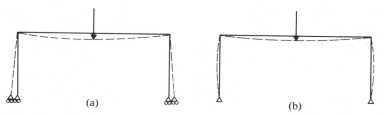

Figure 5.52 Explanation of the behaviour of portal frames in reducing sagging bending moments: (a) if the frame were supported on rollers, the legs would splay out as the beam bends under the effect of the load, the beam being in effect simply supported; (b) with the feet pinned, horizontal forces must develop which cause hogging moments in the beam adjacent to the columns

moments are the result of some form of continuity and there are several methods of achieving this.

A portal frame bridge is shown in Fig. 5.50. Although the legs of the portal frame are below the surface of the water (Fig. 5.51), the curved soffit line with increasing depth to the supports makes it clear that a form of continuity has been adopted. The hogging bending moments at the ends of the beams comes from the horizontal reactions developed at the base of the legs. Figure 5.52 explains how the horizontal reactions occur in portal frames as a result of the bending of the beam under load.

When multiple spans are adopted, hogging moments at the supports are obtained by making the beams continuous (Fig. 5.53). The maximum bending moments at the

Figure 5.53 Waterloo bridge, London. This is a multi-span continuous-beam bridge with variable depth beams (see Fig. 5.54). Again, the high hogging bending moments at the position of the supports require the beams to be deeper there, so leading to the arch-shape (*Source*: British Cement Association)

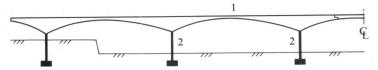

Figure 5.54   Diagram of the Waterloo bridge. Although of arch-type appearance, the super-structure (1) is supported on slender walls (2) (actually hidden from view by the protecting pier shell) which allow longitudinal movement

supports are normally appreciably higher than the maximum sagging bending moments at mid-span, and by proportioning the beams accordingly (somewhat easier in concrete) the pleasing *arch-shaped* beams result. It is important to realize, though, that these are simply continuous beams of variable depth and not a series of structural arches. Arching action implies horizontal forces applied from the supports to the superstructure and there are no such forces in this case. Indeed, the structural engineer has taken care to eliminate the possibility of such forces by supporting the beams on very slender walls (Fig. 5.54), these being hidden from view within the more massive

Figure 5.55   The increased depth often required where beams are continuous over supports has been cleverly used in this bridge to accommodate the slope of the road (*Source*: British Cement Association)

piers. This arrangement allows longitudinal movement of the superstructure and so avoids the development of significant horizontal forces which would cause overturning moments on the piers with undesirable stresses in the foundations.

The greater depth of beam desirable at a support where the beam is continuous, compared with the depth at the support at the end of a beam, has been cleverly used in the bridge shown in Fig. 5.55 to accommodate the inclination of the road in an attractive manner.

5.4 Construction

Construction of concrete structures

At the stage of mixing, concrete behaves essentially as a fluid and must, therefore, be placed within confining formwork (Fig. 5.56) whilst it gradually hardens into the required structural shape. There are essentially two quite distinct methods of construction:

– placing the concrete directly into its final position (Fig. 5.57) in the structure (known as *in-situ* concrete construction);

Figure 5.56 Placing and compaction of the concrete within formwork. In this case the concrete has been transported from the mixer by crane (*Source*: British Cement Association)

Figure 5.57   Concrete can often be transported from the mixer more conveniently by pumping it through pipes, the increased fluidity required for this being obtained by incorporating a Pozzolan cement (*Source*: Pozzolanic Lytag Limited)

Figure 5.58   Formwork for concrete will often require supporting during construction and in major bridges the temporary structures required can be considerable. This shows a concrete bridge under construction but steel girders support the formwork until the concrete has hardened, whilst additional steel girders facilitate the transfer of the formwork from one span to the next (*Source*: British Cement Association)

– placing the concrete into moulds (usually under factory conditions) from which the concrete member is subsequently removed, transported to site and erected in position (known as *precast* concrete construction).

Both systems have their advantages and disadvantages, and it is desirable that students begin to make their own observations on this.

For placing concrete *in-situ*, not only is temporary formwork required but also supporting structures, termed falsework. The amount of such temporary work required can be considerable as shown in Fig. 5.58. In some cases (Fig. 5.59) the design of the supports to the formwork must involve as much structural engineering as the design of the final structure. In multi-storey structures it will be appreciated that *in-situ* concrete floor construction involves an awkward transfer of forms and falsework from one floor to the next.

The use of precast concrete members avoids some of these problems. Large span bridges can be constructed using precast segments of beams (Fig. 5.60a) which are subsequently prestressed together. This may require falsework to support the

Figure 5.59 An attractive concrete bridge in Switzerland built in 1929 (but one which poses a difficult constructional problem for *in-situ* concrete) (*Source*: British Cement Association)

(a)

(b)

Figure 5.61   A precast concrete segmental bridge under construction using the cantilever method (*Source*: British Cement Association)

segments (Fig. 5.60b) whilst *in-situ* concrete joints between the segments are made prior to the prestressing. Alternatively, over difficult terrain, the precast units may be added one by one to give a cantilever form of construction which avoids the need for most of the temporary supports (Fig. 5.61).

Precast concrete is frequently used for multi-storey structures (Figs 5.62 and 5.63). It should, however, be understood that the structural system adopted when using precast concrete members is usually quite different from that which would be adopted when using *in-situ* concrete. Whilst with the latter the monolithic nature of construction ensures flexurally rigid beam–column connections, such connections are much more difficult to effect in precast concrete and speedier erection is obtained if

◄ **Figure 5.60**   Two views of a precast concrete segmental bridge under construction using false-work as temporary supports until the segments have been prestressed together. (a) The precast segments are temporarily stacked on site after delivery from the factory and prior to position-ing on the temporary supports. (b) Units positioned on the temporary supports. Note the posi-tions of the ducts for the prestressing tendons. As the bridge is continuous over several spans, the prestressing tendons will need to be near the top of the units adjacent to a pier but near the bottom in units positioned near mid-span (*Source*: British Cement Association)

Figure 5.62 A precast concrete multi-storey building under construction. In this case, 5-storey columns are initially positioned on the foundations, to be quickly followed by precast beams, floor slabs, shear walls, staircases and cladding (*Source*: Trent Concrete Structures Limited)

simple hinged joints are retained. This can set problems in resisting horizontal forces, particularly as the precast floors do not always give the continuity and stiffness necessary for transferring the wind forces to the braced parts of the structure. The making of reliable joints capable of resisting moments is a difficulty which has not always been satisfactorily overcome (Fig. 5.64). Another disadvantage of precast concrete which affects costs is the considerable weight of the units as larger tower cranes, demanding a greater financial capital outlay, are then required.

The bending and fixing of the reinforcing bars (sometimes referred to as *rebars*) prior to the placing of the concrete in the formwork is an important part of concrete construction (Fig. 5.65). The steelfixer secures the bars in the position required by the designer and in accordance with his drawings. Nevertheless, a certain amount of tolerance applies to the accuracy to which one can expect bars to be positioned. For example, a displacement of 5 mm out of position will normally be acceptable, and for large diameter bars the tolerance could be greater than this as long as the concrete cover is not adversely affected. Another factor is that bent bars cannot have sharp right-angled bends even if drawn that way (Fig. 5.66). This can affect the detailing of a

Figure 5.63   View from below a floor in a reinforced concrete building in which precast concrete units of T or double-T shaped cross-section were used for the construction of the floor (*Source*: Dow-Mac Concrete Limited)

congested area. In this connection it should also be noted that the deformations on a bar increase the overall size of the bar by at least 10 per cent of its *nominal* diameter (i.e., the diameter determining the effective cross-sectional area of the bar).

In detailing the reinforcement, the engineer must always have an understanding of how the bars will be assembled. At a beam–column junction, for example, whilst it would be usual for the lower column reinforcement cage to pass through to lap with the upper column cage above the junction, the reinforcement cages for the beams would stop just short (Fig. 5.67). This will allow the beam cages to be prefabricated and lowered into position in the formwork. It is, of course, essential that reinforcing bars pass from the beams into the column, but this can be effected more easily by subsequently adding additional 'loose bars' to the cages.

It will be recalled that the chemical reaction between cement and water proceeds over a long period of time and the strength of the concrete increases correspondingly. However, this only occurs as long as moist conditions prevail within the concrete. The most rapid gain in strength occurs during the first few days after casting, and during

Figure 5.64   A precast concrete multi-storey building which became unstable during construction and collapsed (*Source*: Building Research Establishment)

this critical period it is essential that the moist conditions within the concrete are maintained. If the concrete is allowed to lose its water by evaporation, the hydration of the cement cannot proceed. The preservation of the moist conditions within the concrete during this early period is known as *curing*—and is an important step in the construction of concrete structures.

Different methods of curing concrete are available. The surface of the concrete can be kept moist by periodic spraying, often using hessian or similar absorbent material spread over the concrete to retain this water. Polythene sheets are often additionally used to reduce the loss of this water by evaporation. An alternative method used on large flat surfaces is to seal the surface of the concrete by means of a membrane applied by spraying.

Construction in steelwork

Steel-framed buildings have the great advantage that the erection of the steel components can be carried out very rapidly, often by lifting from the delivery vehicle

Figure 5.65 Reinforcement fixed in position within formwork prior to placing the concrete (*Source*: R. Ward, Tarmac Construction Limited)

directly into position. However, in order to maintain this advantage it is essential that the floors, inevitably in concrete, are also constructed efficiently and quickly. This requirement therefore precludes the construction of floors using *in-situ* concrete in conjunction with temporary formwork and falsework. However, *in-situ* concrete can be used in conjunction with profiled steel sheeting which acts as both permanent formwork and reinforcement for the concrete (Fig. 5.68). Precast concrete floor units are also a means of maintaining fast construction (Fig. 5.69). However, it is then much more difficult to effect composite T-beam action between steel beam and floor, and such action is usually disregarded.

Figure 5.66 Sharp right-angled bends in reinforcement are not possible. This must be borne in mind when detailing

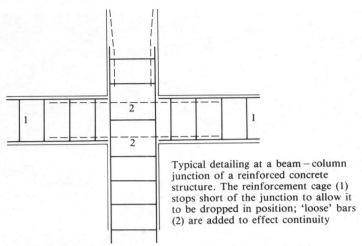

Typical detailing at a beam – column junction of a reinforced concrete structure. The reinforcement cage (1) stops short of the junction to allow it to be dropped in position; 'loose' bars (2) are added to effect continuity

Figure 5.67 Typical detailing at a beam–column junction of a reinforced concrete structure

Figure 5.68 Steel sheeting can be used as both the support to the concrete during construction of a floor and as part of the reinforcement of the hardened concrete. It can also form useful temporary stacking areas on a crowded site (*Source*: British Steel Corporation)

Figure 5.69   Construction of floors in a steel-framed building using precast concrete hollow rectangular flooring units (*Source*: Bison Limited)

Steelwork erection is often preceded by the construction of those vertical reinforced concrete features (e.g., staircases or cores) which serve to stiffen the structure. The erection of the steelwork may be storey-by-storey or bay-by-bay or a combination of the two, depending on the shape of the building under construction (Fig. 5.70). For those multi-storey buildings in which the outer structural steelwork is suspended from the internal core, erection commences at the top floor and proceeds downward (Fig. 5.71). It is of interest to note that, as construction proceeds, the upper suspension members become increasingly stressed in tension with a resulting deflection of the floors relative to the core. This must be allowed for.

The choice between steelwork and concrete

Both steel and concrete are incorporated in most civil engineering structures. There would be few concrete structures without the steel used for reinforcing or prestressing the concrete; all steelwork structures utilize concrete foundations and most steel-framed buildings require concrete floors and frequently concrete shear walls and concrete circulation cores. In this sense then, the two materials are mutually dependent.

Nevertheless, there is an interesting tussle between the steelwork and structural

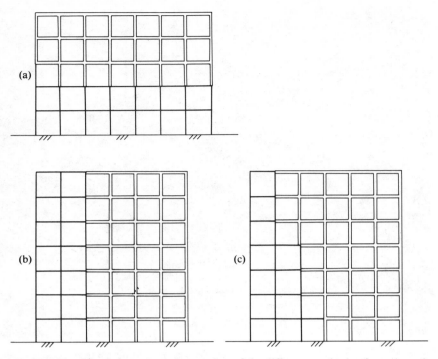

Figure 5.70   Diagrammatic representation of the different methods of erection of steel-framed multi-storey buildings: (a) storey by storey, (b) bay by bay, (c) stepped

concrete interests to dominate those load-carrying systems for which either could be adopted. It is not possible to suggest general rules which determine when it is appropriate to use one material or the other. However the basic influences are that:

- the cost of the *materials* for a concrete structure would be less than that for a structure in steelwork;
- the cost of *erection* of a steelwork structure is less than that for the construction of an *in-situ* concrete structure;
- the cost of obtaining the required *fire resistance* of steelwork structures is greater than that for concrete structures;
- the period required for construction is less for steelwork structures with a consequential reduction in the costs related to *interest charges*.

It is the overall cost from such considerations which largely determines the structural material to be preferred for a given situation.

5.5 Philosophy of design
The aim of design is clearly to obtain a satisfactory structure. There are several ways

Figure 5.71   Construction of the multi-storey steel building shown in Fig. 5.16 using the procedure of 'building from the top' (*Source*: Centre Belgo-Luxembourgeois d'Information de l'Acier)

in which a structure could become unsatisfactory. For example, there is a stage when the sag of a floor resulting from the deflection of supporting beams would be unacceptable. Similarly, there is a stage when the widths of cracks in reinforced concrete structures would be unacceptable. In such cases of deflection and crack widths, when the acceptable limits are exceeded, the structure is said to have become unserviceable. The *limits of serviceability* are determined largely as a result of experience, and the process of design aimed at achieving a satisfactory structure is termed *limit state* design.

Needless to say, for a structure to remain satisfactory it must not collapse, a condition which is termed the *ultimate limit state*. Clearly then one part of the design problem is to ensure that:

| The load which would cause the structure to reach its ultimate limit state | > | The maximum load which is actually applied to the structure |

The right-hand side of this inequality is not known exactly and the designer can only estimate the probable maximum load. Thus, in order to ensure the safety of the structure, this estimated load is multiplied by a suitable *load factor*.

Similarly, the left-hand side of the inequality relating to the strength of the structure cannot be determined exactly. It can only be estimated and for this the designer needs to make assumptions regarding behaviour of the structure and assumptions regarding the strength of the materials to be used in the structure. Again, in order to be on the safe side, safety factors are included in the calculation. The calculated strength is determined assuming that the strengths of the materials are equal to the specified strengths divided by appropriate *material safety factors*.

The inequality therefore used for design would be of the form:

$$\frac{\text{Estimated actual strength}}{K_1} \geqslant K_2 \times \text{Estimated maximum load}$$

where K_1 and K_2 are factors greater than unity.

This is only intended to be indicative of the philosophy behind design. The values of the factors and the way they are introduced into the calculations differ with the type of loading and with the particular structural material under consideration. Such details

Figure 5.72   Structural codes of practice (*Source*: Manchester University Engineering Department)

will be covered in second year studies. It will suffice here to indicate that the overall safety margin against collapse ($K_1 \times K_2$) for most building structures is of the order of 1.7.

A similar approach is adopted to the design for serviceability. The difference is that the limit states of deflection and cracking are obviously not so critical as the ultimate limit states, and therefore much lower safety factors are included in the calculations.

Structural engineers are guided in their designs by *codes of practice* and *manuals* (Fig. 5.72). These are documents which synthesize the knowledge, judgement and experience of the profession. In general, the codes give fairly specific methods for the analysis and calculations, and also state the factors of safety to be adopted in design. Although presented as recommendations, in practice they often become mandatory as a result of a reference to them in local Building Regulations. In this sense then, codes of practice can be considered as a basis of protection of the public against undesirable and dangerous undercutting of standards of design and construction.

Nevertheless it must be made clear that a code of practice deals only in general terms with design and construction. There are no rules which determine the best structural arrangement for a particular situation. Decisions on structural arrangements are made entirely by the structural engineer, and such decisions form the most important part of the design. The subsequent proportioning and analysis of the structure to conform to the appropriate codes of practice ensure the safety and serviceability of the structure but have no bearing on the overall efficiency of the structure. Indeed, the degree of success of a structural engineer will depend more on his imagination and creative capacity than his knowledge of codes of practice.

Appendix A: Answers to problems

Chapter 1

1.1 Compression in member 5; tension in member 6.

1.2 No, the members could not be removed.

1.3 Pratt truss.

1.4 The moment is $\dfrac{wb}{2}\left(\dfrac{2b}{3} + a\right)$, or $\dfrac{wb}{2}\left(\dfrac{2b}{3} - a\right)$ if A is within BC.

1.5 12, 4, -6 kN; -24 kN m.

1.6 Yes. The moment tending to overturn the crane is 5370 kN m. The righting moment from the dead load of the crane is 7000 kN m. The ratio of the two values gives a safety factor less than 1.4.

1.7 $2W\left(1 - \dfrac{r}{R}\right).$

1.8 $R_{ya} = \dfrac{W \cos\theta_1 \sin\theta_2}{\sin(\theta_1 + \theta_2)}$ $\qquad R_{yb} = \dfrac{W \cos\theta_2 \sin\theta_1}{\sin(\theta_1 + \theta_2)}$

$R_{xa} = -\dfrac{W \cos\theta_1 \cos\theta_2}{\sin(\theta_1 + \theta_2)}$ $\quad R_{xb} = \dfrac{W \cos\theta_1 \cos\theta_2}{\sin(\theta_1 + \theta_2)}$

$P_{ac} = \dfrac{W \cos\theta_1}{\sin(\theta_1 + \theta_2)}$ $\qquad P_{bc} = \dfrac{W \cos\theta_2}{\sin(\theta_1 + \theta_2)}.$

1.9 $R_{ya} = 8$ kN $\qquad R_{yb} = 24$ kN

$R_{xa} = 12.4$ kN $\qquad R_{xb} = -12.4$ kN.

1.10 There is no single answer regarding the external force (or forces) acting on the supported face. The important point is to ensure that all the forces described as acting on the block are in equilibrium. Check by taking moments about the centre of the supported face. The possible movements are overturning and sliding.

Chapter 2

2.1 The two vertical members nearest the two supports.

2.2 $F_1 = 30\text{ kN};$ $F_2 = 30\sqrt{2}\text{ kN};$ $F_3 = -60\text{ kN}.$

2.3 $F_1 = 43.7\text{ kN};$ $F_2 = -51.5\text{ kN};$ $F_3 = -20\text{ kN};$ $F_4 = -45.1\text{ kN}.$

2.4 $F_1 = 3.75\text{ kN};$ $F_2 = 15.9\text{ kN};$ $F_3 = -15.9\text{ kN};$ $F_4 = -3.75\text{ kN}.$

2.5 $F_2 = -150\text{ kN}.$

2.6

$$
\begin{bmatrix}
-\dfrac{1}{\sqrt{5}} & \dfrac{2}{\sqrt{5}} & & & & & & \dfrac{7}{\sqrt{53}} & \\[6pt]
-\dfrac{2}{\sqrt{5}} & \dfrac{1}{\sqrt{5}} & & & & & & -\dfrac{2}{\sqrt{53}} & \\[6pt]
& -\dfrac{2}{\sqrt{5}} & 1 & & & \dfrac{5}{\sqrt{34}} & & & \\[6pt]
& -\dfrac{1}{\sqrt{5}} & & & & -\dfrac{3}{\sqrt{34}} & & & \\[6pt]
& & -1 & \dfrac{2}{\sqrt{5}} & & & & & -\dfrac{5}{\sqrt{34}} \\[6pt]
& & & -\dfrac{1}{\sqrt{5}} & & & & & -\dfrac{3}{\sqrt{34}} \\[6pt]
& & & -\dfrac{2}{\sqrt{5}} & \dfrac{1}{\sqrt{5}} & & & -\dfrac{7}{\sqrt{53}} & \\[6pt]
& & & \dfrac{1}{\sqrt{5}} & -\dfrac{2}{\sqrt{5}} & & & -\dfrac{2}{\sqrt{53}} & \\[6pt]
& & & & -\dfrac{1}{\sqrt{5}} & -\dfrac{5}{\sqrt{34}} & -\dfrac{7}{\sqrt{53}} & &
\end{bmatrix}
\begin{bmatrix}
F_1 \\ F_2 \\ F_3 \\ F_4 \\ F_5 \\ F_6 \\ F_7 \\ F_8 \\ F_9
\end{bmatrix}
= -
\begin{bmatrix}
P_{xb} \\ P_{yb} \\ P_{xc} \\ P_{yc} \\ P_{xd} \\ P_{yd} \\ P_{xe} \\ P_{ye} \\ P_{xf}
\end{bmatrix}
$$

Alternatively, expressing the member forces in terms of tension coefficients (see the solution to Problem 2.3 in Part 2), the equations are:

$$
\begin{bmatrix}
-1 & 2 & & & & & 7 & & \\
-2 & 1 & & & & & -2 & & \\
& & -2 & 2 & & & 5 & & \\
& & -1 & & & & -3 & & \\
& & & -2 & 2 & & & & -5 \\
& & & & -1 & & & & -3 \\
& & & -2 & 1 & & -7 & & \\
& & & 1 & -2 & & -2 & & \\
& & & & -1 & -5 & -7 & &
\end{bmatrix}
\begin{bmatrix}
t_1 \\ t_2 \\ t_3 \\ t_4 \\ t_5 \\ t_6 \\ t_7 \\ t_8 \\ t_9
\end{bmatrix}
= -
\begin{bmatrix}
P_{xb} \\ P_{yb} \\ P_{xc} \\ P_{yc} \\ P_{xd} \\ P_{yd} \\ P_{xe} \\ P_{ye} \\ P_{xf}
\end{bmatrix}
$$

2.7 20.4 mm.

2.8 $\dfrac{29.3PL}{AE}$.

2.9 835 mm^2.

2.10 25 mm.

2.11 2.4 mm; -6.8 mm; 30 kN; 40 kN.

2.12 $25\Delta_{xb} - 43\Delta_{xc} + 9\Delta_{xd} + 12\Delta_{yd} + 9\Delta_{xe} = -15$.

2.13 Members DG and DF; members EC and EB.

2.14 FB: 53.4 kN; FC: -53.4 kN; FE: -62.5 kN; FG: zero.

2.15 BE and CE: -20.4 kN; DE: 26.8 kN; BD and CD: -11.0 kN; AD: 33.5 kN.

2.16 3.1 mm.

2.18 $\dfrac{vwL}{4s}$.

2.19 3760 kN.

2.21 Maximum bending moment $= 691$ kN m.

2.22 353 kN m.

2.23 Both reactions are 5 kN; $M_x = 5x - 10[x - 2] + 10[x - 2]^\circ$.

2.24 (a) -28 kN m; (b) 84 kN m; (c) 36 kN m.

2.25 7 m from the left-hand end of the beam.

2.26 Maximum shear force in the cantilever spans is $0.414wL$ and in the internal span is $0.586wL$. Maximum bending moment is $0.086wL^2$.

2.27 Shear force in AB at A is -67.3 kN and at B is 92.7 kN; shear force in BC at B is -87.3 kN and at C is 72.7 kN. The maximum sagging bending moment in AB is 112 kN m and in BC is 161 kN m; the hogging moment at B is 104 kN m and at C is 204 kN m.

2.28 Maximum shear force is $\dfrac{wL}{4}$; maximum bending moment is $\dfrac{wL^2}{12}$. New position of maximum moment at $x = \dfrac{L}{\sqrt{12}}$; maximum bending moment is $\dfrac{wL^2}{36\sqrt{3}}$.

2.29 $M_a = M_e = $ zero; $M_b = 54.4$ kN m sagging; $M_c = 53.1$ kN m sagging; $M_d = 6.7$ kN m hogging.
In AB: $S_a = -24$ kN; $S_b = -3.2$ kN.
In BC: $S_b = -3.7$ kN; $S_c = 5.3$ kN.
(a) none; (b) reduced.
Increased.

2.30 (a) $\dfrac{Wa^2(L + a)}{3EI}$; (b) $\dfrac{5wL^4}{768EI}$; (c) $\dfrac{wL^4}{128EI}$.

2.31 $\dfrac{3L}{5}$.

2.33 The deflection of the tapered beam is one-half of that of the uniform beam.

2.34 (a) $\dfrac{wL^2a}{16EI}$ upward; (b) $\dfrac{WL^2}{16EI}$ anticlockwise; (c) $\dfrac{5WL^3}{48EI}$ downward.

2.35 $\dfrac{25Ph^3}{3072EI}$.

2.36 $\dfrac{wL^3}{12EI}$.

2.37 (a) $M_b = M_d = 64$ kN m (tension on the outside);
(b) $M_b = 32$ kN m (tension on the inside); $M_d = 16$ kN m (tension on the outside).
For the combined loading: $M_b = 32$ kN m (tension on the outside); $M_d = 80$ kN m (tension on the outside).

2.40 $M_b = 20$ kN m (hogging); $M_c = 30$ kN m (hogging); $M_d = 25$ kN m (hogging).

mid-span BC $= 6.3$ kN m (sagging);
mid-span CD $= 17.5$ kN m (sagging);
mid-span DE $= 18.8$ kN m (sagging).

For the statically determinate system:
mid-span   BC = 21.3 kN m;   mid-span   CD = 45 kN m;   mid-span   DE = 31.3 kN m.

2.41 $\dfrac{5P}{32}$.

2.42 $\Delta_x = 1.65$ mm; $\Delta_y = -4.68$ mm.
$F_1 = 20.7$ kN; $F_2 = 65.6$ kN; $F_3 = 27.5$ kN.

Chapter 3

3.2   109 N/mm², − 109 N/mm².

3.3 $\tau = 0.062\left(1200 + 20y_1 - \dfrac{2}{3}y_1^2\right)$.

3.4   (i) $\dfrac{2G_s}{G_c}$.   (ii) $\dfrac{8G_s}{G_c}$.

3.5 $F_p = \begin{bmatrix} 190.1 & 0 & 0 \\ 0 & 9.9 & 0 \\ 0 & 0 & 0 \end{bmatrix}$ N/mm²   with respect to axes at 28.2° anticlockwise from xy.

3.6 $F_{T1} = \begin{bmatrix} 100 & 90.1 & 0 \\ 90.1 & 100 & 0 \\ 0 & 0 & 0 \end{bmatrix}$ N/mm²   with respect to axes at 16.9° clockwise from xy.

$F_{T2} = \begin{bmatrix} 100 & -90.1 & 0 \\ -90.1 & 100 & 0 \\ 0 & 0 & 0 \end{bmatrix}$ N/mm²   with respect to axes at 73.1° anticlockwise from xy.

3.7   (i) $S_p = \begin{bmatrix} 460.6 & 0 & 0 \\ 0 & -260.6 & 0 \\ 0 & 0 & \varepsilon_z \end{bmatrix} \times 10^{-6}$ with respect to axes at 16.85° anticlockwise from xy.

$S_{T1} = \begin{bmatrix} 100 & 360.6 & 0 \\ 360.6 & 100 & 0 \\ 0 & 0 & \varepsilon_z \end{bmatrix} \times 10^{-6}$ with respect to axes at 28.15° clockwise from xy.

(ii) $S_p = \begin{bmatrix} 482.8 & 0 & 0 \\ 0 & -82.8 & 0 \\ 0 & 0 & \varepsilon_z \end{bmatrix} \times 10^{-6}$ with respect to axes at 67.5° clockwise from xy.

$S_{T1} = \begin{bmatrix} 200 & 282.8 & 0 \\ 282.8 & 200 & 0 \\ 0 & 0 & \varepsilon_z \end{bmatrix} \times 10^{-6}$ with respect to axes at 67.5° anticlockwise from xy.

(iii) The circle reduces to a point at $\varepsilon = 200 \times 10^{-6}$.
See full solution in Part Two.

(iv) $S_p = \begin{bmatrix} -117.2 & 0 & 0 \\ 0 & -682.8 & 0 \\ 0 & 0 & \varepsilon_z \end{bmatrix} \times 10^{-6}$ with respect to axes at 22.5° anticlockwise from xy.

$S_{T1} = \begin{bmatrix} -400 & 282.8 & 0 \\ 282.8 & -400 & 0 \\ 0 & 0 & \varepsilon_z \end{bmatrix} \times 10^{-6}$ with respect to axes at 22.5° clockwise from xy.

3.8 10.9° anticlockwise.

3.11 $\sigma_1 = 63$ N/mm² at 72° clockwise from the direction of x.
$\sigma_2 = -73$ N/mm² at 18° anticlockwise from the direction of x.
$\varepsilon_1 = 0.8126 \times 10^{-3}$, $\varepsilon_2 = -0.8876 \times 10^{-3}$.

3.12 237.5 N/mm².

3.14 469 kN.

3.15 k = 7.78 mm.

3.16 (i) 1.53 m,   (ii) 3 mm.

3.17 62.7 mm, 3.8 mm.

3.18 1.25.

3.19 (a) 70.6 N/mm², -70.6 N/mm² at 45° anticlockwise and 45° clockwise respectively from the direction of x; 70.6 N/mm² with respect to xy.
(b) 147.5 N/mm² in a direction 19.1° anticlockwise from x.
-17.7 N/mm² in a direction 70.9° clockwise from x.
82.6 N/mm² with respect to axes 25.9° clockwise from xy.

3.20 $S_p = \begin{bmatrix} 1.75 & 0 & 0 \\ 0 & -0.75 & 0 \\ 0 & 0 & \varepsilon_z \end{bmatrix} \times 10^{-3}$ with respect to axes at 18.4° anticlockwise from xy.

3.21 27.7 kN.

$$F_p = \begin{bmatrix} -92.3 & 0 & 0 \\ 0 & -27.7 & 0 \\ 0 & 0 & 0 \end{bmatrix} \text{N/mm}^2 \text{ with respect to axes } xy.$$

3.22 $72.56 \times 10^3 \text{ N/mm}^2$;
331.6 N/mm^2;
0.31;
$27.7 \times 10^3 \text{ N/mm}^2$;
$63.6 \times 10^3 \text{ N/mm}^2$.

3.23 $\dfrac{e_z}{312.3} + \dfrac{e_y}{756.3} = 1.$

Chapter 4

4.2 545 kN.

4.3 Top: $2.27 \times 10^6 \text{ mm}^3$; bottom: $1.38 \times 10^6 \text{ mm}^3$;
top flange about the y–y axis: $0.264 \times 10^6 \text{ mm}^3$.

4.4 $b_r d_n^2 + 2(d_s[b - b_r] + mA_s)d_n - [b - b_r]d_s^2 - 2mA_s d = 0.$

Appendix B: Guidance on the procedure for solving selected problems

Chapter 1

1.1 The answers are obtained by considering equilibrium at joint C. It is necessary to acknowledge the nature of the forces in members 2 and 3 as determined in the earlier discussion.

1.2 Use the fact of symmetry of both frame and loading.

1.3 The relationship between the strength of a compression member and its length is relevant here.

1.4 For ease in determining the value of the integration it will be better to position the origin at the point of zero load, distance a from A.

1.5 Let the length of AB represent the 14 kN force. The lengths of the components of AB are easily obtained from the coordinates of A and B. The moment is determined by assuming that the components act at either A or B—the same answer is obtained whichever point is used.

1.6 Firstly, calculate the horizontal and vertical components of the force in the cable required to lift the bridge. Now consider the effect of these forces on the crane—they will tend to overturn the crane about its front support. Remember that the components of the tensile force can be considered to act at any convenient position along the line of action of the force. It is better to use the two components of the force rather than the single force, as their perpendicular distances about the point of rotation are easier to determine.

1.7 The tube will tend to overturn about the edge below the point of contact with the upper ball. Note that the tube has no bottom and therefore the only forces from the balls to the tube are applied horizontally at the points of contact. The first step is to find these horizontal forces by considering equilibrium of the balls.

1.8 For this structure it is possible to isolate a part of the structure, say the part to the right of C, and, knowing that C is a pin and incapable of resisting a moment, a fourth equation in terms of the reactive forces can be obtained.

1.9 As in problem 1.8, it is possible to isolate a part of the structure about a pin to obtain the fourth equation. Note that although the shape of the arch was defined as 'parabolic', it does not enter into this problem.

1.10 There are several different ways of describing the forces (and stresses) acting on the supported face. In other words, there are several force systems which are equivalent to one another. The student should attempt to describe as many as possible before referring to those described in the solutions in Part Two.

Chapter 2

2.1 In Chapter 1, consideration of equilibrium for a system of concurrent forces established a number of corollaries. One of these stated: 'a concurrent 3 force system can only be in equilibrium if the three forces lie in the same plane and are such that no two of the forces have the same line of action.' It follows from this that, when three members meet at an unloaded pin-joint and two of the members have the same line of action, the force in the third member must be zero. This enables the members of the frame having zero force to be identified.

2.2 With a cantilever frame of this kind, it is not necessary to determine the reactive forces. In using the method of sections one can always work on that part of the frame where all the external forces are known.

2.3 Use the method of joints. To facilitate the determination of the component of the member forces in the x and y directions, obtain the coordinates of the joints and the lengths of the members. (See the 'Use of joint coordinates' in Section 2.4.)

2.4 As the section through the panel in question cuts through four members, four equations are required. The simple use of the method of sections generates three equations, but a fourth can be obtained from a consideration of horizontal equilibrium at the joint C.

2.5 Assemble the column matrix for the particular loading arrangement and then multiply this with the appropriate line of the matrix solution for this frame given in Section 2.1.

2.6 There are nine members and hence nine equilibrium equations are required. They are obtained by considering horizontal and vertical equilibrium at the four joints B, C, D and E, and horizontal equilibrium at joint F. Utilize the joint coordinates to obtain the horizontal and vertical components of the member forces—see the 'Use of joint coordinates' in Section 2.4. (The equations are more easily expressed in terms of tension coefficients—see the solution to Problem 2.3 in Part 2.)

2.7 Use the method of joints to determine the forces F_0 in all the members caused by the specific loads. Similarly, for the forces f_1 in the case of a unit load, $X_1 = 1$

applied at D. Symmetry reduces the arithmetic involved. In checking the deflection using the matrix method, you will need to assemble the column matrix for the specified loads—as in Problem 2.5.

2.8 The unit load method for calculating deflections usually requires the determination of the forces in all members for two systems of loading, namely, the original loading and the imaginary unit load. In this particular example, however, the position of the imaginary load is such that, for this load, there is zero force in the majority of the members. The first step is to identify these members because it does mean that there is no need to determine the forces in these particular members caused by the original loads.

2.9 Symmetry of the frame and the loading ensure that the maximum deflection will occur at the mid-span position. Use the unit load method to determine this deflection in terms of the unknown cross-sectional area A of the members. Hence, set up an inequality concerning the deflection constraint. The maximum member force obtained during this calculation will enable an inequality to be set up concerning the stress constraint. The more stringent of the two conditions will give the minimum value of A.

2.10 In contrast to Problem 2.9, the members of this frame have different cross-sectional areas, so arranged to give the stated strains. Since

$$\frac{\text{Force in a member}}{\text{Cross-sectional area} \times \text{Young's Modulus}} = \text{Strain in member}$$

the unit load expression for deflection can be re-written as

$$\Delta_1 = \sum \frac{F_0 f_1 L}{AE} = \sum \varepsilon_0 f_1 L$$

where ε_0 is the strain caused by the original load. The value of ε_0 in a particular member depends on the type of force in the member caused by the original load, but note that these types of force will coincide with those caused by the unit load $X_1 = 1$.

2.11 Utilize the expression obtained for the force in a general member of a frame in the part of Section 2.2 dealing with joint movements using matrices.

2.12 As in the previous example, utilize the expression for the tensile force in a general member from which the forces acting at joint C from the members 2, 5 and 6 can be expressed in terms of the unknown displacements at joints B, C, D and E. Obtaining the x-component of these forces enables the corresponding equilibrium equation of joint C to be obtained. The most likely sources of errors in developing the equation are the signs of $\cos \theta_i$ and $\sin \theta_i$. For example, $\cos \theta_5 = -\frac{3}{5}$ and $\sin \theta_5 = -\frac{4}{5}$.

2.13 Refer to the corollaries regarding equilibrium at joints for space frames and select the one appropriate to the case for determining the two members which have zero force. Prior to proving that the member BC also has zero force, use simple logic to determine the other two members which would then also have zero force. This observation then helps in deciding the route to take in determining the force in BC.

2.14 There are four members meeting at F, so that one cannot start the analysis at this joint. However, the use of simple logic at another joint shows that the force in one of the members meeting at F has zero force. Hence, one can come immediately back to joint F and solve for the remaining forces using simple equilibrium at that joint. (The arithmetic is slightly easier if one uses tension coefficients as demonstrated in Problems 2.3 and 2.6 in Part 2.)

2.15 Acknowledge symmetry of the frame in designating the unknown forces. There are four unknown forces and these can be determined by the method of joints at E and D. (Again the arithmetic is simplified by the use of tension coefficients.)

2.16 The forces F_0 in the members caused by the applied loads were determined in an example covered in Section 2.4 and these can be utilized in this problem. The main part of the problem therefore reduces to finding the member forces f_1 due to the unit load $X_1 = 1$ applied at the joint D in the x direction. The use of tension coefficients, whilst not essential, would simplify the arithmetic.

2.17 As the bending moment in the cable is zero at every point, consider any typical point on the cable (x, y) and obtain a relationship by taking moments about that point for all forces to one side. Eliminate the vertical reaction from the relationship by obtaining its value in terms of W and H from the moment equilibrium equation about the opposite support. A rearrangement of the relationship will then prove the proposition.

2.18 Find expressions for the two vertical reactions at the supports by considering moment equilibrium about both supports. The horizontal reactions H can be eliminated from these expressions by the use of the proposition proved in the previous problem.

2.19 The horizontal component of the tensile force is constant throughout the cable and is obtained by the use of the proposition in Problem 2.17. The maximum tensile force in the cable will therefore occur at the position where the vertical component of the cable force is a maximum, and this occurs adjacent to the support at B—see solution to Problem 2.18. The horizontal and maximum vertical components are then combined to give the maximum force in the cable.

2.20 A curve is said to have a catenary shape when its gradient at any point P is proportional to the distance along the curve from the point P to the lowest point of the curve. To prove this for a suspended cable, consider the equilibrium of a part of the cable of length s measured from the lowest point.

2.21 The first step is to calculate the reaction at the left-hand support. This then enables a general equation for the shear force at a section distance x from this support to be obtained, and hence the shear force diagram to be drawn. Similarly for the bending moment. The position of zero shear force determines the position of maximum bending moment.

2.22 We know from the previous analysis of this beam summarized in Fig. 2.48 that the maximum bending moment will be in the region of sagging bending. To find the position of maximum sagging moment, determine the position of zero shear force.

2.23 From a knowledge of the reactive forces at the supports, the shear force and bending moment diagrams can be built up by considering the two regions of the beam. For example, calculate the shear force at a section on the left of the bracket by considering forces to the *left* of the section, whilst at a section on the right of the bracket by considering forces to the *right* of the section. The single equation for the bending moment throughout the beam requires a knowledge of the forces applied to the beam at the interface of bracket and beam. These forces are best obtained by considering initially the equilibrium of the bracket.

2.24 Following the calculation of the reactive forces, the shear force diagram is developed by initially visualizing the equilibrium of a region between the left-hand end of the beam and a nearby section, and subsequently moving this section along the beam to the right, the shear force on the section changing as the loads are passed. The bending moment diagram is best obtained by obtaining the single general equation for bending moment at a section distance x from the left-hand end. Substituting appropriate values of x (at positions of the concentrated forces and at, say, 1 m intervals in the region of the distributed load) enables the bending moment diagram to be plotted.

2.25 From the general equation for the bending moment at any section distance x from the left-hand end, obtain that part relevant to the region of the distributed load. The square brackets can also be eliminated by observing that the maximum bending moment in question cannot occur in the cantilever part of the beam. The remaining expression can then be differentiated to find the position of the maximum bending moment.

2.26 Moving the supports towards the centre of the beam will have the effect of increasing the hogging bending moments over the supports (because of the increased span of the cantilevers) and decreasing the maximum sagging moments between the supports, and vice versa. It follows that the lowest possible value of the bending moment is when the maximum hogging bending moment equals the maximum sagging moment. Note that the maximum sagging moment will be given by

$$\frac{w \times \text{span}^2}{8} - \text{hogging moment at the supports}$$

2.27 The first step is to find the unknown reactive forces at A and C. To plot the bending moment diagram for the span AB, obtain the bending moment equation for this span and so obtain the bending moment values at, say 1 m intervals. As the diagram for BC consists of straight lines, only one additional value of bending moment is required to complete the diagram—the value at the position of the concentrated load. This value is best obtained by considering the moment of all forces to the *right* of the load, namely the applied moment and the vertical reaction at C.

2.28 Find the reactions, and also the intensity of load at a position distance x from the left-hand support but within the left-hand half. Hence, establish the shear force and bending moment equations in this half. For this one can use either first principles or the relationships:

$$w = \frac{\mathrm{d}S}{\mathrm{d}x} = -\frac{\mathrm{d}^2 M}{\mathrm{d}x^2}$$

A concentrated load P applied at the mid-span of a beam causes a bending moment at that position equal to $PL/4$. Hence, calculate the required upward force. The new values of the reactions can then be determined, and so also the new expression for bending moment at a position distance x from the support. This is analysed to determine the position of maximum bending moment.

2.29 Despite the inclination of the structure, the orientation of the roller support is such that the reactive forces acting on the structure are vertical. Find these reactions by first calculating the total weights of the sections of the structure (AB, BC, etc.) and the distances of the lines of action of these gravity forces from the supports. Hence, determine the bending moment at the required positions by considering the forces to one side—either the right or left whichever has the fewer forces to take into account. For the shear force diagram, note that the shear force at a particular position on a member is that acting on a section perpendicular to the axis of the member. Hence, for AB you need to obtain the appropriate components of the reaction and the distributed loading. In considering the effects of the additional loading on bending moments, for two of the cases it is first necessary to consider how the additional loading affects the reaction at A.

2.30 The restrictions pertaining to the strain energy method immediately determine the case for which this method was used. Of the two remaining cases, only one can generate the standard bending moment diagrams which enable the tabulated coefficients to be used.

2.31 Use the standard expression for a cantilever to obtain the deflection at the position of the load. Obtain an expression for the deflection at the end of the beam using the unit load method. This will involve the product integration of a triangle and a quadrilateral. As Table 2.7 does not give a coefficient for the product integration involving a quadrilateral, split the quadrilateral into two triangles and complete the product integration in two parts. (See the solution to this problem for an alternative way of using the unit load method.)

2.32 Starting from the known expression for the intensity of load at any point on the beam, integrate this to obtain an expression for the shear force at any point. The constant of integration can be determined from a known value of the shear force at a particular point, namely, that adjacent to a support. A series of similar integrations lead to expressions for bending moment (and hence curvature), slope and deflection.

2.33 Use the standard expression for the deflection of a cantilever subjected to a uniformly distributed load to obtain the deflection at the end of the uniform beam. Obtain the deflection of the non-uniform beam by using the differential equation of bending, but allowing for the variation in the second moment of area along the beam. The solution of the equation requires the integration of $\log x$ which is $x \log x - x$.

2.34 These problems are very quickly solved as long as the bending moment diagrams can be immediately established. For all three problems, the M_0 and m_1 bending moment diagrams are simple standard cases and it is imperative that students can obtain the diagrams immediately without having to revert to fundamental calculations. A knowledge of the standard cases depicted in Table 2.6 and of the case of a moment applied at one end of a beam is therefore essential.

2.35 The unit load method should be used although the product integration of the M_0 and m_1 diagrams is somewhat awkward. Nevertheless, the diagrams can be split up into regions involving the integration of triangles, so permitting the use of the coefficients. The integration must, of course, allow for the change of section.

2.36 As the two beams and their loads are identical, the angular discontinuity at B is equal to twice the rotation at the end of one of the beams. Determine this rotation using the unit load method.

2.37 In order to draw the bending moment diagrams it is only necessary to calculate the bending moments at B and D. To obtain these bending moments, the first step for each case is to find the reactive forces at A and E. The four reactive forces are found using the three equations of equilibrium for the whole frame, together with an equation obtained from the knowledge that the bending moment at C is zero.

2.38 The first step is to obtain the reactive forces and then to use these to obtain an expression for the bending moment at a general point of coordinates x, y. For the parabolic relationship between x and y, the expression for bending moment becomes zero.

2.39 Although a two-pinned arch is statically indeterminate, a knowledge of the horizontal reactive forces caused by a concentrated load allows the forces caused by the uniformly distributed load to be determined. A short length (dx) of the load can be considered as equivalent to a concentrated load. Hence, the horizontal forces caused by the full distributed load can be obtained by integration. An expression for bending moment can now be obtained which, as in Problem 2.38, becomes zero.

2.40 Students should note that this structure is 'twice' statically indeterminate—for vertical loads only there are initially four unknown reactive forces but only two available equilibrium equations. However, since we are given two of the unknown forces (in this case, one of the vertical reactive forces and one of the hogging bending moment values), it is possible to obtain all the forces (reactions, shear forces, bending moments) relevant to the structure. In this problem, only the bending moments are required.

The information available allows the hogging moments at the three supports B, C and D to be quickly obtained. The 'free' maximum bending moments for the spans BC, CD and DE (i.e., the bending moments assuming simply supported spans) are also quickly obtained. These six values enable the bending moment diagram to be sketched.

2.41 As the load P is applied, the downward movement of the left-hand cantilever will cause a downward force (say R) to be applied to the right-hand cantilever at the position of the pin. Action and reaction being equal and opposite, there will be an upward force R applied to the left-hand cantilever. Since the cantilevers are connected, the two cantilevers (one with the downward force R only and the other with the load P and an upward force R) must have equal end deflections. These deflections can be obtained in terms of P and R using the given expression, and hence the value of R can be obtained.

2.42 Use the expression for the tensile force in a general member i in terms of x and y components of the displacement at the ends of the member as developed in Section 2.2 (see also Fig. 2.25). Determine the values of $\sin \theta_i$, $\cos \theta_i$, and $(AE/L)_i$ for the three members, and so obtain the expressions for the forces in the members in terms of Δ_x and Δ_y at the loaded joint. The components of these forces acting *in the positive direction of the axes* are obtained by multiplying by the appropriate values of $\cos \theta_i$ and $\sin \theta_i$.

Chapter 3

3.1 Consider a vertical strip of the triangle of width dz and distance z from the vertical principal axis Gy. Calculate the area of the strip as a function of z and integrate the second moment of area of the strip about Gy between the limits $z = B/2$ and $z = 0$. Double this value to get I_y.

3.2 The maximum stress will occur at the cross-section where the bending moment has its maximum value and at the extreme fibre of the cross-section, that is, when $M = M_{max}$ and $y = y_{max}$ in Eq. (3.8).

The bending moment diagram is the addition of that due to the self-weight of the beam (a uniformly distributed load) and the imposed point load of 400 kN.

Decide where the maximum value of the bending moment will occur as the load moves slowly across the beam and calculate its value.

3.3 Use Eq. (3.27). Read the defintiions of $\bar{A}$ and $\bar{y}$ very carefully.

3.4 Use the double Eq. (3.37). Because the ends of the tubes are fixed to rigid plates, the inside tube will rotate through the same angle as the outside tube, i.e., (θ/L), is the same for both tubes. This equality allows the solution to be found.

3.5 This question involves the straightforward use of Mohr's circle of stress.

3.6 This question too can be solved by Mohr's circle of stress.

3.7 Use Mohr's circle of strain to represent the state of strain in each case. Remember to move 2θ on the circle when you move through θ on the element.

3.8 Draw Mohr's circle of strain for the given state of strain. The rectangle was drawn in an orthogonal set of axes for which $\gamma_{\bar{x}\bar{y}}$ is zero.

3.9 Use Eqs (3.121) and (3.121a).

3.10 The given expressions for σ_x, σ_y and σ_z are obtained from the manipulation of the expressions for ε_x, ε_y and ε_z in Eqs (3.123) and (3.124).

3.11 The question itself outlines the procedure. To convert principal stresses to principal strains, use Eq. (3.121).

3.12 Draw the bending moment and shear force diagrams for the loaded beam. The maximum tensile stress in the web due to bending alone will occur at the point where the web meets the flange at the section of maximum bending moment. From the shear force diagram it will be evident that the section to consider is *just* to the right (or left) of the middle of the beam. Combine the tensile stress and the shear stress using Mohr's circle to find the maximum (i.e., principal) stress.

3.13 (i) Use Eq. (3.142) and find the limiting value of e_z by putting $\sigma_x = 0$ at the opposite end of a diameter. (ii) Again, use Eq. (3.142). The only thing that changes from part (i) of the question is I.

3.14 Transfer the 50 kN load to the vertical axis of the column accompanied by a moment $M_y = (50 \times 0.15)$ kN m. Now use Eq. (3.144) and put σ_x to its limiting value of -150 N/mm² when $z = -75$ mm. This equation will yield the required value of $\bar{P}$.

3.15 Find the maximum magnitude of the bending moment by considering the load applied at different points across the beam. Express the second moment of area of the beam cross-section in terms of k. Use Eq. (3.8) and put σ_x to its limiting value with $y = 10k$. This equation will give the required value of k.

3.16 (i) Find I by first finding J, remembering that $J = I_z + I_y$ and that, for a circular cross-section, $I_z = I_y = I$. (ii) Replace the 6 kN load by its horizontal and vertical components acting at the top of the mast. The vertical component is an axial compressive load acting through the entire mast. The horizontal component causes a bending moment which increases linearly with distance from the top of the mast. Use Eq. (3.144) to consider, firstly, the point on tube A which experiences the highest compressive stress (this will give the allowable length of tube A) and, secondly, the point on tube B which will experience the highest compressive stress (this will give the minimum wall thickness of tube B).

3.17 The solution is found by the use of Eq. (3.37) and the fact that both tubes must twist by the same amount because they are fixed together at their ends. That is, (θ/L) is the same for both tubes.

3.18 Draw Mohr's circle of stress for the given stress state. Find the maximum shear stress. The ratio of 135 N/mm^2 divided by the existing maximum shear stress is the factor required.

3.19 Use Eqs (3.8) and (3.27) respectively to find σ_x and τ at the points of interest in the web, having first determined the values of the shear force and the bending moment at the section 0.5 m from the end of the cantilever.

At each of the two points in turn, use Mohr's circle to find the directions and values of the principal stresses and the maximum shear stress.

3.20 From the information given, deduce the values of the x-direction strain ε_x, the y-direction strain ε_y, and the shear strain γ_{xy} on the plate. ε_x and ε_y are found from two simultaneous equations derived by using the given percentage changes in area and perimeter. Remember to ignore second-order quantities.

Having found ε_x, ε_y and γ_{xy}, draw the Mohr's circle of strain and hence find ε_1 and ε_2 (the principal strains) and their directions relative to the xy axes.

3.21 ε_x and ε_y are principal strains and $\varepsilon_x = 0$. σ_x and σ_y are principal stresses. Use Eq. (3.121) to express ε_y in terms of σ_y. Draw the Mohr's strain circle in terms of σ_y and find σ_y by knowing the strain in the direction m. Everything else will follow.

3.22 Use Eq. (3.145) to find an equation in e_z and e_y based upon the stress at e being equal to -250 N/mm^2. This will define the position of line QQ. When defining the entire area within which the load must act, remember that the section is doubly symmetrical so that there are *seven* other corners like e to be protected from a stress less than -250 N/mm^2.

Chapter 4

4.1 Combine Eqs (4.1), (4.2) and (4.3) to give Eq. (4.4) by eliminating those parameters which do not appear in Eq. (4.4). For example, commence with Eqs (4.2) and (4.3), eliminating A_s by putting it equal to *rbd*. The parameter which does not appear in Eq. (4.4) is now *mr*. Obtain a value for this from Eq. (4.1). (In manipulating Eq. (4.1), rearrange to give the square root only on one side of the equation. Now square both sides and simplify to find *mr*.)

4.2 Two modes of failure are possible: the failure of the bolts by shearing on a plane perpendicular to their axis, or the failure by crushing of the bolts or plate as a result of high bearing stresses. After calculating the shear capacity and the bearing capacity of a bolt, the smaller value will determine the design shear force for the connection. Regarding the bearing strength, as the beam end plate is thinner than the column flange, this will be the more critical of the two and so the thickness of the flange will not enter into the calculation. The bearing capacity of a single bolt is

$$\text{diameter} \times \text{plate thickness} \times \text{maximum bearing stress}$$

4.3 The positions of the y–y and z–z axes of the beam and channel are given in Fig. 4.65. For the compound girder, note that the y–y axis of one section is parallel to the z–z axis of the other. Although the width of the flange of the beam has not been given, it is possible to calculate its I_{yy} from the data given.

4.4 Use the transformed section method. The problem then simply resolves into a matter of taking moments of areas to find the position of the centroid of that section.

PART TWO: Solutions to Problems

In publishing the solutions to the problems set in Part One, the authors are very much aware of the danger that this presents. The thought processes involved in solving problems, or indeed in attempting to solve problems, form a very important part in the development of one's understanding of a subject, and it is essential that this set of solutions does not eliminate that activity. We therefore urge that reference to our solutions should only be made after completing your own, or when you have reached a stage where you cannot proceed further. On that premise, we are sure that this Section will present a valuable source of further instruction in the fundamentals of analysis of civil engineering structures.

Problem 1.1

Use simple logic to determine the nature of the forces in members 5 and 6 of the pin-jointed truss in Fig. S1.1a.

Simple logic applied to the equilibrium of joints can often (but not always) determine the nature of the forces in the members of pin-jointed frames. The discussion in Part One (page 29) established that both members 2 and 3 had tensile forces.

The situation at joint C at this stage is therefore that shown in Fig. S1.1b—both members 2 and 3 are pulling on the joint, but the types of force exerted by members 5 and 6 are not yet known. In order that the joint is able to resist the vertical component of the pull from member 3, the member 5 must push on the joint—member 5 is therefore in *compression*.

The known situation is now that in Fig. S1.1c. As the forces from both members 3 and 5 have horizontal components acting from right to left, and so in the same direction as the force from member 2, the force from member 6 must act from left to right in order to balance these. In other words, member 6 must be pulling on the joint—member 6 is therefore in *tension*. The situation is summarized in Fig. S1.1d.

Following the study of the bending of *beams*, the forces in the upper and lower chords of a truss will often be intuitively obvious. For example, for a truss supported

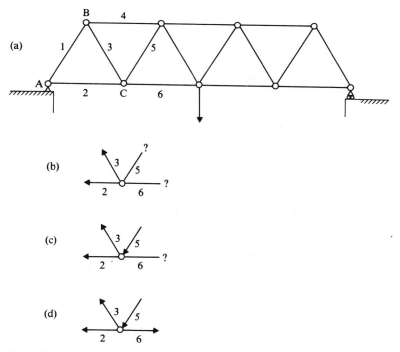

Figure S1.1

at its ends and loaded vertically (as in Fig. S1.1a), the members forming the end posts and the top chord will always be in compression whilst the members forming the bottom chord will always be in tension. The internal members will be either tension or compression, depending on the position of the loads.

Problem 1.2

Use simple logic to show that the force in each of the two inclined internal members of the symmetrical truss in Fig. S1.2a is zero. Could these members therefore be removed?

We are told that the frame is symmetrical, and we can observe that the loading is also symmetrical. It follows that the forces in the members of a corresponding pair must be equal in magnitude and of the same type. In the case of the pair of internal inclined members, the forces in the members must be zero, otherwise it would not be possible to achieve equilibrium at the central joint (Fig. S1.2b). For example, if both inclined members were subject to tensile forces, they would both have upward-acting components and there is no force available to resist these components.

It is perhaps instructive to pursue this problem further by considering the case when only one of the loads is acting (Fig. S1.2c). The use of simple logic (starting from the right-hand support) establishes that the types of forces acting on the central joint would be as shown in Fig. S1.2d—one of the inclined internal members is in tension and pulling on the joint, whilst the other inclined internal member is in compression and pushing on the joint. For the case when the other load W is acting alone, the situation would be reversed for the inclined members. It follows that when both loads are acting together, their effects on the two internal members would cancel out.

It is not possible to remove these internal members simply because they have zero force. Whilst it is conceivable that for the structure shown in Fig. S1.2e equilibrium could momentarily be achieved, the merest change from perfect symmetry—either in the configuration of the frame or in the loads—would cause such a frame to collapse. For structures in which the members are connected through pins, a triangulated arrangement of the members is essential.

Problem 1.3

For vertical loads applied at the joints along the bottom chord, which of the two trusses, Pratt or Howe, is to be preferred? What is the particular advantage of the K-truss?

In the case of the Pratt truss (Fig. S1.3a), simple logic will show that, when the loads are applied at each of the joints of the bottom chord, the internal inclined members (termed the 'diagonals', the end inclined members being termed the 'end posts') act in tension whilst the forces in the vertical members are either tension or compression, or in one case zero. The nature of the forces in the internal members is indicated on the diagram. In order to use the procedure of simple logic to establish the nature of the member forces, it is necessary to observe that the central vertical member has zero force—at the joint with the top chord there is no means of resisting a force from this vertical member. It then follows from a consideration of equilibrium at the bottom

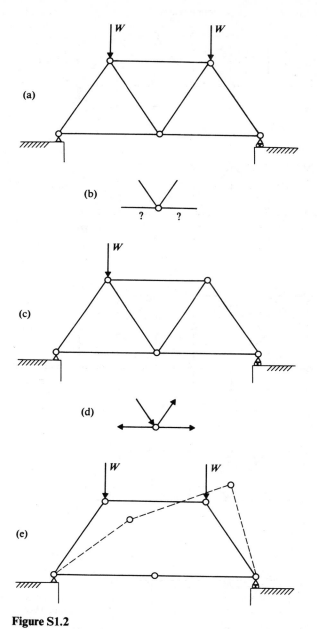

Figure S1.2

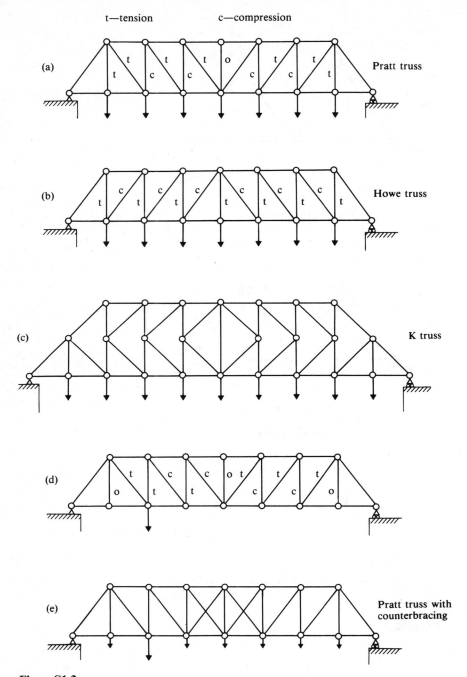

Figure S1.3

chord that the two central diagonals must be in tension; the tensile forces from these two members at their junction with the top chord must be resisted by compressive forces from the vertical members meeting there. And so on, until at the positions of the two outer vertical members, equilibrium at the bottom chord establishes that these members must be in tension.

In the case of the Howe truss (Fig. S1.3b), the situation is largely reversed with all the diagonals acting in compression and all the vertical members acting in tension.

For such loading, the Pratt system is preferred since its compression members (the verticals) are shorter than those (the diagonals) of the Howe truss. The shorter the better for compression members from the standpoint of resisting the tendency to bend out-of-line of the applied force.

It is for this reason that the K-truss (Fig. S1.3c) is sometimes adopted when a very deep truss is required. The effective length of the verticals, for example, has been reduced as a result of the K-type diagonals, and this reduces their bending out-of-line in the plane of the truss.

It is here that one might point out the reason for incorporating the central vertical of the Pratt truss, which equilibrium tells us will always have zero force. Its value is in reducing the effective length in the plane of the truss of the central part of the upper chord.

A further point to note is that the previous discussion related to the case of symmetrical loading along the bottom chord, and this will apply to the dead load of the structure. For asymmetrical imposed loading of the Pratt truss (as in Fig. S1.3d, for example), some of the diagonals would be stressed in compression. In those panels where such compressive forces due to the imposed load overcome the tensile forces due to the dead load, the buckling tendency of these diagonals would need to be considered in design. Alternatively, the use of counter-bracing (see Part One, page 379) in these panels could be adopted (Fig. S1.3e).

Problem 1.4

Figure S1.4a shows a distributed load which varies linearly from zero at B to w kN/m at C. Show that for moments about A, the statically equivalent force system is a concentrated load of magnitude equal to the total load acting at the centroid of the distributed load. Show that this also applies if A is within BC.

Consider an element of the load (Fig. S1.4b) positioned x from B.

Intensity of load at this position $= \dfrac{x}{b} w$ kN/m

Value of the elemental load $\quad = \dfrac{x}{b} w \, dx$ kN

Distance of elemental load from A $= a + x$

Moment of elemental load about A $= (a + x)\dfrac{x}{b} w \, dx$

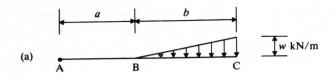

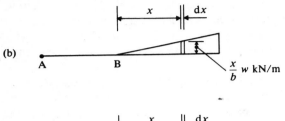

Figure S1.4

Moment of total load about A $= \displaystyle\int_{0}^{b} (a + x)\frac{x}{b}\, w\, dx$

$$= \frac{w}{b}\left(\frac{b^3}{3} + \frac{ab^2}{2}\right)$$

$$= \frac{wb}{2}\left(a + \frac{2b}{3}\right)$$

Thus moment = total load × distance of centroid of load from A

For the case when A is positioned within BC, see Fig. S1.4c.

Distance of elemental load from A $= x - a$

Moment of total load about A $= \displaystyle\int_{a}^{b} (x - a)\frac{x}{b}\, w\, dx$

$$= \frac{wb}{2}\left(\frac{2b}{3} - a\right)$$

Thus moment = total load × distance of centroid of load from A.

Problem 1.5

The line of action of a force of magnitude 14 kN is from A to B, where A and B have coordinates in an orthogonal system of axes x, y, z of 0, 2, 3 and 6, 4, 0, the distance units being metres. Calculate the components of the force in the directions of the axes and also the moment of the force about the x-axis.

$$\text{Distance from A to B} = \sqrt{((6-0)^2 + (4-2)^2 + (0-3)^2)}$$

$$= \sqrt{49} = 7\,\text{m}$$

Let the force of 14 kN be represented by the line AB. The scale is then $1\,\text{m} = 2\,\text{kN}$.

Projected length of AB in the x direction $= 6 - 0 = 6\,\text{m}$

Projected length of AB in the y direction $= 4 - 2 = 2\,\text{m}$

Projected length of AB in the z direction $= 0 - 3 = -3\,\text{m}$

Component of the force in the x direction $= 6 \times 2 = 12\,\text{kN}$

Component of the force in the y direction $= 2 \times 2 = 4\,\text{kN}$

Component of the force in the z direction $= -3 \times 2 = -6\,\text{kN}$

Consider that the components act at A as in Fig. S1.5a. Only the components in the y

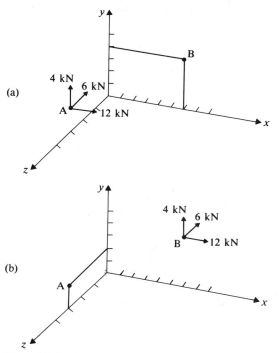

(a)

(b)

Figure S1.5

and z directions have a moment about the x-axis. Considering clockwise moments as positive when viewed in the direction of the x-axis:

$$\text{Moment about the } x\text{-axis} = -(4 \times 3) - (6 \times 2) = \underline{-24\,\text{kN m}}$$

Alternatively, if we consider the components to act at B as in Fig. S1.5b, only the component in the z direction has a moment about the x-axis:

$$\text{Moment about the } x\text{-axis} = -(6 \times 4) = \underline{-24\,\text{kN m}}$$

Problem 1.6

The roller supporting the bridge truss at B in Fig. S1.6a requires maintenance. For this, the truss is to be rotated slightly about A using the crane as shown. The members of the truss weigh 2 kN/m. Determine whether extra weights need to be added to the crane to ensure a factor of safety against overturning of 1.4.

Each member of the truss is of length 10 m

$$\text{Weight of each member} = 10 \times 2 = 20\,\text{kN}$$

$$\text{Total weight of truss} = 11 \times 20 = 220\,\text{kN}$$

As the truss is lifted off the roller by the crane, for purposes of considering equilibrium of the truss, the external forces acting on the truss are those shown in Fig. S1.6b where

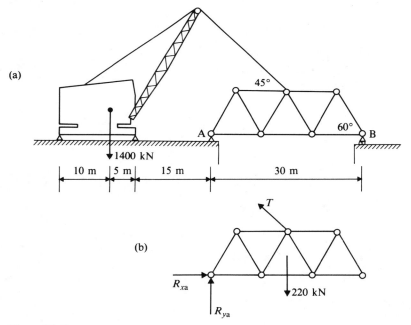

Figure S1.6

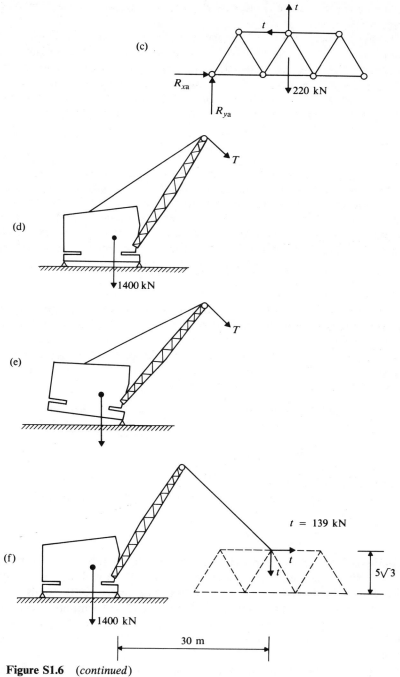

(c)

R_{xa}

R_{ya}

220 kN

(d)

T

1400 kN

(e)

T

(f)

$t = 139$ kN

$5\sqrt{3}$

1400 kN

30 m

Figure S1.6   (*continued*)

T is the tensile force in the cable, and R_{xa} and R_{ya} are the resulting reactive forces at the support A.

By taking moments about A, the value of T can be obtained without the need to calculate the values of the reactions. However, rather than calculate T, it will be preferable to obtain its vertical and horizontal components as their perpendicular distances to A can be obtained more readily. As the cable is inclined at 45°, the horizontal and vertical components of T will be equal. Let their value be t. The external forces acting on the truss are therefore as in Fig. S1.6c.

Using $\sum M_a = 0$ $\qquad\qquad (t \times 15) + (t \times 5\sqrt{3}) - (220 \times 15) = 0$

$$t = 139 \text{ kN}$$

Consider now the forces acting on the crane (Fig. S1.6d). Clearly, the problem is that the crane may tip up about its front wheels (Fig. S1.6e). To ensure that this does not happen, the engineer requires that the righting moment due to the weight of the crane shall be 1.4 times the estimated overturning moment due to the tensile force in the cable. It is, therefore, a matter of comparing these moments about the front wheels.

Righting moment due to dead load of crane = $1400 \times 5 = 7000 \text{ kN m}$

In order to determine the overturning moment due to the force from the cable, it will again be preferable to use the vertical and horizontal components calculated previously. But where should these be positioned? It should be noted that we are not actually told the position of the junction of the cable and the crane jib. We can, of course, consider the components to act at any convenient position along the line of action of the cable. The most convenient position for our calculation will be at the junction of the cable and the truss (Fig. S1.6f).

Overturning moment due to force in cable = $(139 \times 30) + (139 \times 5\sqrt{3})$

$$= 5374 \text{ kN m}$$

$$\text{Ratio} \frac{\text{Righting moment}}{\text{Overturning moment}} = \frac{7000}{5374} = 1.3$$

As this is less than the required 1.4, extra weights would need to be added to the crane.

Problem 1.7

Two balls, each of radius r and weight W are placed in a tube of radius R as shown in Fig. S1.7a. Find the minimum weight of the tube which will ensure stability.

● Initially, let the horizontal distance between the centres of the balls be h, and the vertical distance be v. (This is not essential, but it will simplify the terms for the horizontal and vertical components of the forces acting at the positions of contact of the balls. The values of h and v can be determined later in terms of r and R when required.)

The external forces acting on the upper ball are shown in Fig. S1.7b where P is the

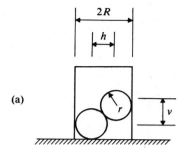

(a)

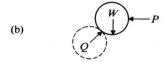

(b)

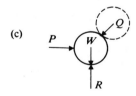

(c)

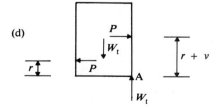

(d)

Figure S1.7

reaction acting from the tube to the ball and Q is the reaction from the lower ball to the upper ball. These forces must be in equilibrium, hence:

Using $\sum P_y = 0$

$$\frac{v}{2r} Q - W = 0$$

$$Q = \frac{2r}{v} W$$

Using $\sum P_x = 0$

$$\frac{h}{2r} Q - P = 0$$

$$P = \frac{h}{2r} Q$$

$$= \frac{h}{v} W$$

The force acting from the upper ball to the lower ball will, of course, also be of value Q (Fig. S1.7c). Clearly then, the horizontal force between the lower ball and the tube will also be of the same value P.

The horizontal forces acting on the tube are therefore as summarized in Fig. S1.7d. The difference in height of these equal forces means that the tube will tend to tip about the edge beneath the point of contact with the higher ball. At the stage at which tipping just commences, the reactive force from the ground to the tube will be concentrated at the point A. If W_t is this limiting weight of the tube, taking moments about A gives

$$(W_t \times R) + \left(\frac{h}{v} W \times r\right) = \frac{h}{v} W(r + v)$$

Note that the vertical forces due to the weights of the balls do not enter into this equation as they are conveyed directly to the ground.

Hence

$$W_t = \frac{hW}{R}$$

As

$$h = 2R - 2r = 2(R - r)$$

$$W_t = \frac{2(R - r)W}{R}$$

Problem 1.8

The two bar pin-jointed structure in Fig. S1.8a supports a load W. By considering only external forces, calculate the horizontal and vertical reactive forces at the supports A and B. Explain how it is possible to obtain these four reactive forces when only three equations of equilibrium apply to the complete system of forces on a planar structure. Show that the resultant of the vertical and horizontal forces at a particular support is

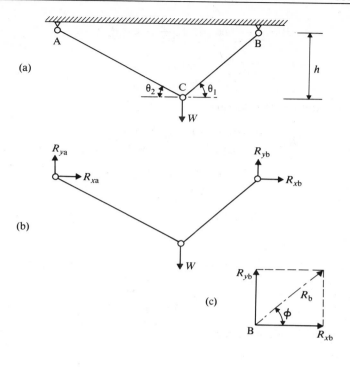

Figure S1.8

at the same angle as the member at that support, and explain why this must be the case. Also obtain expressions for the forces in the members by considering equilibrium at the loaded joint.

Let the horizontal and vertical reactive forces at the two pins A and B be as shown in Fig. S1.8b.

The lengths AC, BC and AB in terms of h, θ_1 and θ_2 are:

$$AC = \frac{h}{\sin \theta_2}$$

$$BC = \frac{h}{\sin \theta_1}$$

$$AB = \frac{h}{\sin \theta_1} \cos \theta_1 + \frac{h}{\sin \theta_2} \cos \theta_2$$

$$= \frac{h \sin (\theta_1 + \theta_2)}{\sin \theta_1 \sin \theta_2}$$

Using $\sum M_a = 0$ for the whole structure:

$$R_{yb} \frac{h \sin (\theta_1 + \theta_2)}{\sin \theta_1 \sin \theta_2} - W \frac{h \cos \theta_2}{\sin \theta_2} = 0$$

$$R_{yb} = \frac{W \sin \theta_1 \cos \theta_2}{\sin (\theta_1 + \theta_2)}$$

Using $\sum P_y = 0$ for the whole structure:

$$R_{ya} + R_{yb} - W = 0$$

$$R_{ya} = W - R_{yb}$$

$$= \frac{W \sin \theta_2 \cos \theta_1}{\sin (\theta_1 + \theta_2)}$$

Using $\sum P_x = 0$ for the whole structure:

$$R_{xa} + R_{xb} = 0$$

$$R_{xa} = -R_{xb}$$

Using $\sum M_c = 0$ for the forces to the right-hand side of C:

$$R_{yb} \frac{h \cos \theta_1}{\sin \theta_1} - R_{xb}h = 0$$

$$R_{xb} = R_{yb} \frac{\cos \theta_1}{\sin \theta_1}$$

$$= \frac{W \cos \theta_1 \cos \theta_2}{\sin (\theta_1 + \theta_2)}$$

Thus
$$R_{xa} = -\frac{W \cos \theta_1 \cos \theta_2}{\sin (\theta_1 + \theta_2)}$$

Obtaining the four reactive forces for this structure required four equilibrium equations. In addition to the three equilibrium equations for the structure as a whole, the fourth equation was obtained by isolating a part of the structure and using a known condition.

The horizontal and vertical reactive forces at B are shown in Fig. S1.8c. These two forces can be combined into a single force R_b. If this resultant acts at an angle ϕ:

$$\tan \phi = \frac{R_{yb}}{R_{xb}} = \frac{W \sin \theta_1 \cos \theta_2}{\sin (\theta_1 + \theta_2)} \times \frac{\sin (\theta_1 + \theta_2)}{W \cos \theta_1 \cos \theta_2}$$

$$= \tan \theta_1$$

Clearly then, this line of action of the resultant of the horizontal and vertical forces at the support B is along the line BC. This must, of course, be the case in order that the moment of this resultant about the joint C is zero.

Let the tensile forces in the members be T_{ac} and T_{bc}. Consider the equilibrium of the forces at C (Fig. S1.8d):

Using $\sum P_x = 0$ $\qquad\qquad T_{bc} \cos \theta_1 - T_{ac} \cos \theta_2 = 0$

Using $\sum P_y = 0$ $\qquad\qquad T_{bc} \sin \theta_1 + T_{ac} \sin \theta_2 = W$

Solving these two simultaneous equations gives

$$T_{ac} = \frac{W \cos \theta_1}{\sin (\theta_1 + \theta_2)}$$

$$T_{bc} = \frac{W \cos \theta_2}{\sin (\theta_1 + \theta_2)}$$

Problem 1.9

Explain why, for a body in equilibrium under the action of three forces only, the three forces must be concurrent or parallel. Calculate the horizontal and vertical components of the reactions at the supports of the three-pinned parabolic arch shown in Fig. S1.9a. Combine graphically the horizontal and vertical components into their resultants and show that these, with the applied load, satisfy the three-force condition. Show also that the resultant at A acts along a line passing through the pin at C and explain why this must be the case.

Let the horizontal and vertical reactive forces at A and B be as shown in Fig. S1.9b.

Using $\sum M_b = 0$ for the whole structure $(32 \times 7) - (R_{ya} \times 28) = 0$

$$R_{ya} = 8 \text{ kN}$$

Using $\sum P_y = 0$ for the whole structure $\qquad R_{ya} + R_{yb} - 32 = 0$

$$R_{yb} = 24 \text{ kN}$$

Using $\sum P_x = 0$ for the whole structure $\qquad R_{xa} + R_{xb} = 0$

$$R_{xb} = -R_{xa}$$

Using $\sum M_c = 0$ for the part of the structure to the left of C

$$(R_{xa} \times 9) - (R_{ya} \times 14) = 0$$

$$R_{xa} = 12.4 \text{ kN}$$

Thus $\qquad\qquad\qquad\qquad\qquad\qquad\qquad R_{xb} = -12.4 \text{ kN}$

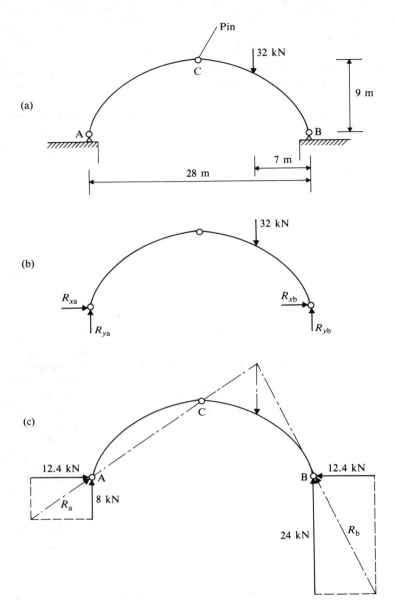

Figure S1.9

The horizontal and vertical components are combined graphically in Fig. S1.9c. By drawing the arch to scale, it can be seen that the lines of action of these resultants and the line of action of the load are concurrent. Moreover, the line of action of the resultant reaction at A passes through the pin at C. This must be the case as the resultant reaction is the only force to the left of C. In order that the moment of this force about C is zero, its line of action must pass through C.

Problem 1.10

A block of wood of mass M lies on a flat surface as in Fig. S1.10a. A horizontal force P acts at the position shown, although this does not cause it to move. Describe all the external forces acting on the block. If P gradually increases, what types of movement are possible?

The external forces acting on the block are:

– the horizontal force P;
– the vertical gravitational force Mg;
– the reactive force (or forces) at the supported face.

Clearly, the system of forces is in equilibrium and hence the reactive forces (Fig. S1.10b) can be determined. Using simple logic the forces at the supported face must include:

– a horizontal force P to counter the effect of the applied load;
– a vertically upward force Mg to counter the gravitational force.

These forces are summarized in Fig. S1.10c, but they cannot be the only ones on the supported face as the block would not then be in equilibrium. This is clear by taking moments about the midpoint of the baseline for such a system of forces, when it is seen that there is an unbalanced moment equal to Ph. It follows that the system of forces at the supported face must include a counteracting moment of value Ph as in Fig. S1.10d.

The forces at the supported face are the summation of stresses distributed over the face. For example, the shear force P is the result of shear stresses distributed over the face. The direct force Mg and the moment Ph are both the result of distributed direct stresses. In order to understand the development of the moment, it is desirable to look closer into this distribution of direct stresses, although it may be preferable to delay this particular study until Chapter 3 has also been covered.

The direct force Mg acting on the face would require a uniform distribution of stress as in Fig. S1.10e, whereas the moment would require a triangular distribution of stress as in Fig. S1.10f. The tensile stresses implied by the latter distribution of stress do not, of course, occur—indeed, they could not occur as the block is simply placed on the table. Provided the uniform compressive stresses due to the force Mg are greater than the implied maximum tensile stress due to the moment, the resulting combination would be the trapezoidal distribution of stress shown in Fig. S1.10g.

Such a distribution of stress also has a *single* resultant Mg, but this is now located eccentrically at the centroid of the area. The single eccentric force is, of course, equivalent to the combination of the axial force and the moment, and is simply an

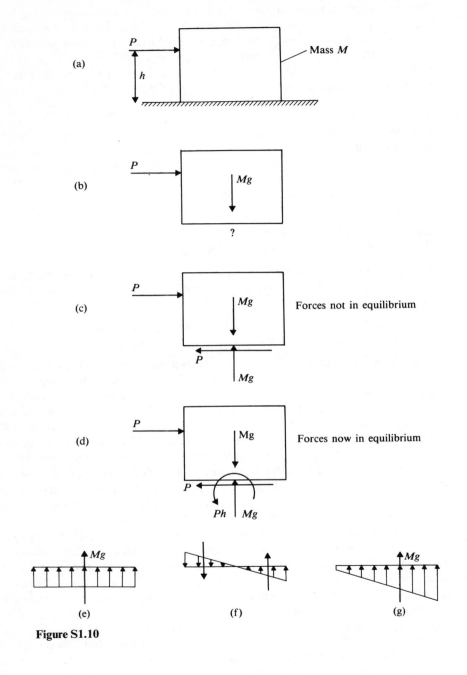

(a)

(b)

(c)

(d)

(e)

(f)

(g)

Figure S1.10

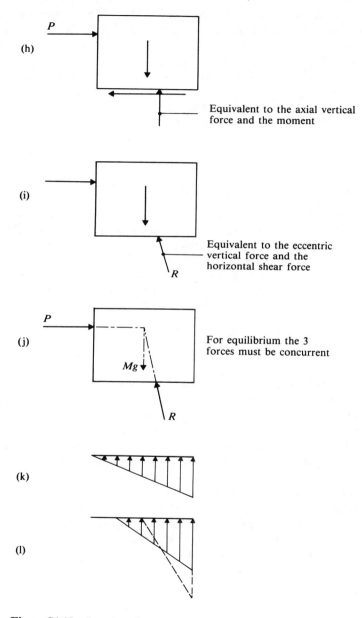

(h)

Equivalent to the axial vertical force and the moment

(i)

Equivalent to the eccentric vertical force and the horizontal shear force

(j)

For equilibrium the 3 forces must be concurrent

(k)

(l)

Figure S1.10   *(continued)*

alternative way of viewing the forces. A summary of this alternative system of external forces on the block is shown in Fig. S1.10h. One could alter this further by combining the two forces on the supported face into their resultant as in Fig. S1.10i. This single reactive force now represents the previous shear force P, the central vertical force Mg, and the moment Ph. As in this representation there are now only three external forces acting on the block, they must be concurrent (Fig. S1.10j) in order to establish equilibrium.

If the applied force P is gradually increased, there are two possible types of movement which would occur. The block would slide if the shearing force P exceeds the maximum possible frictional forces which can be developed at the interface. It will start to overturn about its edge when the overturning moment Ph exceeds the righting moment due to the dead load.

A study of the change in the distribution of the direct stresses immediately prior to overturning is instructive. It should be noted that as P increases, the uniform stresses due to the weight of the block remain constant but the stresses due to the moment effect will increase. At some stage during the increase in P, the trapezoidal distribution of stress will become triangular (Fig. S1.10k). Further increase in P will reduce the base length of the triangle of compressive stresses (Fig. S1.10*l*)—the total area will remain the same (the stresses must still sum to the value of Mg) and the centroid of the triangle will move across (corresponding to the higher moment). Eventually, a stage is reached when the triangle will disappear and the block will start to overturn.

Problem 2.1

In the pin-jointed frame in Fig. S2.1a, which of the members must have a zero force?

Members 1 and 2 (Fig. S2.1b) will have zero force for loading applied only at the joints of the upper chord. This will be clear by considering equilibrium at the joints connecting these members with the bottom chord. For example, of the three members meeting at joint C, only member 1 has a component of its length in the y direction. As the joint C has no externally applied load, it follows that member 1 must have zero force.

It is important to realize that this recognition of members with zero force is not restricted to a consideration of equilibrium in the x or y directions. For example, Fig. S2.1c shows three members meeting at an unloaded pin-joint. As members 3 and 4 are collinear, the force in member 5 must be zero as is seen by considering equilibrium in the direction *perpendicular* to members 3 and 4.

Problem 2.2

For the frame in Fig. S2.2a, use the method of sections to determine the forces in members 1, 2 and 3. Check the values using an alternative method.

Consider the section XX as shown in Fig. S2.2b. The forces acting on the part of the

(a)

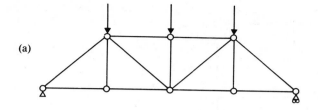

(b)

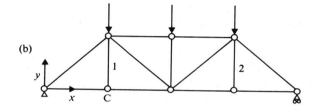

(c)

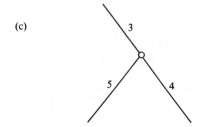

Figure S2.1

structure to the right of XX are shown in Fig. S2.2c, where F_1, F_2 and F_3 are the forces to be determined.

Using $\sum P_y = 0$ $\qquad \dfrac{F_2}{\sqrt{2}} - 10 - 10 - 10 = 0$

$$F_2 = \underline{30\sqrt{2}\,\text{kN}}$$

Using $\sum M_e = 0$ $\qquad (F_1 \times 2) - (10 \times 2) - (10 \times 4) = 0$

$$F_1 = \underline{30\,\text{kN}}$$

(It should be recognized that the joint E was adopted for the consideration of

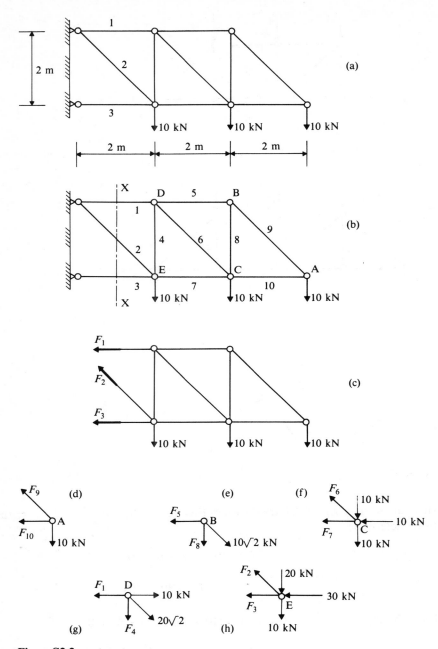

Figure S2.2

rotational equilibrium in order to eliminate F_2 and F_3 from the equation and so obtain F_1 directly.)

Using $\sum P_x = 0$

$$-F_1 - \frac{F_2}{\sqrt{2}} - F_3 = 0$$

$$F_3 = -30 - 30$$

$$= \underline{-60\ kN}$$

The method of joints will be used to check these values. This could be done either by first determining the reactive forces at the positions of the supports and then the method of joints at these positions, or alternatively starting at joint A and then at successive joints along the frame. We will use the latter. Thus at A, the forces at the start of the analysis are as shown in Fig. S2.2d.

Using $\sum P_y = 0$

$$\frac{F_9}{\sqrt{2}} - 10 = 0$$

$$F_9 = 10\sqrt{2}\ kN$$

Using $\sum P_x = 0$

$$-\frac{F_9}{\sqrt{2}} - F_{10} = 0$$

$$F_{10} = -10\ kN$$

At B, the known and unknown forces are summarized in Fig. S2.2e. The force in member 9 is acknowledged as being in tension and therefore *pulling* on the joint.

Using $\sum P_y = 0$

$$-F_8 - \frac{10\sqrt{2}}{\sqrt{2}} = 0$$

$$F_8 = -10\ kN$$

Using $\sum P_x = 0$

$$\frac{10\sqrt{2}}{\sqrt{2}} - F_5 = 0$$

$$F_5 = 10\ kN$$

At C, the forces are as in Fig. S2.2f. The forces in members 8 and 10 are acknowledged as being in compression and therefore *pushing* on the joint.

Using $\sum P_y = 0$

$$-10 - 10 + \frac{F_6}{\sqrt{2}} = 0$$

$$F_6 = 20\sqrt{2}\ kN$$

Using $\sum P_x = 0$

$$-\frac{F_6}{\sqrt{2}} - F_7 - 10 = 0$$

$$F_7 = -30\ kN$$

At D, the forces are as in Fig. S2.2g.

$$\text{Using } \Sigma P_y = 0 \qquad -F_4 - \frac{20\sqrt{2}}{\sqrt{2}} = 0$$

$$F_4 = -20 \text{ kN}$$

$$\text{Using } \Sigma P_x = 0 \quad 10 + \frac{20\sqrt{2}}{\sqrt{2}} - F_1 = 0$$

$$F_1 = \underline{30 \text{ kN}}$$

At E, the forces are as in Fig. S2.2h.

$$\text{Using } \Sigma P_y = 0 \qquad \frac{F_2}{\sqrt{2}} - 20 - 10 = 0$$

$$F_2 = \underline{30\sqrt{2} \text{ kN}}$$

$$\text{Using } \Sigma P_x = 0 \qquad -\frac{F_2}{\sqrt{2}} - F_3 - 30 = 0$$

$$F_3 = \underline{-60 \text{ kN}}$$

These confirm the previous values for F_1, F_2 and F_3.

Problem 2.3

Obtain the forces in all four members of the cantilever frame shown in Fig. S2.3a.

To facilitate the determination of the components of the member forces in the x and y directions, the coordinates of the joints are initially determined. These are summarized in Fig. S2.3b. The lengths of the members are:

$$\text{AB} = \sqrt{((10-0)^2 + (10-8)^2)} = \sqrt{104}$$
$$\text{AC} = \sqrt{((10-4)^2 + (10-6)^2)} = \sqrt{52}$$
$$\text{BC} = \sqrt{((4-0)^2 + (6-8)^2)} \quad = \sqrt{20}$$
$$\text{CD} = \sqrt{((4-0)^2 + (6-0)^2)} \quad = \sqrt{52}$$

For equilibrium at joint A (Fig. S2.3c):

$$\text{Using } \Sigma P_y = 0 \qquad -F_1 \frac{(10-8)}{\sqrt{104}} - F_2 \frac{(10-6)}{\sqrt{52}} - 20 = 0$$

$$\text{Using } \Sigma P_x = 0 \qquad -F_1 \frac{(10-0)}{\sqrt{104}} - F_2 \frac{(10-4)}{\sqrt{52}} = 0$$

Solving these two simultaneous equations gives:

$$F_1 = \underline{43.7 \text{ kN}} \qquad F_2 = \underline{-51.5 \text{ kN}}$$

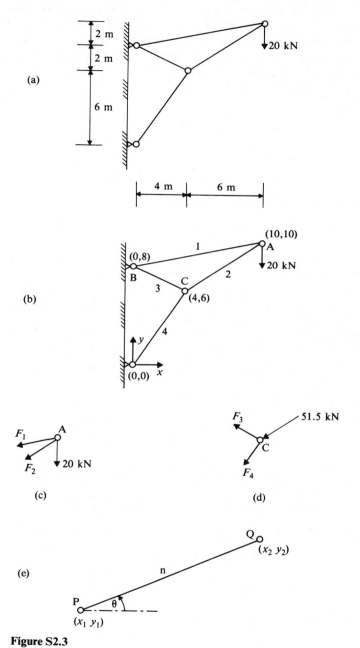

Figure S2.3

Member 1 is therefore in tension and member 2 in compression. (Students should confirm that these are in accordance with what they would expect to be the case. Such a cantilever frame subjected only to gravity forces would always sustain tension in the top members and compression in the bottom members. The opposite would be the case for a frame simply supported at its ends.)

For equilibrium at joint C (Fig. S2.3d):

Using $\sum P_y = 0$ $F_3 \dfrac{(8-6)}{\sqrt{20}} - F_4 \dfrac{(6-0)}{\sqrt{52}} - 51.5 \dfrac{(10-6)}{\sqrt{52}} = 0$

Using $\sum P_x = 0$ $-F_3 \dfrac{(4-0)}{\sqrt{20}} - F_4 \dfrac{(4-0)}{\sqrt{52}} - 51.5 \dfrac{(10-4)}{\sqrt{52}} = 0$

Solving these two simultaneous equations gives:

$$F_3 = -20 \text{ kN} \qquad F_4 = -45.1 \text{ kN}$$

The use of electronic calculators allows the solutions of these pairs of simultaneous equations to be obtained with relative ease. However, it is of interest to note that prior to calculators and computers it was important to develop techniques which reduced the arithmetical tedium. One such technique was used in the solution of such equilibrium equations. For example, solving the simultaneous equations relative to joint A initially for the values $F_1/\sqrt{104}$ and $F_2/\sqrt{52}$, and then subsequently determining F_1 and F_2 undoubtedly simplifies the arithmetic. Such a technique was known as tension coefficients. Although now only of historical interest, the procedure is worth demonstrating.

The tension coefficient of a particular member is defined as

$$\text{Tension coefficient} = \frac{\text{Force in member}}{\text{Length of member}}$$

Its use is illustrated by reference to the bar n in Fig. S2.3e. Its joints P and Q are defined by x and y coordinates. Let the tension coefficient of member n be t_n.

$$\text{Force in member n} = t_n \times \text{Length of member}$$

The horizontal component of the force exerted by the bar on the joint P

$$= t_n \times \text{Length of member} \times \cos\theta$$

$$= t_n \times \text{Length of member} \times \frac{(x_2 - x_1)}{\text{Length of member}}$$

$$= t_n(x_2 - x_1)$$

Similarly, the vertical component of the force exerted by the bar on the joint P

$$= t_n(y_2 - y_1)$$

Applying this technique to the previous problem, at joint A:

Using $\sum P_y = 0$ $\qquad$ $t_1(8 - 10) + t_2(6 - 10) - 20 = 0$

Using $\sum P_x = 0$ $\qquad$ $t_1(0 - 10) + t_2(4 - 10) = 0$

Solving these simultaneous equations gives

$$t_1 = \frac{30}{7} \, \text{kN/m} \qquad t_2 = -\frac{50}{7} \, \text{kN/m}$$

Hence the forces in the members are

$$F_1 = \frac{30}{7} \sqrt{104} = \underline{43.7 \, \text{kN}}$$

$$F_2 = -\frac{50}{7} \sqrt{52} = \underline{-51.5 \, \text{kN}}$$

The student is encouraged to use the method of tension coefficients to consider equilibrium at joint C to obtain the forces in members 3 and 4.

Problem 2.4

Check that the frame in Fig. S2.4a is statically determinate. Use the method of sections in conjunction with a consideration of equilibrium at joint C to find the forces in members 1, 2, 3 and 4. Why is it not possible to use the method of sections alone?

Number of members $m = 25$

Number of reactive forces $r = 2 + 1 = 3$

Number of joints $j = 14$

As $m + r = 2j$, the frame is statically determinate.

The simple use of the method of sections with a section passing through the panel containing members 1, 2, 3 and 4 would not suffice to obtain the member forces. Only three equations would be generated from a consideration of the part of the frame to one side of the section. A fourth equation is therefore necessary and this can be obtained from a consideration of horizontal equilibrium of the joint at C.
   Initially, the reactions at A (see Fig. S2.4b) are determined. As the applied loads are all vertical, clearly the reaction R_{xa} will be zero. The value of R_{ya} is obtained by taking moments about B:

$$(10 \times 9) + (10 \times 6) + (10 \times 3) - (10 \times 3) - (10 \times 6) - (R_{ya} \times 12) = 0$$

$$R_{ya} = 7.5 \, \text{kN}$$

The part of the structure to the left of the section XX is shown in Fig. S2.4c.

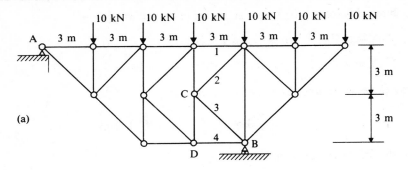

(a)

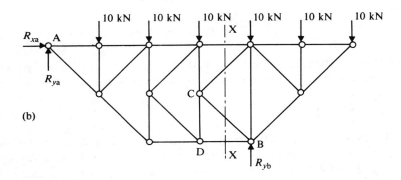

(b)

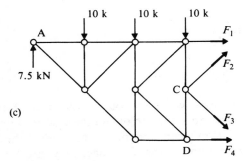

(c)

Figure S2.4

Using $\sum P_x = 0$ for the joint at C:

$$\frac{F_2}{\sqrt{2}} + \frac{F_3}{\sqrt{2}} = 0$$

$$F_2 = -F_3$$

Consider now the equilibrium of the part of the structure shown in Fig. S2.4c.

Using $\sum P_y = 0$ $\dfrac{F_2}{\sqrt{2}} - \dfrac{F_3}{\sqrt{2}} + 7.5 - 10 - 10 - 10 = 0$

$$F_2 = 15.9 \text{ kN}$$

$$F_3 = -15.9 \text{ kN}$$

Using $\sum M_d = 0$ $(10 \times 6) + (10 \times 3) - (7.5 \times 9) - (F_1 \times 6) = 0$

(Note that the moments due to the forces F_2 and F_3 cancel each other out.)

$$F_1 = 3.75 \text{ kN}$$

Using $\sum P_x = 0$ $F_1 + F_4 + \dfrac{F_2}{\sqrt{2}} + \dfrac{F_3}{\sqrt{2}} = 0$

$$F_4 = -3.75 \text{ kN}$$

Problem 2.5

For the frame in Fig. S2.5a used to illustrate the matrix method of joints (in Section 2.1 of Part One), use the matrix solution previously obtained for the general loading case to calculate the force in member 2 for the loading case of vertical loads of 100 kN at each of the joints B, C and D. Check the answer using the method of sections.

The matrix solution for this frame given in Part One for the general case of loading shown in Fig. S2.5a is

$$
\begin{bmatrix} F_1 \\ F_2 \\ F_3 \\ F_4 \\ F_5 \\ F_6 \\ F_7 \end{bmatrix}
= -
\begin{bmatrix}
-0.417 & -0.938 & -0.417 & -0.313 & 0 & -0.625 & 0 \\
0.5 & -0.375 & -0.5 & -0.375 & 0 & -0.75 & 0 \\
0.417 & -0.313 & 0.417 & 0.313 & 0 & 0.625 & 0 \\
-0.75 & 0.563 & -0.75 & 0.188 & -1 & 0.375 & -1 \\
-0.417 & 0.313 & -0.417 & -0.313 & 0 & 0.625 & 0 \\
0.417 & -0.313 & 0.417 & -0.937 & 0 & -0.625 & 0 \\
-0.25 & 0.188 & -0.25 & 0.563 & 0 & 0.375 & -1
\end{bmatrix}
\begin{bmatrix} P_{xb} \\ P_{yb} \\ P_{xc} \\ P_{yc} \\ P_{xd} \\ P_{yd} \\ P_{xe} \end{bmatrix}
$$

The column matrix of loads for the case shown in Fig. S2.5b is

$$
\begin{bmatrix} 0 \\ -100 \\ 0 \\ -100 \\ 0 \\ -100 \\ 0 \end{bmatrix}
$$

Hence $F_2 = -(37.5 + 37.5 + 75) = \underline{-150 \text{ kN}}$

The force in member 2 is, therefore, compressive and has a value of 150 kN.

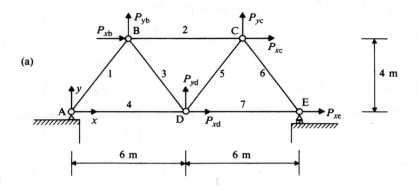

(a)

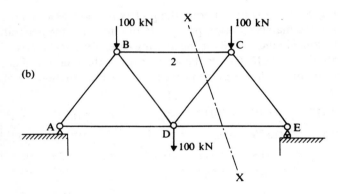

(b)

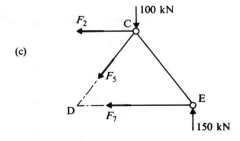

(c)

Figure S2.5

To check this answer using the method of sections, it will be necessary to calculate the reaction(s) at one of the supports. We will calculate the vertical reaction at the roller support at E.

Using $\sum M_a = 0$ for the whole frame:

$$(R_{ye} \times 12) - (100 \times 3) - (100 \times 6) - (100 \times 9) = 0$$

$$R_{ye} = 150 \text{ kN}$$

(This value is obtained more easily by observing the symmetry of the frame and the loading. The vertical reaction at E must therefore equal half the total load.)

Consider the equilibrium of the part of the frame to the right of section XX (Fig. S2.5c). Although there are three unknown forces acting on this part of the frame, we are only required to find F_2. This can be obtained directly by considering the rotational equilibrium about D, since the lines of action of unknown forces F_5 and F_7 pass through D and therefore do not feature in this equation.

Using $\sum M_d = 0$ $\quad (F_2 \times 4) + (150 \times 6) - (100 \times 3) = 0$

$$F_2 = -150 \text{ kN}$$

This confirms the previous value.

Problem 2.6

Show that the frame in Fig. S2.6a is statically determinate. Explain why the simple hand methods of sections and joints cannot be used directly to obtain the forces in the members. Assemble in matrix form the equilibrium equations which would enable the member forces to be calculated for the generalized loading system and assuming the following (x, y) coordinates of the joints: A (0, 0); B (1, 2); C (3, 3); D (5, 3); E (7, 2); F (8, 0).

Number of members $\quad\quad\quad m = 9$

Number of reactive forces $\quad\quad r = 2 + 1 = 3$

Number of joints $\quad\quad\quad\quad j = 6$

As $m + r = 2j$ (i.e., the number of unknowns equals the number of available equilibrium equations), the frame is statically determinate. However, the simple method of joints cannot be used to obtain the member forces—there are only two equilibrium equations available for the analysis of a joint but there is no joint at which only two members meet. Hence one cannot start the process. Moreover, the method of sections cannot be used as every section would cut either more than three members or three members meeting at a point. In the latter case, the method of sections reduces to the analysis of a joint for which, as stated, only two equilibrium equations are available.

The series of interconnected equilibrium equations relating the member forces and the loads shown in Fig. S2.6b is obtained by considering horizontal and vertical

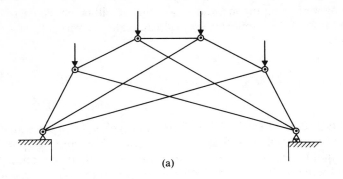

(a)

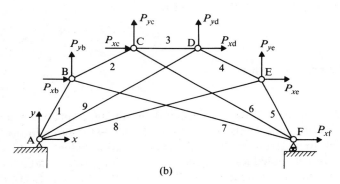

(b)

Figure S2.6

equilibrium at the four joints B, C, D and E and horizontal equilibrium at the support F. This leads to nine simultaneous equations which can be solved for the nine member forces.

As the shape of the frame is defined by x, y coordinates, it will be advantageous to present the equations in terms of tension coefficients (see the earlier discussion in relation to Problem 2.3). Let the tension coefficients for the nine members be $t_1 \, t_2 \, t_3 \ldots t_9$.

Consider the equilibrium at joint B:

Using $\sum P_x = 0$ $t_1(0 - 1) + t_2(3 - 1) + t_7(8 - 1) + P_{xb} = 0$

$$-t_1 + 2t_2 + 7t_7 = -P_{xb}$$

Using $\sum P_y = 0$ $t_1(0 - 2) + t_2(3 - 2) + t_7(0 - 2) + P_{yb} = 0$

$$-2t_1 + t_2 - 2t_7 = -P_{yb}$$

Consider the equilibrium at joint C:

Using $\sum P_x = 0$ $t_2(1 - 3) + t_3(5 - 3) + t_6(8 - 3) + P_{xc} = 0$

$$-2t_2 + 2t_3 + 5t_6 = -P_{xc}$$

Using $\sum P_y = 0$ $t_2(2 - 3) + t_3(3 - 3) + t_6(0 - 3) + P_{yc} = 0$

$$-t_2 - 3t_6 = -P_{yc}$$

Similarly, five other equations are obtained by considering the equilibrium at joints D, E and F. The resulting nine simultaneous equations assembled in matrix form are

$$\begin{bmatrix} -1 & 2 & & & & & 7 & & \\ -2 & 1 & & & & & -2 & & \\ & & -2 & 2 & & & 5 & & \\ & & -1 & & & & -3 & & \\ & & & -2 & 2 & & & -5 & \\ & & & -1 & & & & -3 & \\ & & & & -2 & 1 & & -7 & \\ & & & & 1 & -2 & & -2 & \\ & & & & & -1 & -5 & -7 & \end{bmatrix} \begin{bmatrix} t_1 \\ t_2 \\ t_3 \\ t_4 \\ t_5 \\ t_6 \\ t_7 \\ t_8 \\ t_9 \end{bmatrix} = - \begin{bmatrix} P_{xb} \\ P_{yb} \\ P_{xc} \\ P_{yc} \\ P_{xd} \\ P_{yd} \\ P_{xe} \\ P_{ye} \\ P_{xf} \end{bmatrix}$$

Using a computer to invert the square matrix allows equations for the values of the tension coefficients t to be obtained. These can then be translated to member forces by multiplying by the appropriate member length.

Length of members 1 and 5 $= \sqrt{((1 - 0)^2 + (2 - 0)^2)} = \sqrt{5}$ m

Length of members 2 and 4 $= \sqrt{((3 - 1)^2 + (3 - 2)^2)} = \sqrt{5}$ m

Length of member 3 $= 5 - 3 = 2$ m

Length of members 6 and 9 $= \sqrt{((5 - 0)^2 + (3 - 0)^2)} = \sqrt{34}$ m

Length of members 7 and 8 $= \sqrt{((7 - 0)^2 + (2 - 0)^2)} = \sqrt{53}$ m

Problem 2.7

Use the unit load method to calculate the vertical deflection at joint D of the frame in Fig. S2.7a. Check the value using the matrix inverse for this frame given in Section 2.2 of Part One.

To determine the vertical deflection at D caused by the applied loads, a unit load $X_1 = 1$ is applied at D as shown in Fig. S2.7b. The forces f_1 caused in the members by this unit load are summarized in Table S2.7. (The forces are determined by the successive use of the method of joints. Note that symmetry of the frame and the loading reduces the arithmetic involved.)

The member forces F_0 caused by the original 100 kN loads are also summarized in Table S2.7. Again symmetry reduces the arithmetic.

Young's Modulus for all members   = 200 kN/mm^2
Cross-sectional area for members 1,3,5 and 6 = 625 mm^2
Cross-sectional area for members 2,4 and 7   = 750 mm^2

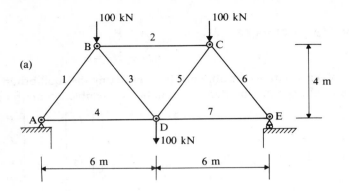

(a)

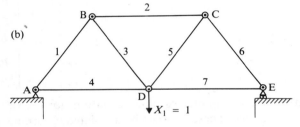

(b)

Figure S2.7

Table S2.7

Member	F_0 kN	f_1	L mm	A mm^2	$\dfrac{F_0 f_1 L}{A}$ kN/mm
1	−187.5	−0.625	5 × 10^3	625	937.5
2	−150	−0.75	6 × 10^3	750	900
3	62.5	0.625	5 × 10^3	625	312.5
4	112.5	0.375	6 × 10^3	750	337.5
5	62.5	0.625	5 × 10^3	625	312.5
6	−187.5	−0.625	5 × 10^3	625	937.5
7	112.5	0.375	6 × 10^3	750	337.5

$$\sum \frac{F_0 f_1 L}{A} = 4075 \text{ kN/mm}$$

Hence

$$\Delta_{yd} = \frac{\sum \dfrac{F_0 f_1 L}{A}}{E} = \frac{4075}{200}$$

$$= \underline{20.4 \text{ mm}} \quad \text{(to 3 significant figures)}$$

To check this value using matrices, reference must be made to the matrix inverse for this frame given in Section 2.2 of Part One for the general loading case. For convenience, this is repeated here:

$$
\begin{bmatrix} \Delta_{xb} \\ \Delta_{yb} \\ \Delta_{xc} \\ \Delta_{yc} \\ \Delta_{xd} \\ \Delta_{yd} \\ \Delta_{xe} \end{bmatrix} = -0.01
\begin{bmatrix}
-6.28 & 2.62 & -4.28 & 1.88 & -3.00 & 3.00 & -4.00 \\
2.62 & -6.66 & 1.12 & -2.97 & 2.25 & -5.38 & 3.00 \\
-4.28 & 1.12 & -6.28 & 0.38 & -3.00 & 0 & -4.00 \\
1.88 & -2.97 & 0.38 & -6.66 & 0.75 & -5.38 & 3.00 \\
-3.00 & 2.25 & -3.00 & 0.75 & -4.00 & 1.50 & -4.00 \\
3.00 & -5.38 & 0 & -5.38 & 1.50 & -9.63 & 3.00 \\
-4.00 & 3.00 & -4.00 & 3.00 & -4.00 & 3.00 & -8.00
\end{bmatrix}
\begin{bmatrix} P_{xb} \\ P_{yb} \\ P_{xc} \\ P_{yc} \\ P_{xd} \\ P_{yd} \\ P_{xe} \end{bmatrix}
$$

For our loading case, the load matrix is

$$
\begin{bmatrix} 0 \\ -100 \\ 0 \\ -100 \\ 0 \\ -100 \\ 0 \end{bmatrix}
$$

Hence
$$\Delta_{yd} = -(5.38 + 5.38 + 9.63)$$

$$= -20.4 \text{ mm (to 3 significant figures)}$$

This confirms the previous answer. (The positive sign in the unit load method means that the deflection corresponds to the direction of the unit load, namely, downward. The negative sign in the matrix method means that the deflection is in the opposite direction to the positive direction of the y-axis, i.e., the deflection is downward.)

Problem 2.8

In the pin-jointed cantilever frame in Fig. S2.8a, all the members have a cross-sectional area A and a modulus of elasticity E. Use the unit load method to determine the vertical deflection at D and check this using the graphical procedure. Why cannot the strain energy method be used? (Take care not to do unnecessary calculations in the unit load method.)

The caution against unnecessary calculations is because the unit load method for calculating deflections usually requires the determination of the forces in all members for two systems of loading, namely, the original loads and the imaginary unit load. In this particular case, however, the position of the imaginary load ($X_1 = 1$ at D in Fig.

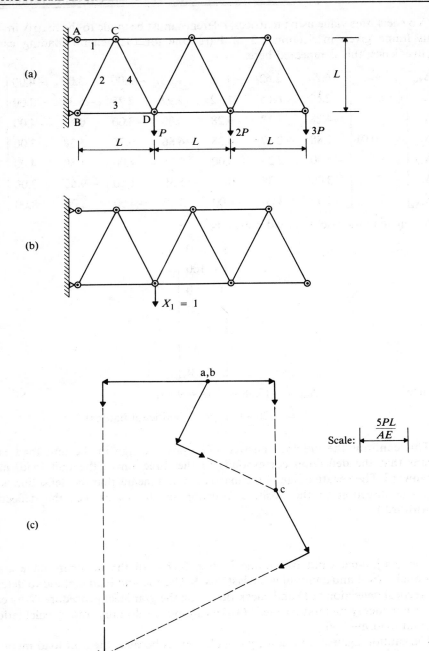

Figure S2.8

S2.8b) is such that, for this load, there is zero force in the majority of members—namely the eight undesignated members. (This can be seen by considering the equilibrium of the part of the frame to the right of a section passing through any three of the undesignated members.) It follows that there is also no need to calculate the member forces caused by the original loads (Fig. S2.8a) for these same eight members.

The forces F_0 in members 1, 2, 3 and 4 due to the original loads are summarized in the table below. Clearly, the method of sections will be preferable for calculating these forces. The forces f_1 caused by $X_1 = 1$ are also given in Table S2.8.

Table S2.8

Member	F_0	f_1	Length	$F_0 f_1 \times Length$
1	$14P$	1	$\dfrac{L}{2}$	$7.00PL$
2	$-3\sqrt{(5)}P$	$-\dfrac{\sqrt{5}}{2}$	$\dfrac{\sqrt{(5)}L}{2}$	$8.39PL$
3	$-11P$	$-\dfrac{1}{2}$	L	$5.50PL$
4	$3\sqrt{(5)}P$	$\dfrac{\sqrt{5}}{2}$	$\dfrac{\sqrt{(5)}L}{2}$	$8.39PL$

$$\sum (F_0 f_1 \times Length) = 29.28PL$$

Hence
$$\Delta_{yd} = \frac{29.28PL}{AE}$$

$$= \frac{29.3PL}{AE}$$

To check this using the graphical procedure requires the calculation of the changes in length of members 1, 2, 3 and 4.

$$\text{Extension of member 1} = \frac{14P}{AE} \times \frac{L}{2} = \frac{7PL}{AE}$$

$$\text{Extension of member 2} = -\frac{3\sqrt{(5)}P}{AE} \times \frac{\sqrt{(5)}L}{2} = -\frac{7.5PL}{AE}$$

$$\text{Extension of member 3} = -\frac{11P}{AE} \times L = -\frac{11PL}{AE}$$

$$\text{Extension of member 4} = \frac{3\sqrt{(5)}P}{AE} \times \frac{\sqrt{(5)}L}{2} = \frac{7.5PL}{AE}$$

The construction of the Williot diagram is shown in Fig. S2.8c. This confirms the previous value for the vertical deflection of D. It also indicates a horizontal component of the movement of D equal to $11PL/AE$, this being, of course, equal to the shortening of member 3.

The strain energy method can only be used when a single load is applied to the structure, and then only when the deflection of the frame is required at the point of application of the load and in the direction of the load. In this problem, loads are applied at three positions and the energy equation would involve three unknown deflections.

Problem 2.9

The steel pin-jointed truss in Fig. S2.9a is to be designed to support the loads shown over the span of 32 m. Some of the constraints on design are that (a) all members must be of the same cross-section, (b) the stress in any member shall not exceed 150 N/mm², and (c) the maximum deflection below the supports shall not exceed 40 mm. Determine the minimum cross-sectional area of the members to satisfy these requirements. Assume that E is 200 kN/mm².

The maximum deflection will clearly occur at mid-span. The unit load method can be used to determine this deflection in terms of the cross-sectional area of the members used for the frame. Symmetry of the frame and the loading will reduce the arithmetic involved.

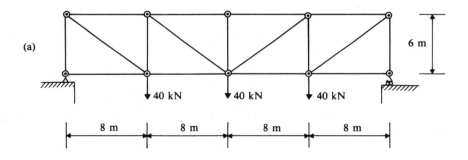

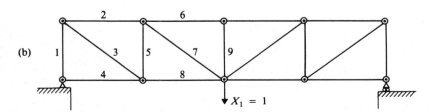

Figure S2.9

The forces F_0 in the members in the left-hand half of the frame caused by the applied loads are determined using the method of joints and are summarized in the table below. The forces f_1 in these members caused by the unit load $X_1 = 1$ applied at mid-span (Fig. S2.9b) are also summarized in Table S2.9.

Table S2.9

Member	L m	F_0 kN	f_1	$F_0 f_1 L$ kN m
1	6	−60	−0.5	180
2	8	−80	−0.667	426.9
3	10	100	0.833	833
4	8	0	0	0
5	6	−20	−0.5	60
6	8	−106.7	−1.33	1135.3
7	10	33.3	0.833	277.4
8	8	80	0.667	426.9
9	6	0	0	0

$\sum F_0 f_1 L$ for the whole frame $= 6679$ kN m

To satisfy the deflection condition:

$$\frac{6679 \times 10^3}{A \times 200} \leqslant 40$$

Hence
$$A \geqslant 835 \text{ mm}^2$$

To satisfy the stress condition:

$$\text{Maximum member force} = 106.7 \text{ kN}$$

Hence
$$\frac{106.7 \times 10^3}{A} \leqslant 150$$

$$A \geqslant 711 \text{ mm}^2$$

Thus to satisfy both conditions minimum cross-sectional area $= \underline{835 \text{ mm}^2}$

Problem 2.10

The designer of the pin-jointed frame in Fig. S2.10a proportioned the members so that, under the design load P, the strain developed in the tension members is 750×10^{-6} whilst in the compression members it is 250×10^{-6}. Explain why he has differentiated in this way between the two types of member. Calculate the vertical deflection of joint E caused by the load P.

Because of the problem of buckling, the load-carrying capacity of compression members can be appreciably less than that of tension members of comparable size. In

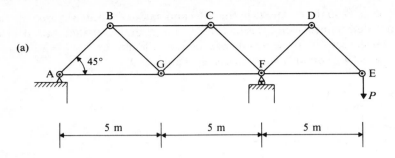

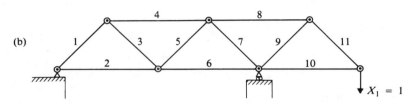

Figure S2.10

design, this reflects itself in the maximum stresses (and hence strains) which can be permitted under the applied load, those in compression members being lower than in tension members.

The members of this frame have different cross-sectional areas, so arranged to give the stated strains. Since

$$\frac{\text{Force in a member}}{\text{Cross-sectional area} \times \text{Young's Modulus}} = \text{strain in member}$$

the unit load expression for deflection can be re-written as:

$$\Delta_1 = \sum \frac{F_o f_1 L}{AE} = \sum \varepsilon_0 f_1 L$$

where ε_0 is the strain in the member caused by the original load.

The forces f_1 due to the unit load $X_1 = 1$ applied at E (Fig. S2.10b) are found using the method of joints, and these are given in the table below. The signs of these forces (i.e., whether the members are in compression or tension) indicate also the signs of the forces in the members under the original load P since both P and $X_1 = 1$ are applied at the joint E and in the same direction. This enables the values of ε_0 to be decided for each member and these are summarized in Table S2.10.

Table S2.10

Member	f_1	ε_0	$\dfrac{L}{mm}$	$\dfrac{\varepsilon_0 f_1 L}{mm}$
1	$\dfrac{1}{\sqrt{2}}$	750×10^{-6}	$\dfrac{5}{\sqrt{2}} \times 10^3$	1875×10^{-3}
2	$-\dfrac{1}{2}$	-250×10^{-6}	5×10^3	625×10^{-3}
3	$-\dfrac{1}{\sqrt{2}}$	-250×10^{-6}	$\dfrac{5}{\sqrt{2}} \times 10^3$	625×10^{-3}
4	1	750×10^{-6}	5×10^3	3750×10^{-3}
5	$\dfrac{1}{\sqrt{2}}$	750×10^{-6}	$\dfrac{5}{\sqrt{2}} \times 10^3$	1875×10^{-3}
6	$-\dfrac{3}{2}$	-250×10^{-6}	5×10^3	1875×10^{-3}
7	$-\dfrac{1}{\sqrt{2}}$	-250×10^{-6}	$\dfrac{\sqrt{5}}{2} \times 10^3$	625×10^{-3}
8	2	750×10^{-6}	5×10^3	7500×10^{-3}
9	$-\sqrt{2}$	-250×10^{-6}	$\dfrac{5}{\sqrt{2}} \times 10^3$	1250×10^{-3}
10	-1	-250×10^{-6}	5×10^3	1250×10^{-3}
11	$\sqrt{2}$	750×10^{-6}	$\dfrac{5}{\sqrt{2}} \times 10^3$	3750×10^{-3}

$$\sum \varepsilon_0 f_1 L = 25\,000 \times 10^{-3}$$

Hence deflection at joint E = 25 mm

Problem 2.11

Both members of the frame in Fig. S2.11a are of steel (Young's Modulus 200 kN/mm^2), but member 1 has a cross-sectional area of 125 mm^2 whilst member 2 has a cross-sectional area of 250 mm^2. The application of the 50 kN load causes joint A to undergo displacements Δ_{xa} and Δ_{ya}. Obtain expressions for the forces in the members in terms of these displacements. Use the expressions to obtain the equilibrium equations at joint A and solve the simultaneous equations for Δ_{xa} and Δ_{ya}. Hence obtain the member forces. Check the forces using the method of joints directly, and check the displacements using the graphical method.

It was shown in Part One, Section 2.2, that the tensile force in a general member i in terms of the joint displacements at its two ends as in Fig. S2.11b is given by

$$F_i = \left(\frac{AE}{L}\right)_i ((\Delta_{xn} - \Delta_{xm}) \cos \theta_i + (\Delta_{yn} - \Delta_{ym}) \sin \theta_i)$$

In the case of members 1 and 2 of the structure in Fig. S2.11a:

$$\cos \theta_1 = -\frac{4}{5} \qquad \sin \theta_1 = \frac{3}{5}$$

$$\cos \theta_2 = \frac{3}{5} \qquad \sin \theta_2 = \frac{4}{5}$$

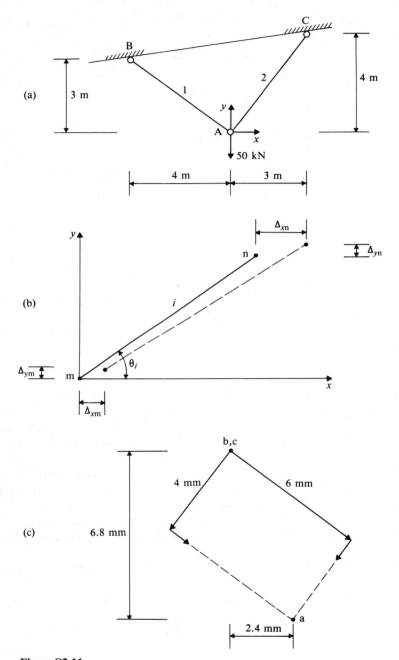

Figure S2.11

$$\left(\frac{AE}{L}\right)_1 = \frac{125 \times 200}{5 \times 10^3} = 5 \text{ kN/mm}$$

$$\left(\frac{AE}{L}\right)_2 = \frac{250 \times 200}{5 \times 10^3} = 10 \text{ kN/mm}$$

Hence the forces in members 1 and 2 in terms of the displacements at A are

$$F_1 = 5\left((0 - \Delta_{xa})\left(-\frac{4}{5}\right) + (0 - \Delta_{ya})\left(\frac{3}{5}\right)\right)$$

$$= 4\Delta_{xa} - 3\Delta_{ya}$$

$$F_2 = 10\left((0 - \Delta_{xa})\left(\frac{3}{5}\right) + (0 - \Delta_{ya})\left(\frac{4}{5}\right)\right)$$

$$= -6\Delta_{xa} - 8\Delta_{ya}$$

To determine the two values of Δ, these forces are translated into their horizontal and vertical components in order to obtain the horizontal and vertical equilibrium equations at joint A. Thus:

$$F_{x1} = F_1 \cos \theta_1 = -\frac{16}{5}\Delta_{xa} + \frac{12}{5}\Delta_{ya}$$

$$F_{y1} = F_1 \sin \theta_1 = \frac{12}{5}\Delta_{xa} - \frac{9}{5}\Delta_{ya}$$

$$F_{x2} = F_2 \cos \theta_2 = -\frac{18}{5}\Delta_{xa} - \frac{24}{5}\Delta_{ya}$$

$$F_{y2} = F_2 \sin \theta_2 = -\frac{24}{5}\Delta_{xa} - \frac{32}{5}\Delta_{ya}$$

Using $\sum P_x = 0$ at joint A:

$$-\frac{16}{5}\Delta_{xa} + \frac{12}{5}\Delta_{ya} - \frac{18}{5}\Delta_{xa} - \frac{24}{5}\Delta_{ya} = 0$$

$$17\Delta_{xa} + 6\Delta_{ya} = 0 \tag{2.11a}$$

Using $\sum P_y = 0$ at joint A:

$$\frac{12}{5}\Delta_{xa} - \frac{9}{5}\Delta_{ya} - \frac{24}{5}\Delta_{xa} - \frac{32}{5}\Delta_{ya} - 50 = 0$$

$$12\Delta_{xa} + 41\Delta_{ya} = -250 \tag{2.11b}$$

Solving Eqs (2.11a) and (2.11b) gives:

$$\Delta_{xa} = \underline{2.4 \text{ mm}}$$

$$\Delta_{ya} = \underline{-6.8 \text{ mm}}$$

Substituting these values in the earlier equations for F_1 and F_2 gives:

$$F_1 = \underline{30 \text{ kN}}$$

$$F_2 = \underline{40 \text{ kN}}$$

Although the above is a tedious hand method of determining member forces in a pin-jointed frame, it does form the basis of most computer methods. Needless to say, for simple cases like this particular structure, the calculation of the member forces can be determined by hand quite easily by considering simple equilibrium at joint A:

Using $\sum P_y = 0$
$$\frac{3}{5} F_1 + \frac{4}{5} F_2 - 50 = 0$$

$$3F_1 + 4F_2 = 250 \tag{2.11c}$$

Using $\sum P_x = 0$
$$-\frac{4}{5} F_1 + \frac{3}{5} F_2 = 0$$

$$-4F_1 + 3F_2 = 0 \tag{2.11d}$$

Solving Eqs. (2.11c) and (2.11d) verifies the previous values of F_1 and F_2.

To use the graphical method of determining the movement at A, it is first necessary to calculate the changes of length of the two members:

$$\delta_1 = \frac{30 \times 5 \times 10^3}{125 \times 200} = 6 \text{ mm}$$

$$\delta_2 = \frac{40 \times 5 \times 10^3}{250 \times 200} = 4 \text{ mm}$$

The Williot diagram is shown in Fig. S2.11c. This confirms the previous values of the displacements at A.

Problem 2.12

Obtain the equilibrium equation corresponding to the x direction for joint C for the frame in Fig. S2.12 in terms of joint displacements. Check the answer with the appropriate equation of the complete set of equilibrium equations given in matrix form on page 103 of Part One.

As in the previous problem, the expression for the tensile force in a general member will be used, namely

$$F_i = \left(\frac{AE}{L}\right)_i ((\Delta_{xn} - \Delta_{xm}) \cos \theta_i + (\Delta_{yn} - \Delta_{ym}) \sin \theta_i)$$

For member 2:

$$A_2 = 750 \text{ mm}^2$$

$$L_2 = 6 \times 10^3 \text{ mm}$$

Young's Modulus for all members   = 200 kN/mm^2
Cross-sectional area for members 1,3,5 and 6 = 625 mm^2
Cross-sectional area for members 2,4 and 7   = 750 mm^2

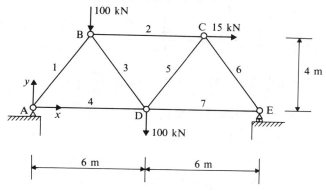

Figure S2.12

$$\left(\frac{AE}{L}\right)_2 = \frac{750 \times 200}{6 \times 10^3} = 25 \text{ kN/mm}$$

$$\cos \theta_2 = -1 \qquad \sin \theta_2 = 0$$

$$F_2 = 25((\Delta_{xb} - \Delta_{xc})(-1) + (\Delta_{yb} - \Delta_{yc})0)$$

$$= -25\Delta_{xb} + 25\Delta_{xc}$$

Hence
$$F_{x2} = F_2 \cos \theta_2 = 25\Delta_{xb} - 25\Delta_{xc}$$

For member 5:

$$A_5 = 625 \text{ mm}^2$$

$$L_5 = 5 \times 10^3 \text{ mm}$$

$$\left(\frac{AE}{L}\right)_5 = \frac{625 \times 200}{5 \times 10^3} = 25 \text{ kN/mm}$$

$$\cos \theta_5 = -\frac{3}{5} \qquad \sin \theta_5 = -\frac{4}{5}$$

$$F_5 = 25\left((\Delta_{xd} - \Delta_{xc})\left(-\frac{3}{5}\right) + (\Delta_{yd} - \Delta_{yc})\left(-\frac{4}{5}\right)\right)$$

$$= -15(\Delta_{xd} - \Delta_{xc}) - 20(\Delta_{yd} - \Delta_{yc})$$

Hence
$$F_{x5} = F_5 \cos \theta_5$$

$$= 9(\Delta_{xd} - \Delta_{xc}) + 12(\Delta_{yd} - \Delta_{yc})$$

For member 6:

$$\left(\frac{AE}{L}\right)_6 = 25 \text{ kN/mm}$$

$$\cos \theta_6 = \frac{3}{5} \qquad \sin \theta_6 = -\frac{4}{5}$$

$$F_6 = 25\left((\Delta_{xe} - \Delta_{xc})\frac{3}{5} + (0 - \Delta_{yc})\left(-\frac{4}{5}\right)\right)$$

$$= 15(\Delta_{xe} - \Delta_{xc}) + 20\Delta_{yc}$$

Hence $\qquad F_{x6} = 9(\Delta_{xe} - \Delta_{xc}) + 12\Delta_{yc}$

At joint C:

Using $\sum P_x = 0$ $\qquad\qquad F_{x2} + F_{x5} + F_{x6} + 15 = 0$

Substituting the previous expressions for F_{x2}, F_{x5} and F_{x6} gives

$$25\Delta_{xb} - 43\Delta_{xc} + 9\Delta_{xd} + 12\Delta_{yd} + 9\Delta_{xe} = -15$$

Problem 2.13

The space frame in Fig. S2.13a is in the form of a cube. It is loaded by the twin forces P acting along the line of the diagonal. Which two members of the frame must clearly be unstressed under this loading? Prove that the force in member BC also has zero force. Which additional two members will therefore also be unstressed?

There are four members meeting at the unloaded joint G. However, as three of these members are in the same plane (BGHE), it follows that the force in the fourth member (DG) must be zero. A similar consideration at joint F shows that member DF must also have zero force.

If it can be proved that the force in member BC has a zero force, it will follow from a consideration of equilibrium at C in the z direction (Fig. S2.13b) that the force in member CE must also be zero—C is an unloaded joint and members CB and CE are the only ones meeting at that joint having a component of length in the z direction. Similarly, the force in member BE will also be zero.

This observation suggests that one should aim to determine the force in member BC from a consideration of equilibrium in the x direction at joint C. This will require a knowledge of the forces in members CA and CD, and so the first steps of the analysis will aim to determine these forces.

If the length of the side of the cube is considered as unity, the length of the diagonal AH is $\sqrt{3}$. Hence the tensile forces in members AB, AC and AD are all equal to $P/\sqrt{3}$, as a consideration of equilibrium at joint A clearly indicates. (For example, the component of P in the x direction must be resisted by the force in AC.)

Now, considering equilibrium at D whilst acknowledging that the forces in DG and DF are zero:

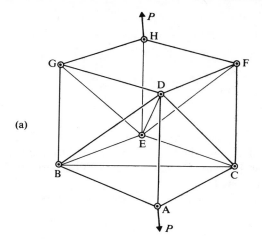

(a)

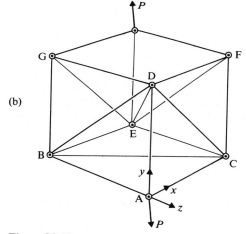

(b)

Figure S2.13

Using $\sum P_z = 0$
$$-\frac{F_{DB}}{\sqrt{2}} - \frac{F_{DE}}{\sqrt{3}} = 0$$

Using $\sum P_y = 0$
$$-F_{AD} - \frac{F_{CD}}{\sqrt{2}} - \frac{F_{DB}}{\sqrt{2}} - \frac{F_{DE}}{\sqrt{3}} = 0$$

Hence
$$-\frac{P}{\sqrt{3}} - \frac{F_{CD}}{\sqrt{2}} = 0$$

$$F_{CD} = -\frac{\sqrt{(2)}P}{\sqrt{3}}$$

Now considering joint C:

Using $\sum P_x = 0$ $\qquad -F_{CA} - \dfrac{F_{CB}}{\sqrt{2}} - \dfrac{F_{CD}}{\sqrt{2}} = 0$

Hence $\qquad\qquad\qquad -\dfrac{P}{\sqrt{3}} - \dfrac{F_{CB}}{\sqrt{2}} + \dfrac{P}{\sqrt{3}} = 0$

$$F_{CB} = 0$$

Problem 2.14

The pin-jointed cantilever space frame in Fig. S2.14a is supported from a vertical wall at A, B, C, D. Planes ADGE and BCF are horizontal and EFG is vertical. Calculate the forces in the members meeting at F.

If D is considered as the origin of coordinates, the coordinates of the various joints are

A	0, 0, 3	E	4, 0, 3
B	0, 2, 3	F	4, 2, 1.5
C	0, 2, 0	G	4, 0, 0
D	0, 0, 0		

As there are four members meeting at F, one cannot start the analysis at this joint. However, consideration of equilibrium in the y-direction at the unloaded joint G shows that the force in member FG must be zero. Hence, one can return immediately to joint G to find the forces in the remaining three members. Tension coefficients (explained in Problem 2.3) will be used as this slightly simplifies the arithmetic. The members meeting at F have been designated 1, 2, 3 and 4 as in Fig. S2.14b.

Using $\sum P_x = 0$ $\qquad t_1(0 - 4) + t_2(0 - 4) = 0$

Using $\sum P_y = 0$ $\qquad t_3(0 - 2) - 50 = 0$

Using $\sum P_z = 0$ $\qquad t_1(0 - 1.5) + t_2(3 - 1.5) + t_3(3 - 1.5) = 0$

Solving these gives $\qquad t_1 = -12.5 \text{ kN/m}$

$$t_2 = 12.5 \text{ kN/m}$$

$$t_3 = -25 \text{ kN/m}$$

Length of member 1 $\qquad = \sqrt{(4^2 + 1.5^2)} = 4.27 \text{ m}$

member 2 $\qquad = 4.27 \text{ m}$

member 3 $\qquad = \sqrt{(2^2 + 1.5^2)} = 2.5 \text{ m}$

Hence $\qquad\qquad F_1 = -12.5 \times 4.27 = \underline{-53.4 \text{ kN}}$

$$F_2 = \underline{53.4 \text{ kN}}$$

$$F_3 = -25 \times 2.5 = \underline{-62.5 \text{ kN}}$$

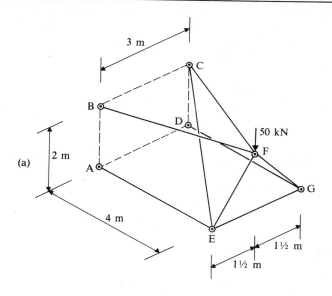

(a)

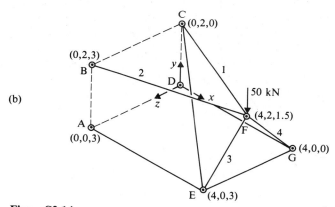

(b)

Figure S2.14

Problem 2.15

A pin-jointed space frame is shown in elevation and plan in Fig. S2.15a. Calculate the forces in the members caused by the load of 20 kN applied at E.

Although there are six members of the frame, symmetry reduces the number of unknown forces to four and this is acknowledged in designating the members in Fig. S2.15b. (Note that this observation is equivalent to using $\sum P_z = 0$ at the joints D and E.) Tension coefficients will again be used in order to slightly simplify the arithmetic involved in the method of joints.

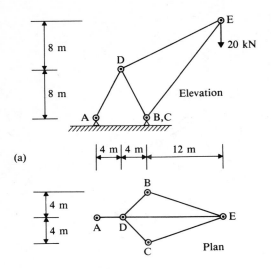

(a)

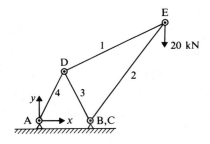

(b)

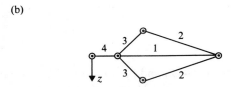

Figure S2.15

Adopting joint A as the origin of coordinates, the coordinates of the various joints are

$$\begin{array}{llll} A & 0, 0, 0 \\ C & 8, 0, 4 \\ D & 4, 8, 0 \\ E & 20, 16, 0 \end{array}$$

For the joint E:

Using $\sum P_x = 0$ $t_1(4 - 20) + t_2(8 - 20) \times 2 = 0$

$$2t_1 + 3t_2 = 0 \qquad (2.15a)$$

Using $\sum P_y = 0$ $t_1(8 - 16) + t_2(0 - 16) \times 2 - 20 = 0$

$$2t_1 + 8t_2 = -5 \qquad (2.15b)$$

Solving the two simultaneous Eqs (2.15a) and (2.15b) gives

$$t_1 = 1.5 \text{ kN/m}$$

$$t_2 = -1 \text{ kN/m}$$

For the joint D:

Using $\sum P_x = 0$ $t_1(20 - 4) + t_3(8 - 4) \times 2 + t_4(0 - 4) = 0$

$$2t_3 - t_4 = -6 \qquad (2.15c)$$

Using $\sum P_y = 0$ $t_1(16 - 8) + t_3(0 - 8) \times 2 + t_4(0 - 8) = 0$

$$2t_3 + t_4 = 1.5 \qquad (2.15d)$$

Solving the two simultaneous Eqs (2.15c) and (2.15d) gives

$$t_3 = -1.125 \text{ kN/m}$$

$$t_4 = 3.75 \text{ kN/m}$$

Length of member 1 $= \sqrt{((20 - 4)^2 + (16 - 8)^2)} = 17.9 \text{ m}$

member 2 $= \sqrt{((20 - 8)^2 + (16 - 0)^2 + (4 - 0)^2)} = 20.4 \text{ m}$

member 3 $= \sqrt{((8 - 4)^2 + (8 - 0)^2 + (4 - 0)^2)} = 9.8 \text{ m}$

member 4 $= \sqrt{((4 - 0)^2 + (8 - 0)^2)} = 8.94 \text{ m}$

Hence

$$F_1 = 1.5 \times 17.9 = \underline{26.8 \text{ kN}}$$

$$F_2 = -1 \times 20.4 = \underline{-20.4 \text{ kN}}$$

$$F_3 = -1.125 \times 9.8 = \underline{-11.0 \text{ kN}}$$

$$F_4 = 3.75 \times 8.94 = \underline{33.5 \text{ kN}}$$

Problem 2.16

The loads applied to the joint D of the tripod-shaped frame in Fig. S2.16a cause a displacement of the joint. Use the unit load method to calculate the component of that movement in the x direction. (The cross-sectional area of all three members is 250 mm^2 and Young's Modulus is 200 kN/mm^2.)

The unit load method requires the determination of member forces F_0 due to the original load and the determination of the member forces f_1 due to the unit load applied at the particular joint in the required direction. For this structure, the forces F_0 were determined in an example covered in Section 2.4 of Part One and these calculations will not be repeated here. The forces F_0 are summarized in Table S2.16.

Application of the unit load is shown in Fig. S2.16b. Consideration of equilibrium at joint D allows the forces f_1 to be determined. The use of tension coefficients will be adopted to simplify the arithmetic. The coordinates of the joints are given in Fig. S2.16b.

Using $\sum P_x = 0$ $\qquad t_1(0 - 2) + t_2(6 - 2) + t_3(3 - 2) + 1 = 0$

$$-2t_1 + 4t_2 + t_3 = -1 \qquad (2.16a)$$

Using $\sum P_y = 0$ $\qquad t_1(0 - 5) + t_2(0 - 5) + t_3(0 - 5) = 0$

$$t_1 + t_2 + t_3 = 0 \qquad (2.16b)$$

Using $\sum P_z = 0$ $\qquad t_1(0 - 2) + t_2(1 - 2) + t_3(5 - 2) = 0$

$$-2t_1 - t_2 + 3t_3 = 0 \qquad (2.16c)$$

The solutions of the three simultaneous Eqs (2.16a), (2.16b) and (2.16c) give

$$t_1 = \frac{4}{27} \qquad t_2 = -\frac{5}{27} \qquad t_3 = \frac{1}{27}$$

The lengths of the members are

$$L_1 = \sqrt{33} \text{ m} \qquad L_2 = \sqrt{42} \text{ m} \qquad L_3 = \sqrt{35} \text{ m}$$

Hence the forces due to $X_1 = 1$ are

$$F_1 = 0.851 \qquad F_2 = -1.2 \qquad F_3 = 0.219$$

Table S2.16

Member	F_0 (kN)	f_1	L (m)	$F_0 f_1 L$ (kN m)
1	1.92	0.851	$\sqrt{33}$	9.39
2	−17.28	−1.2	$\sqrt{42}$	134.38
3	7.89	0.219	$\sqrt{35}$	10.22

$$\sum F_0 f_1 L = 154 \text{ kN m}$$

Hence $\qquad\qquad \Delta_{xd} = \dfrac{\sum F_0 f_1 L}{AE} = \dfrac{154 \times 10^3}{250 \times 200} = \underline{3.1 \text{ mm}}$

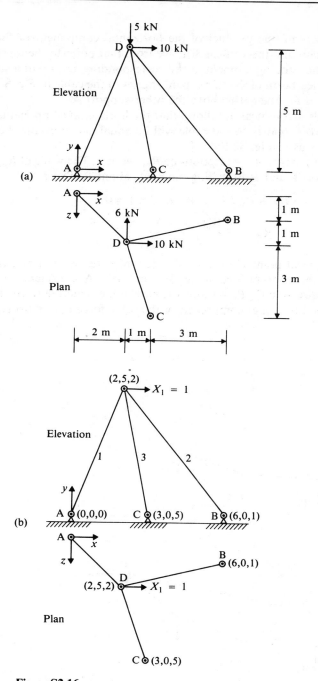

Figure S2.16

Problem 2.17

Show that the proposition 'the product of the horizontal component of the cable tensile force at any point and the vertical distance from that point to the line joining the supports equals the bending moment at the corresponding section of a similarly loaded simply supported beam of the same span' applies to the cable in Fig. S2.17a in which one support B is at a height v above the other support A.

As the applied loads are vertical, it follows that the horizontal components of the reactions from the two supports to the cable will be equal and opposite. Let these horizontal forces be H as in Fig. S2.17b.

Let the vertical components of the reactions be R_{ya} and R_{yb}. The value of R_{ya} can be determined in terms of W and H by taking moments about B. Thus:

Using $\sum M_b = 0$ $\qquad$ $Hv + R_{ya}L - \sum W_n(L - x_n) = 0$

$$R_{ya} = \frac{\sum W_n(L - x_n)}{L} - \frac{Hv}{L}$$

Consider now any typical point C on the cable (coordinates x, y) and in this case positioned between the concentrated loads W_1 and W_2. A consideration of the equilibrium in the x direction ($\sum P_x = 0$) for the part of the structure to the left of C clearly shows that the horizontal component of the cable force at C must equal H.

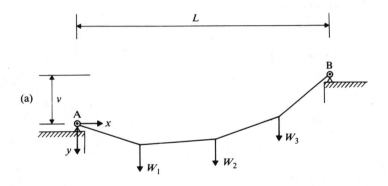

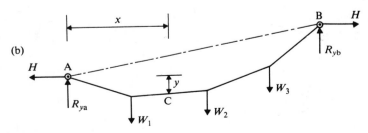

Figure S2.17

As the bending moment in the cable at every point is zero, taking moments of all the forces to the left of C gives

$$R_{ya}x - Hy - W_1(x - x_1) = 0$$

Substituting the previous value of R_{ya} and rearranging gives

$$H\left(y + \frac{vx}{L}\right) = \frac{x \sum W_n(L - x_n)}{L} - W_1(x - x_1)$$

A study of this equation shows that the left-hand side is the product of the horizontal component of the cable tensile force at C and the vertical distance from C to the line joining the supports, whilst the right-hand side is the bending moment at the corresponding section of a similarly loaded, simply-supported beam of the same span.

Problem 2.18

Calculate the difference between the vertical reactions at the supports for the cable in Fig. S2.18a.

Let the reactions at A and B be as shown in Fig. S2.18b.

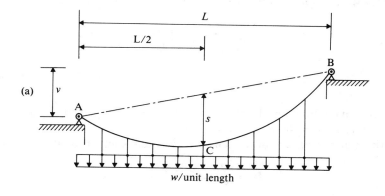

(a)

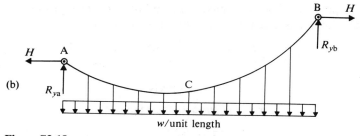

(b)

Figure S2.18

Using $\sum M_b = 0$

$$Hv + R_{ya}L - \frac{wL^2}{2} = 0$$

$$R_{ya}L = \frac{wL^2}{2} - Hv \tag{2.18a}$$

Using $\sum M_a = 0$

$$Hv - R_{yb}L + \frac{wL^2}{2} = 0$$

$$R_{yb}L = \frac{wL^2}{2} + Hv \tag{2.18b}$$

Using the proposition proved in Problem 2.17 for the mid-span section:

$$Hs = \frac{wL^2}{8}$$

$$H = \frac{wL^2}{8s} \tag{2.18c}$$

Combining Eqs (2.18a), (2.18b) and (2.18c) gives

$$R_{yb} - R_{ya} = \frac{vwL}{4s}$$

Problem 2.19

Calculate the maximum tensile force in the cable arrangement in Fig. S2.18 if $v = 10$ m, $s = 12$ m, $L = 48$ m, $w = 100$ kN/m.

The horizontal component of the tensile force is constant throughout the cable at

$$H = \frac{wL^2}{8s} = \frac{100 \times 48^2}{8 \times 12}$$

$$= 2400 \text{ kN}$$

The maximum tensile force in the cable occurs at the position where the vertical component of the tensile force is a maximum, and this occurs adjacent to the support at B. The value of R_{yb} is given by (see solution to Problem 2.18):

$$R_{yb} = \frac{wL}{2} + \frac{Hv}{L} = \frac{100 \times 48}{2} + \frac{2400 \times 10}{48}$$

$$= 2900 \text{ kN}$$

Hence

$$T_{max} = \sqrt{((2900)^2 + (2400)^2)}$$

$$= 3760 \text{ kN}$$

Problem 2.20

Show that a suspended uniform cable acting under its own weight only will adopt a catenary shape.

The cable in Fig. S2.20a is hanging freely under its own weight. That part of the cable between the lowest point O and any point P on the cable at distance s from O is in equilibrium under the forces acting on it, namely:

- the horizontal tensile force H at O;
- the inclined tensile force T at P;
- the distributed load of w per unit length of cable.

The resultant of the distributed load is ws acting through the centroid of the length of the cable OP. The equilibrium of the three forces acting on the cable OP (Fig. S2.20b) is represented by the triangle of forces in Fig. S2.20c.

From the triangle of forces, it follows that the inclination of the force T (and which, of course, is tangential to the cable at point P) is given by

$$\tan \phi = \frac{ws}{H}$$

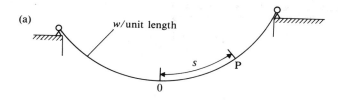

(a)

w/unit length

s

P

0

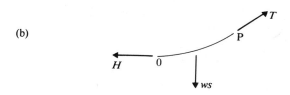

(b)

T

P

H   0

ws

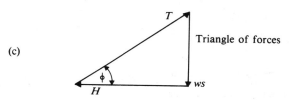

(c)

T

Triangle of forces

φ

H

ws

Figure S2.20

This equation determines the shape of the cable. That it is of catenary shape follows from the fact that the gradient of the curve at any point is proportional to the distance along the curve from the point to the lowest point of the curve.

Problem 2.21

Obtain the shear force and bending moment diagrams for the beam in Fig. S2.21a and determine the maximum value of the bending moment.

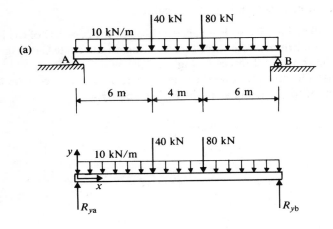

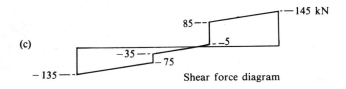

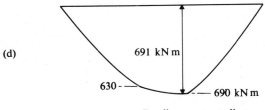

Figure S2.21

The first step is to find the reactive forces (Fig. S2.21b):

Using $\sum M_b = 0$ $16R_{ya} - (40 \times 10) - (80 \times 6) - \dfrac{10 \times 16^2}{2} = 0$

$$R_{ya} = 135\,\text{kN}$$

Using $\sum P_y = 0$ $R_{ya} + R_{yb} - 40 - 80 - (10 \times 16) = 0$

$$R_{yb} = 145\,\text{kN}$$

(It is not essential to know the value of R_{yb} in order to obtain the shear force and bending moment diagrams, but it gives a useful check on the arithmetic relating to the shear force diagram.)

The equation representing the shear force at any section distance x from the left-hand end of the beam is

$$S_x = -135 + 10x + 40[x - 6]^\circ + 80[x - 10]^\circ$$

where the values of the square bracket terms become zero if the expression within the bracket is negative. (And note that, since the bracket is raised to the power zero, the value of the term will be unity if the expression within the bracket is positive.)

This general equation is built up by considering firstly a section in the left-hand region (i.e., $0 \leqslant x \leqslant 6$) and subsequently moving to the right to the successive regions bounded by the concentrated forces. Thus:

$$0 \leqslant x \leqslant 6 \qquad S_x = -135 + 10x$$

$$6 \leqslant x \leqslant 10 \qquad S_x = -135 + 10x + 40$$

$$10 \leqslant x \leqslant 16 \qquad S_x = -135 + 10x + 40 + 80$$

These three equations may be combined into the single general equation above. Students should adopt the particular system, i.e., the single general equation or the multiple simpler equations, which they find the most convenient. In these solutions, we shall adopt the single general equation.

In order to draw the diagram, values of shear force are determined only at the positions of concentrated force, as we know that between these positions the value of the shear force varies linearly. Thus:

at $x = 0$ $\qquad\qquad\qquad\qquad\qquad\qquad\qquad\qquad$ $S = -135\,\text{kN}$

at $x = 6$ (left of the concentrated load) $\qquad\qquad$ $S = -75\,\text{kN}$

at $x = 6$ (right of the concentrated load) $\qquad\quad$ $S = -35\,\text{kN}$

at $x = 10$ (left of the concentrated load) $\qquad\quad$ $S = 5\,\text{kN}$

at $x = 10$ (right of the concentrated load) $\qquad$ $S = 85\,\text{kN}$

at $x = 16$ $\qquad\qquad\qquad\qquad\qquad\qquad\qquad\quad$ $S = 145\,\text{kN}$

The calculated value of S at $x = 16$ is confirmed by the previously calculated value of the reaction at B. The shear force diagram is given in Fig. S2.21c.

The general equation for the bending moment at a section distance x from the left-hand support is

$$M_x = 135x - \frac{10x^2}{2} - 40[x - 6] - 80[x - 10]$$

As this equation involves x^2, in order to draw the diagram reasonably accurately it will be necessary to obtain the values of the bending moment at more frequent intervals than was the case for the shear force. Note, however, that there is now no sudden change in value as we pass from one side of a concentrated load to the other. Thus:

at $x = 0$	$M = 0$
at $x = 2$	$M = 250 \text{ kN m}$
at $x = 4$	$M = 460 \text{ kN m}$
at $x = 6$	$M = 630 \text{ kN m}$
at $x = 8$	$M = 680 \text{ kN m}$
at $x = 10$	$M = 690 \text{ kN m}$
at $x = 12$	$M = 500 \text{ kN m}$
at $x = 14$	$M = 270 \text{ kN m}$
at $x = 16$	$M = 0$

The resulting diagram is shown in Fig. 2.21d. There are several points to note:

1. The line representing the bending moments is curving throughout, but with discontinuities in the curve at the positions of the concentrated loads. It should be recalled that the value of the shear force at a particular section represents the rate of change of bending moment at that section, and hence the slope of the bending moment diagram at that position. As there are sudden jumps in the shear force diagram (at the position of the concentrated loads), it follows that the slope of the bending moment diagram changes abruptly at these positions, i.e., there are discontinuities in the curve representing the bending moment.

2. It follows from the relationship between shear force and bending moment that the position of maximum bending moment occurs at the position of zero shear force.

3. The bending moment in the beam at the position of the supports is zero. This will always be the case for beams simply supported at their ends when there is no external applied moment at those positions.

From the shear force diagram, zero shear force occurs at $6 + \dfrac{35}{10} = 9.5$ m from the support A. Hence:

$$\text{Maximum bending moment} = \underline{691 \text{ kN m}}$$

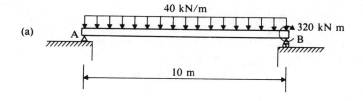

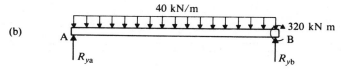

Figure S2.22

Problem 2.22

Calculate the maximum bending moment in the beam in Fig. S2.22a.

If we assume no prior knowledge of the analysis given in Fig. 2.48 of Part One, we must acknowledge that the maximum bending moment in the beam could be either in the region of hogging bending or in the region of sagging bending. The maximum hogging bending moment will, in fact, be at the support B and have the value of the applied moment 320 kN m. The maximum sagging bending moment will be within the span at the position of zero shear force. The latter can easily be found by first calculating the reaction at A (Fig. S2.22b).

Using $\sum M_b = 0$ $\qquad 10R_{ya} + 320 - \dfrac{40 \times 10^2}{2} = 0$

$$R_{ya} = 168 \text{ kN}$$

Distance of position of zero shear force from A $= \dfrac{168}{40} = 4.2$ m

Hence maximum sagging bending moment $= (168 \times 4.2) - \dfrac{(40 \times 4.2^2)}{2}$

$$= 353 \text{ kN m}$$

The maximum bending moment in the beam is therefore 353 kN m

Problem 2.23

The simply supported beam in Fig. S2.23a is loaded through the bracket. Calculate the reactive forces directly and use these to obtain the shear force and bending moment diagrams. Indicate on separate diagrams: (a) the forces acting on the bracket,

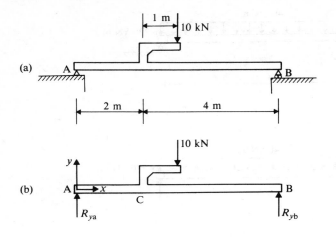

(a)

(b)

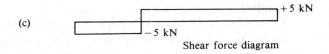

(c)

+5 kN

−5 kN

Shear force diagram

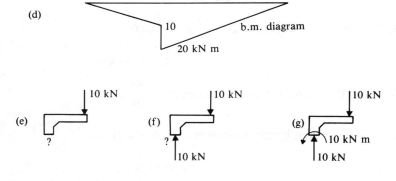

(d)

10

20 kN m

b.m. diagram

(e)

(f)

(g)

(h)

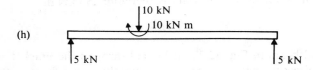

Figure S2.23

(b) the forces acting on the beam. Write down the single equation for the bending moment at a section distance x from the left-hand support and use this to check the bending moment diagram.

By considering the complete structure (Fig. S2.23b):

Using $\sum M_b = 0$

$$6R_{ya} - (10 \times 3) = 0$$

$$R_{ya} = 5 \text{ kN}$$

Using $\sum P_y = 0$

$$R_{ya} + R_{yb} - 10 = 0$$

$$R_{yb} = 5 \text{ kN}$$

(Perhaps it could have been observed without calculation that, since the applied load is midway between the two supports, the two reactions would be equal and each equal to half the applied load.)

Consider a section distance x from support at A, but within the region AC. By considering forces to the *left* of this section:

$$S_x = -5 \text{ kN}$$

i.e., the shear force is constant across the region AC.

$$M_x = 5x$$

When $x = 0$ $M = 0$

When $x = 2$ $M = 10 \text{ kN m}$

Consider a section distance d from the support at B, but within the region BC. By considering forces to the *right* of this section:

$$S_d = 5 \text{ kN}$$

i.e. the shear force is constant across the region BC.

$$M_d = 5d$$

When $d = 0$ $M = 0$

When $d = 4$ $M = 20 \text{ kN m}$

These values are used to draw the shear force and bending moment diagrams (Figs S2.23c and S2.23d).

(a) The bracket is shown in Fig. S2.23e. The bracket is in equilibrium and therefore the forces *from the beam* acting at the interface between beam and bracket must balance the applied load. For vertical equilibrium, it follows that there must be an upward force of 10 kN to balance the downward applied load of 10 kN (Fig. S2.23f).

Although the bracket is now in vertical equilibrium, it is clearly not in rotational equilibrium as there is an unbalanced clockwise moment of $10 \times 1 = 10 \text{ kN m}$. Hence, there must also be an anticlockwise moment of 10 kN m acting at the interface. The forces acting on the bracket are therefore as in Fig. S2.23g.

(b) Action and reaction are equal and opposite. Thus the forces acting from the bracket to the beam are equal and opposite to those acting at the interface in Fig. S2.23g. The forces on the beam are therefore as in Fig. S2.23h.

The general equation for the bending moment at a distance x from support A is

$$M_x = 5x - 10[x - 2] + 10[x - 2]°$$

When $x = 0$ $\qquad\qquad\qquad\qquad\qquad\qquad\qquad\qquad$ $M = 0$

When $x = 2$ (left of the applied moment) $\qquad$ $M = 10\,\text{kN m}$

When $x = 2$ (right of the applied moment) $\quad$ $M = 20\,\text{kN m}$

When $x = 6$ $\qquad\qquad\qquad\qquad\qquad\qquad\qquad\qquad$ $M = 0$

These values confirm the earlier bending moment diagram. The discontinuity in the bending moment diagram is more clearly understood using this second approach. It would be instructive to students to repeat the calculation of the reactions of the beam using the loads on the beam shown in Fig. S2.23h.

Problem 2.24

Obtain the shear force and bending moment diagrams for the beam in Fig. S2.24a. Calculate the bending moment at the position on the beam corresponding to the 40 kN load for the following three cases of loading: (a) the 20 kN load alone, (b) the 40 kN load alone and (c) the 10 kN/m load alone, and then use the principle of superposition to check the value obtained for Fig. S2.24a.

To find the reactive forces, consider the forces on the complete structure as in Fig. S2.24b:

Using $\sum M_b = 0$

$$10R_{ya} - (20 \times 12) - (40 \times 7) - (10 \times 4 \times 3) = 0$$

$$R_{ya} = 64\,\text{kN}$$

Using $\sum P_y = 0$ $\qquad\qquad$ $R_{ya} + R_{yb} - 20 - 40 - (10 \times 4) = 0$

$$R_{yb} = 36\,\text{kN}$$

As a check on this arithmetic, use $\sum M_a = 0$:

$$(40 \times 3) + (10 \times 4 \times 7) - (20 \times 2) - 10R_{yb} = 0$$

$$R_{yb} = 36\,\text{kN}$$

With a knowledge of all the forces acting on the beam, the shear force diagram in Fig. S2.24c is easily obtained by moving along the beam from the left-hand end, changing the ordinates of the shear force diagram as the loads are successively passed.

It will be as well to elaborate on this once and for all. Consider a section of the beam between the left-hand end of the beam and the reactive force at A. By considering the forces to the left of this section, the shear force is seen clearly to be $+20\,\text{kN}$, and this is

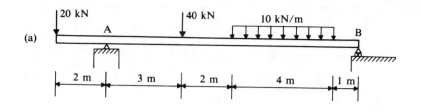

(a)

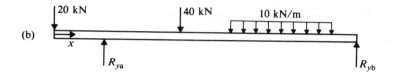

(b)

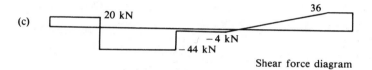

(c)

Shear force diagram

(d)

Building up the s.f. diagram

(e)

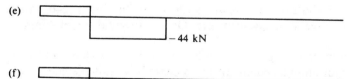

(f)

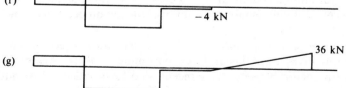

(g)

Figure S2.24

continued

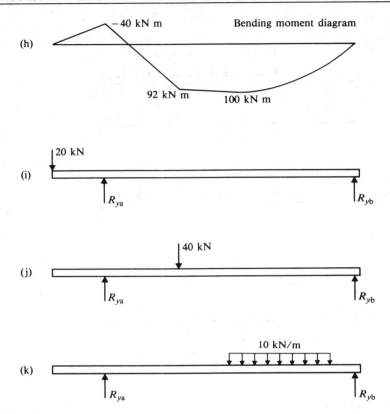

Figure S2.24   (*continued*)

constant over the whole of this region—as illustrated in Fig. S2.24d. (An understanding of the sign convention given in Part One is essential for obtaining shear force diagrams.)

As we now move to the right and past the upward reactive force at A, the shear force will change by an amount equal to the force, 64 kN, and the shear force therefore becomes −44 kN. This will remain constant over the region between A and the 40 kN load as in Fig. S2.24e.

As we now move past this concentrated load, the shear force will change by an amount equal to 40 kN, the shear force becoming −4 kN. Again this remains constant over the region between the concentrated load and the distributed load—Fig. S2.24f.

From this point, as we now move further to the right, the shear force changes continuously at the rate of 10 kN/m over a distance of 4 m—a total change in the shear force of 40 kN, giving a value of 36 kN at the right-hand end of the distributed load as in Fig. S2.24g.

From here the shear force remains constant as we move to the right-hand end of the beam, giving the diagram in Fig. S2.24c. The value of the shear force in this right-hand end region of the beam (36 kN) is confirmed by the reactive force at B found earlier.

The general equation for the bending moment at distance x from the left-hand end of the beam is

$$M_x = -20x + 64[x - 2] - 40[x - 5] - \frac{10[x - 7]^2}{2} + \frac{10[x - 11]^2}{2}$$

The bending moment diagram will, of course, be a straight line over a particular region where the shear force is constant. Over these regions then, it is only necessary to obtain the value of the bending moment at the ends of these regions. Only where the shear force is varying (at the position of the distributed load), will it be necessary to calculate the bending moment at more frequent intervals. Thus:

at $x = 0$ $M = 0$

$x = 2$ $M = -(20 \times 2) = -40$ kN m

$x = 5$ $M = -(20 \times 5) + (64 \times 3) = 92$

$x = 7$ $M = -(20 \times 7) + (64 \times 5) - (40 \times 2) = 100$

$x = 8$ $M = -(20 \times 8) + (64 \times 6) - (40 \times 3) - (5 \times 1^2) = 99$

$x = 9$ $M = -(20 \times 9) + (64 \times 7) - (40 \times 4) - (5 \times 2^2) = 88$

$x = 10$ $M = -(20 \times 10) + (64 \times 8) - (40 \times 5) - (5 \times 3^2) = 67$

$x = 11$ $M = -(20 \times 11) + (64 \times 9) - (40 \times 6) - (5 \times 4^2) = 36$

$x = 12$ $M = -(20 \times 12) + (64 \times 10) - (40 \times 7) - (5 \times 5^2) + (5 \times 1^2) = 0$

Needless to say, one does not really need to substitute $x = 12$ into the general equation, as we know that the bending moment at the right-hand end of the beam must be zero. It is, however, a good check on the equation. Similarly, the value of the bending moment at $x = 11$ can easily be seen to be 36 kN m by considering the moment of all forces to the *right* of this section—there is only one such force, namely the reaction at B.

The bending moment diagram is plotted using these values and is shown in Fig. S2.24h. Note that there is a region of hogging bending extending over the cantilever region and a short way past the support at A.

(a) For the loading shown in Fig. S2.24i, using $\sum M_b = 0$:

$$10R_{ya} - (20 \times 12) = 0$$

$$R_{ya} = 24 \text{ kN}$$

Hence, at $x = 5$ m: $M = -(20 \times 5) + (24 \times 3) = \underline{-28 \text{ kN m}}$

(b) For the load shown in Fig. S2.24j, using $\sum M_b = 0$:

$$10R_{ya} - (40 \times 7) = 0$$

$$R_{ya} = 28 \text{ kN}$$

Hence at $x = 5$ m $\qquad$ $M = 28 \times 3 = \underline{84 \text{ kN m}}$

(c) For the loading shown in Fig. S2.24k, using $\sum M_b = 0$:

$$10R_{ya} - (10 \times 4 \times 3) = 0$$

$$R_{ya} = 12 \text{ kN m}$$

Hence at $x = 5$ m $\qquad$ $M = 12 \times 3 = \underline{36 \text{ kN m}}$

By the principle of superposition, the bending moment at $x = 5$ m for the complete loading is

$$M = -28 + 84 + 36 = \underline{92 \text{ kN m}}$$

This confirms the value previously obtained.

Problem 2.25

Obtain the shear force diagram for the beam in Fig. S2.25a. Write down the expression for the bending moment at any position of the beam distance x from the left-hand end of the beam, and use this to determine the position of the maximum sagging bending moment, given that its position is within the length of the distributed load. Verify the answer from the shear force diagram.

To obtain the reactive forces, consider the equilibrium of the beam shown in Fig. S2.25b.

Using $\sum M_b = 0$ $\quad$ $(R_{ya} \times 8) - (10 \times 10) - (5 \times 8 \times 6) - (10 \times 2) = 0$

$$R_{ya} = 45 \text{ kN}$$

Using $\sum P_y = 0$ $\quad$ $R_{ya} + R_{yb} - 10 - (5 \times 8) - 10 = 0$

$$R_{yb} = 15 \text{ kN}$$

As a check on the arithmetic, using $\sum M_a = 0$:

$$(10 \times 6) + (5 \times 8 \times 2) - (10 \times 2) - (R_{yb} \times 8) = 0$$

$$R_{yb} = 15 \text{ kN}$$

The shear force diagram can now be drawn using the procedure demonstrated in the solution to Problem 2.24. It is shown in Fig. S2.25c, where it can be seen that the position of zero shear force (and hence the position of maximum bending moment) occurs at 5 m from A.

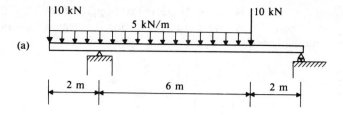

(a)

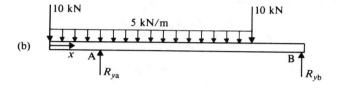

(b)

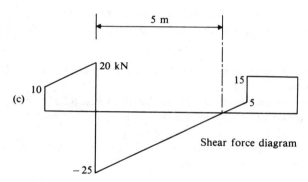

(c)

Shear force diagram

Figure S2.25

The general expression for the bending moment at a section distance x from the left-hand end of the beam is

$$M_x = -10x + 45[x - 2] - \frac{5x^2}{2} - 10[x - 8] + \frac{5[x - 8]^2}{2}$$

The expression relevant to the region of the distributed load reduces to

$$M_x = -10x + 45[x - 2] - \frac{5x^2}{2}$$

Since the position of maximum sagging bending moment occurs within the region of the distributed load, we also know that it must occur within the length $2 \leqslant x \leqslant 8$, as the cantilever region (i.e., $x \leqslant 2$) is clearly subjected to hogging bending. The square brackets can therefore be omitted and the equation for the region containing this maximum bending moment can be written as

$$M_x = -10x + 45(x - 2) - \frac{5x^2}{2}$$

$$= 35x - 90 - \frac{5x^2}{2}$$

$$\frac{\mathrm{d}M_x}{\mathrm{d}x} = 35 - 5x$$

At the position of maximum bending moment, $\dfrac{\mathrm{d}M_x}{\mathrm{d}x} = 0$. Thus:

$$35 - 5x = 0$$

$$x = \underline{7 \text{ m}}$$

This agrees with the earlier finding, namely, 5 m beyond A.

Problem 2.26

The designer of the beam in Fig. S2.26a intends to position the supports symmetrically so that the maximum bending moment is as low as possible. Show that this is when

$$a = (\sqrt{2} - 1)L$$

Draw the shear force and bending moment diagrams for this case and calculate the positions of zero bending moment.

The shape of the bending moment diagram is shown in Fig. S2.26b. If the positions of the supports are moved towards the centre of the span, the maximum hogging moments will increase whilst the maximum sagging moments will decrease. The lowest possible value of the maximum moment will therefore be when the maximum hogging moment equals the maximum sagging moment.

$$\text{Maximum hogging bending moment} = \frac{wa^2}{2}$$

$$\text{Maximum sagging bending moment} = \frac{w(2L - 2a)^2}{8} - \frac{wa^2}{2}$$

$$= \frac{w}{2}(L^2 - 2aL)$$

When these moments are equal:

$$\frac{w}{2}(L^2 - 2aL) = \frac{wa^2}{2}$$

$$a^2 + 2aL - L^2 = 0$$

$$a = \frac{-2L + \sqrt{(4L^2 + 4L^2)}}{2}$$

$$= \underline{(\sqrt{2} - 1)L}$$

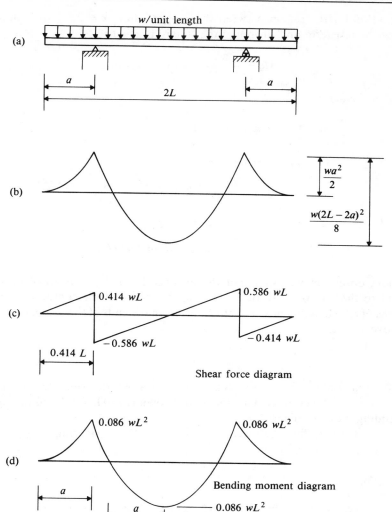

Figure S2.26

The vertical reactions at the two supports is wL. Hence, the shear force diagram is as in Fig. S2.26c. The bending moment diagram is shown in Fig. S2.26d.

The general equation for the bending moment at a section distance x from the left-hand end of the beam is

$$M_x = -\frac{wx^2}{2} + wL[x - 0.414L] + wL[x - 1.586L]$$

The equation for the region between the supports ($a \leqslant x \leqslant 2L - a$) is obtained by omitting the last term and dropping the square brackets:

$$M_x = -\frac{wx^2}{2} + wL(x - 0.414L)$$

For this to be zero:

$$-\frac{wx^2}{2} + wLx - 0.414wL^2 = 0$$

$$x^2 - 2Lx + 0.828L^2 = 0$$

$$x = \frac{2L \pm \sqrt{(4L^2 - (4 \times 0.828L^2))}}{2}$$

$$= L \pm 0.415L$$

$$= \underline{0.585L \text{ or } 1.415L}$$

The more observant will spot that the distance from the sections of zero bending moment to the midspan position equals a, the length of the cantilevers. Hence, the positions of zero bending moment are $L - a$ from the ends of the beam, which confirm the above values.

Problem 2.27

The two-span beam ABC in Fig. S2.27a is statically indeterminate but an analysis indicates that the reaction at B for the loads shown is 180 kN. Obtain the shear force and bending moment diagrams.

The first step is to find the reactions at A and C by considering the equilibrium of the whole beam (Fig. S2.27b):

Using $\sum M_c = 0$

$$16R_{ya} + (180 \times 8) + 204 - (20 \times 8 \times 12) - (160 \times 5) = 0$$

$$R_{ya} = 67.3 \text{ kN}$$

Using $\sum P_y = 0 \qquad R_{ya} + R_{yc} + 180 - (20 \times 8) - 160 = 0$

$$R_{yc} = 72.7 \text{ kN}$$

A check on this arithmetic is obtained by using $\sum M_a = 0$:

$$(20 \times 8 \times 4) + (160 \times 11) + 204 - (180 \times 8) - 16R_{yc} = 0$$

$$R_{yc} = 72.7 \text{ kN}$$

The resulting shear force diagram is shown in Fig. S2.27c.

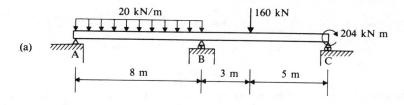

(a)

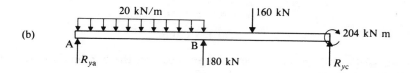

(b)

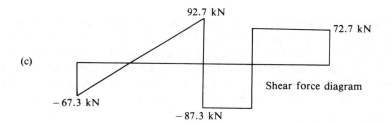

(c)

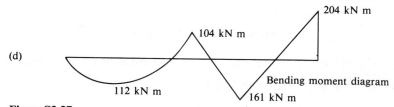

(d)

Figure S2.27

For the span AB, the equation for the moment at distance x from A is

$$M_x = 67x - \frac{20x^2}{2}$$

When

$x = 0$	$M = 0$
$x = 2$	$M = 94 \text{ kN m}$
$x = 3$	$M = 111 \text{ kN m}$
$x = 4$	$M = 108 \text{ kN m}$
$x = 5$	$M = 85 \text{ kN m}$
$x = 6$	$M = 42 \text{ kN m}$
$x = 7$	$M = -21 \text{ kN m}$
$x = 8$	$M = -104 \text{ kN m}$

The position of zero shear force within span AB is $\dfrac{67.3}{20} = 3.36$ m from A. Thus:

$$M_{\mathrm{max}} = (67 \times 3.36) - 10 \times 3.36^2$$

$$= 112 \text{ kN m}$$

The bending moment at the position of the concentrated load in span BC is best obtained by considering the moments of all forces to the *right* of the load. Thus

$$M_{x=11} = (73 \times 5) - 204$$

$$= 161 \text{ kN m}$$

Using the calculated values of the bending moment, the bending moment diagram can now be drawn and is shown in Fig. S2.27d.

Problem 2.28

A simply supported beam of span L is symmetrically loaded with a load that increases linearly from zero at the supports to a maximum of w per unit length at mid-span. Obtain expressions for the shear force and bending moment at a section in the left-hand half of the beam, and sketch the shear force and bending moment diagrams. An upward force is now applied at mid-span and increased until the bending moment at that position is zero. Find the new positions and magnitude of the maximum bending moment.

The total load on the beam (see Fig. S2.28a) $= \dfrac{wL}{2}$

Hence each reaction $= \dfrac{wL}{4}$

The intensity of loading at a distance x from the left-hand end of the beam but within the left-hand half is $\dfrac{2wx}{L}$ per unit length.

The expression for the shear force at this position is

$$S_x = -\frac{wL}{4} + \frac{x}{2}\frac{2wx}{L}$$

$$= -\frac{wL}{4} + \frac{wx^2}{L}$$

Hence when $x = 0$ $\qquad S = -\dfrac{wL}{4}$

$\qquad\qquad x = \dfrac{L}{4} \qquad S = -\dfrac{3wL}{16}$

$\qquad\qquad x = \dfrac{L}{2} \qquad S = 0$

The values are used to sketch the shear force diagram shown in Fig. S2.28b.

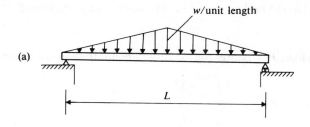

(a)

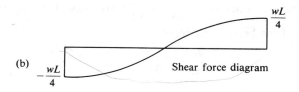

(b)

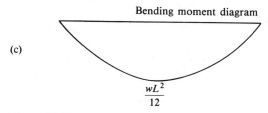

(c)

Figure S2.28

The expression for the bending moment at a distance x from the left-hand end but within the left-hand half is

$$M_x = \frac{wL}{4} x - \frac{wx^2}{L} \frac{x}{3}$$

$$= \frac{wLx}{4} - \frac{wx^3}{3L}$$

Hence when $x = 0$ $M = 0$

$$x = \frac{L}{4} \qquad M = \frac{11wL^2}{192}$$

$$x = \frac{L}{2} \qquad M = \frac{wL^2}{12}$$

These values are used to sketch the bending moment diagram shown in Fig. S2.28c.

The bending moment at the mid-span of a beam caused by a point load P applied at mid-span is $\dfrac{PL}{4}$.

Thus, for an upward load P to cancel the maximum bending moment caused by the distributed load, then

$$\frac{PL}{4} = \frac{wL^2}{12}$$

$$P = \frac{wL}{3}$$

The vertical reactions then become $\dfrac{wL}{4} - \dfrac{wL}{6} = \dfrac{wL}{12}$. The new bending moment equation for the left half of the beam is therefore

$$M_x = \frac{wLx}{12} - \frac{wx^3}{3L}$$

$$\frac{dM_x}{dx} = \frac{wL}{12} - \frac{w}{L}x^2$$

At the position of maximum moment $\dfrac{dM_x}{dx} = 0$

Thus
$$\frac{wL}{12} - \frac{wx^2}{L} = 0$$

$$x = \frac{L}{\sqrt{12}}$$

Hence $$M_{max} = \frac{wL}{12} \times \frac{L}{\sqrt{12}} - \frac{w}{3L} \times \frac{L^3}{12\sqrt{12}} = \frac{wL^2}{36\sqrt{3}}$$

Problem 2.29

The structural form of a flight of reinforced concrete stairs is diagrammatically represented by ABCDE in Fig. S2.29a where DE is a landing cantilevering beyond the

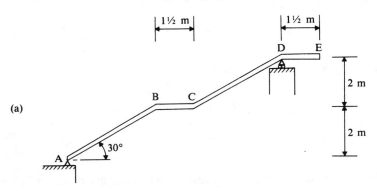

(a)

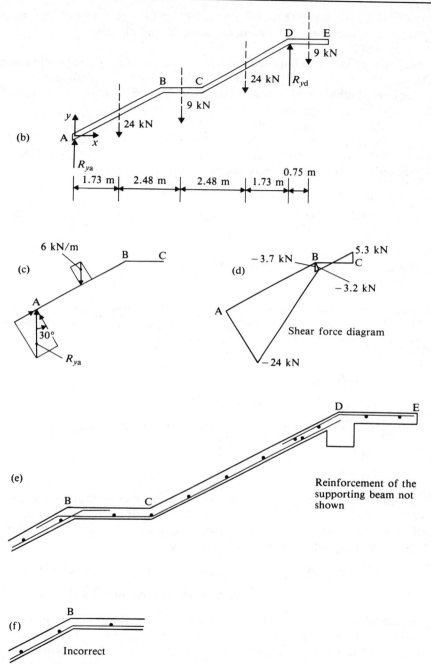

Figure S2.29

support at D. Assuming the dead load is 6 kN per metre run of stairs, obtain for this load the bending moments at A, B, C, D and E, and also the shear force diagram for the part ABC.

State the effect (a) an imposed load on BC would have on the bending moments in DE and (b) an imposed load on DE would have on the bending moments in BC. Determine whether equal and simultaneous distributed loading on BC and DE would cause the bending moment at B to increase or decrease. Sketch how you would arrange the reinforcement in the concrete.

The length of AB is 4 m so that its dead load is 24 kN, as is also that of CD. The length of BC is 1.5 m so that its dead load is 9 kN, as is also that of DE. The forces acting on the structure are therefore as shown in Fig. S2.29b.

Using $\sum M_d = 0$

$$(R_{ya} \times 8.43) - (24 \times 6.7) - (9 \times 4.21) - (24 \times 1.73) + (9 \times 0.75) = 0$$

$$R_{ya} = 27.7 \text{ kN}$$

Using $\sum P_y = 0$ $R_{ya} + R_{yd} - 24 - 9 - 24 - 9 = 0$

$$R_{yd} = 38.3 \text{ kN}$$

The bending moments at a particular position will be obtained by taking moments of all forces to one side, either to the right or to the left, whichever has the fewer forces to take into account.

Thus:

$$M_a = 0$$

$$M_b = (27.7 \times 3.46) - (24 \times 1.73) = 54.4 \text{ kN m}$$

$$M_c = (27.7 \times 4.96) - (24 \times 3.23) - (9 \times 0.75) = 53.1 \text{ kN m}$$

$$M_d = -(9 \times 0.75) = -6.7 \text{ kN m}$$

$$M_e = 0$$

The shear force acting at a position on a member is that acting on a section perpendicular to the axis of the member. Hence, to draw the shear force diagram for AB, it will be necessary to obtain those components of the vertical forces acting perpendicular to AB. Referring to Fig. S2.29c:

Component of R_{ya} perpendicular to AB = $27.7 \cos 30° = 24$ kN

Hence the shear force in AB at B = -24 kN

It is desirable to note here that one cannot use the forces shown in Fig. S2.29b to obtain the shear force diagrams. It was, of course, quite in order to use the total weights of the members acting at their respective centroids when considering the overall equilibrium of the structure, and indeed also for determining the bending moments at the joints. Now that we need to know the distribution of the internal forces along the members AB and BC, we must acknowledge the distribution of the weight of the members. Thus in Fig. S2.29c:

Component of the distributed dead load perpendicular to

$$AB = 6 \cos 30° = 5.2 \text{ kN/m}$$

$$\text{Change in shear force from A to B} = 4 \times 5.2 = 20.8 \text{ kN}$$

$$\text{Hence the shear force in AB at B} = -24 + 20.8 = -3.2 \text{ kN}$$

For the horizontal region BC, it is now the vertical forces which are relevant to the calculation of the shear forces. Thus:

$$\text{shear force in BC at B} = -27.7 + 24 = -3.7 \text{ kN}$$

$$\text{shear force in BC at C} = -3.7 + 9 = 5.3 \text{ kN}$$

The shear force diagram for the region ABC is therefore that shown in Fig. S2.29d.

(a) Since the bending moment at a section equals the sum of the moment of all forces to one side (either to the right or the left), it follows that applying a load to BC can have no influence on the bending moments in DE—the forces to the *right* of a section in DE are not affected by the applied load.

(b) A load imposed on DE will affect both the reactive forces at the supports A and D. As the reactive force at A will be *reduced* by the applied load on DE, so the bending moments in BC will be reduced.

For the effect of equal and simultaneous distributed loads on BC and DE on the bending moment at B, one must determine whether the vertical reaction at A will be increased or decreased. The additional loading on BC will increase R_{ya}, whilst the additional load on DE will decrease R_{ya}. The distances of these extra loads from D indicate that the loading on BC will have the greater effect and so overall the value of R_{ya} will increase. It then follows that the bending moment at B will increase.

In reinforced concrete, the reinforcing bars are positioned to resist the tensile forces caused by the bending moments. In a region of hogging bending (as in the cantilever DE and a short length of CD adjacent to D), tensile stresses are caused in the upper part of the member and so the reinforcing bars are positioned near the upper face. In a region of sagging bending (which applies to the whole of AB and BC and the major part of CD), tensile stresses are caused in the bottom part and so the reinforcing bars are positioned near the lower face. A typical arrangement of bars would be as shown in Fig. S2.29e.

An important detailing point should be noted here. Reinforcing bars acting in tension must *not* be bent to follow the cranked soffit of the stairs at a re-entrant angle as in Fig. S2.29f. It should be clear that such bars acting in tension will tend to straighten and to break through the concrete. For this reason, bars at such re-entrant angles must pass to the opposite face to be anchored in the concrete stressed in compression as in Fig. S2.29e.

Problem 2.30

In order to illustrate the use of three different methods of calculating displacements of beams—the differential equation method, the strain energy method, and the unit load (virtual work) method—a lecturer used the three beams shown in Figs S2.30a, S2.30b and S2.30c, the points marked X being the positions at which the deflections were calculated. Identify which method was used for each beam, noting that in the unit load method the lecturer used tabulated coefficients for the integration of standard bending moment diagrams. Determine the values of the deflections at X in the three beams using the same methods as adopted by the lecturer.

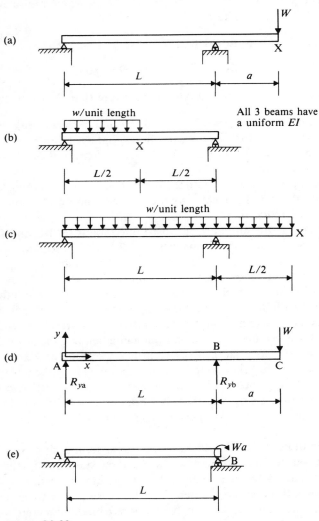

Figure S2.30

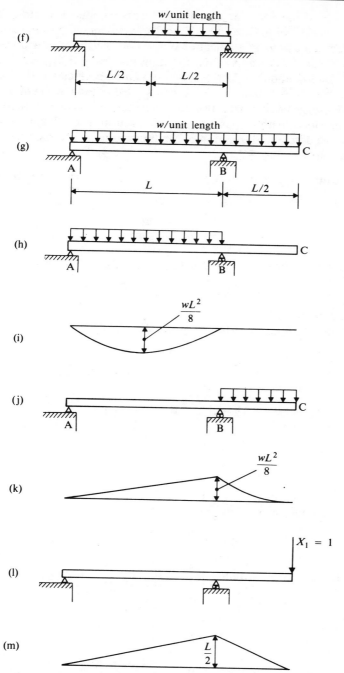

Figure S2.30 (*continued*)

Of the three stated methods for calculating displacements of beams, the strain energy method is restricted to determining the displacements of a beam caused by a concentrated load at the point of application of the load. The lecturer would therefore have illustrated the strain energy method by using the beam in Fig. S2.30a.

As in the unit load method the lecturer used tabulated coefficients, the loading on the beam must have been such as to permit simple standard bending moment diagrams to be drawn. This eliminates the case where the span is only partially loaded, and hence the unit load method was used for the beam in Fig. S2.30c.

It follows that the differential equation of bending was used to obtain the mid-span deflection of the beam in Fig. S2.30b.

(a) The forces acting on the structure of Fig. S2.30a are shown in Fig. S2.30d.

Using $\sum M_b = 0$ $\qquad R_{ya}L + Wa = 0$

$$R_{ya} = -\frac{Wa}{L}$$

Hence the force applied by the support at A to the beam is downward and equal to $\frac{Wa}{L}$.

Consider the part AB.
For an element of length dx at distance x from A:

$$M_x = -\frac{Wax}{L}$$

$$\text{Strain energy in AB} = \int_0^L \frac{M_x^2}{2EI}\, dx = \frac{1}{2EI}\int_0^L \left(-\frac{Wax}{L}\right)^2 dx$$

$$= \frac{W^2a^2L}{6EI}$$

Now consider the part BC.
For an element of length ds at distance s from C:

$$M_s = -Ws$$

$$\text{Strain energy in BC} = \int_0^a \frac{M_s^2}{2EI}\, ds = \frac{1}{2EI}\int_0^a (-Ws)^2\, ds$$

$$= \frac{W^2a^3}{6EI}$$

Thus the total strain energy in the structure is $\dfrac{W^2a^2(L + a)}{6EI}$.

If the vertical displacement at C during the application of the load is Δ, then:

$$\text{Work done by the load} = \frac{W\Delta}{2}$$

By the law of the conservation of energy:

$$\frac{W\Delta}{2} = \frac{W^2 a^2 (L + a)}{6EI}$$

$$\Delta = \frac{W a^2 (L + a)}{3EI}$$

Points to note:

1. As a partial check on such working, it is always desirable to study the units that the final expression would lead to. In this case, if W is in kN, a and L in mm, E in kN/mm^2 and I in mm^4, the expression would give deflections in mm, which of course should be the case.

2. The region of the beam BC is a cantilever. It may be recalled (Section 2.7 of Part One) that the deflection at the end of a comparable cantilever but rigidly fixed at the support is $Wa^3/3EI$. The reason that the point C deflects somewhat more than this value is because the root of the cantilever, namely B, is not rigidly fixed. As a result of the bending of AB, there is a rotation of the beam at B. If the rotation of the beam at B is ϕ, then the *additional* deflection at the end of the cantilever at C is ϕa. It is left to the student to show that the rotation at B for the beam in Fig. S2.30e (and which is equivalent to AB of Fig. S2.30d) is equal to $WaL/3EI$.

(b) The differential equation method of obtaining displacement is always somewhat tedious, and it is therefore tempting to look for ways which make it less so. In this case a *slight* simplification in the equation is obtained by working from the right-hand end of the beam rather than the left. Alternatively, since we are more used to working from the left-hand end of the beam, we can turn the beam around as in Fig. S2.30f. This is the beam we will analyse here.

Using $\sum M_b = 0$
$$R_{ya}L - w \frac{L}{2} \frac{L}{4} = 0$$

$$R_{ya} = \frac{wL}{8}$$

For an element of length dx at distance x from A:

$$M_x = \frac{wL}{8} x - \frac{w}{2} \left[x - \frac{L}{2} \right]^2$$

Hence
$$EI \frac{d^2 v}{dx^2} = \frac{wL}{8} x - \frac{w}{2} \left[x - \frac{L}{2} \right]^2$$

Integrating $\qquad EI\dfrac{dv}{dx} = \dfrac{wL}{16}x^2 - \dfrac{w}{6}\left[x - \dfrac{L}{2}\right]^3 + C_1$

Integrating $\qquad EIv = \dfrac{wL}{48}x^3 - \dfrac{w}{24}\left[x - \dfrac{L}{2}\right]^4 + C_1 x + C_2$

At $x = 0$, $v = 0$; this gives $C_2 = 0$

At $x = L$, $v = 0$; this gives $C_1 = -\dfrac{7wL^3}{384}$

Hence $\qquad EIv = \dfrac{wL}{48}x^3 - \dfrac{w}{24}\left[x - \dfrac{L}{2}\right]^4 - \dfrac{7wL^3}{384}x$

At $x = \dfrac{L}{2}$ $\qquad\qquad EIv = -\dfrac{5wL^4}{768}$

$$\text{Mid-span deflection} = \dfrac{5wL^4}{768EI}\ \text{(downward)}$$

(c) To obtain the standard bending moment diagrams in order to simplify the unit load method, the continuous loading on ABC (Fig. S2.30g) is split into the two cases—the loading on AB (Fig. S2.30h) and the loading on BC (Fig. S2.30j). This gives rise to two sets of standard bending moment diagrams M'_0 and M''_0 as shown in Fig. S2.30i and Fig. S2.30k respectively.

To obtain the deflection at C, the unit load $X_1 = 1$ is applied at that position (Fig. S2.30l), giving rise to the m_1 diagram in Fig. S2.30m.

The coefficients used for the product integration of the bending moment diagrams are obtained from Table 2.7 in Part One.
If Δ_c is the deflection at C, then

$$EI\Delta_c = \int_{AC} M'_0 m_1\ ds + \int_{AC} M''_0 m_1\ ds$$

$$= \int_{AB} M'_0 m_1\ ds + \int_{BC} M'_0 m_1\ ds + \int_{AB} M''_0 m_1\ ds + \int_{BC} M''_0 m_1\ ds$$

$$= -\dfrac{L}{3}\dfrac{wL^2}{8}\dfrac{L}{2} + 0 + \dfrac{L}{3}\dfrac{wL^2}{8}\dfrac{L}{2} + \dfrac{1}{4}\dfrac{L}{2}\dfrac{wL^2}{8}\dfrac{L}{2}$$

$$= \dfrac{wL^4}{128}$$

Hence $\qquad \Delta_c = \dfrac{wL^4}{128EI}$

Problem 2.31

A cantilever beam of length L and uniform cross-section has a concentrated load W applied at a distance d from the support. Determine the value of d which would ensure that the deflection at the end of the beam is twice the deflection at the position of the load.

The beam is shown in Fig. S2.31a. The standard expression for the deflection of a cantilever at the position of a concentrated load gives

$$\Delta_b = \frac{Wd^3}{3EI}$$

An expression for the deflection at C will be found by the unit load method. The M_0 bending moment diagram is shown in Fig. S2.31b. The m_1 diagram corresponding to the force $X_1 = 1$ at C (Fig. S2.31c) is given in Fig. S2.31d.

It will be noted that the bending moments M_0 are zero over the region BC. The moment diagrams to be multiplied together are, therefore, the triangle on AB in Fig. S2.31b and the quadrilateral on AB in Fig. S2.31d. Although coefficients are not given in Table 2.7 of Part One for the product integration involving a quadrilateral, this is easily overcome by splitting the quadrilateral into the two triangles shown in Fig. S2.31e. Thus:

$$EI\Delta_c = \int_{AC} M_0 m_1 \, dx$$

$$= \int_{AB} M_0 m_1 \, dx \quad \left(\text{since} \int_{BC} M_0 m_1 \, dx = 0 \right)$$

$$= \frac{d}{3} WdL + \frac{d}{6} Wd(L - d)$$

$$= \frac{WLd^2}{2} - \frac{Wd^3}{6}$$

As

$$\Delta_c = 2\Delta_b$$

$$\frac{2Wd^3}{3} = \frac{WLd^2}{2} - \frac{Wd^3}{6}$$

$$d = \frac{3L}{5}$$

It is desirable for students to note that, as there is no bending moment in the region BC caused by the applied load W, part of the beam does not bend but remains straight as in Fig. S2.31f. An alternative procedure would, therefore, have been to determine the rotation of the beam at B caused by the load. If this rotation is ϕ, then

$$\phi(L - d) = \Delta_b$$

and the relationship between d and L can be found.

(a)

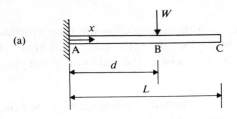

(b)

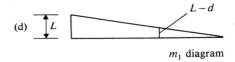

M_0 diagram

(c)

$X_1 = 1$

(d)

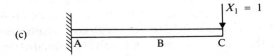

$L - d$

m_1 diagram

(e)

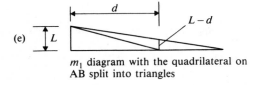

d

$L - d$

m_1 diagram with the quadrilateral on AB split into triangles

(f)

(g)

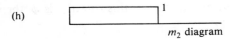

$X_2 = 1$

(h)

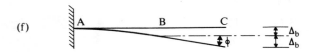

m_2 diagram

Figure S2.31

The rotation can be obtained by using the unit load method. For example, applying a unit moment $X_2 = 1$ at B as in Fig. S2.31g gives the m_2 diagram shown in Fig. S2.31h. The product integration of this rectangle with the M_0 diagram gives

$$EI\phi = \int_{AB} M_0 m_2 \, dx$$

$$= \frac{d}{2} Wd \, 1$$

$$= \frac{Wd^2}{2}$$

Using the above expression now leads to the same answer as before.

Problem 2.32

A beam AB of length L and flexural rigidity EI is simply supported at A and B. It carries a distributed load which at distance x from A has an intensity of $w \sin (\pi x/L)$. Show that each reaction is wL/π. Using the relationships between load intensity and shear force, shear force and bending moment, bending moment and curvature, show that the mid-span deflection is $WL^3/2\pi^3 EI$ where W is the total load on the beam.

The distribution of loading on the beam is illustrated in Fig. S2.32. The loading is symmetrical about mid-span.

$$\text{Total load} = \int_0^L w \sin \frac{\pi x}{L} \, dx = \frac{2wL}{\pi}$$

$$\text{Hence each reaction} = \frac{wL}{\pi}$$

Since

$$w_x = w \sin \frac{\pi x}{L}$$

and

$$\frac{dS_x}{dx} = w_x$$

then

$$S_x = \int w \sin \frac{\pi x}{L} \, dx = -\frac{wL}{\pi} \cos \frac{\pi x}{L} + C_1$$

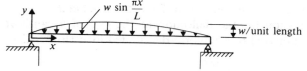

Figure S2.32

At $x = 0$, $S = -\dfrac{wL}{\pi}$ and hence $C_1 = 0$

Since
$$\frac{dM_x}{dx} = -S_x$$

then
$$M_x = \int \frac{wL}{\pi} \cos \frac{\pi x}{L} \, dx = \frac{wL^2}{\pi^2} \sin \frac{\pi x}{L} + C_2$$

At $x = 0$, $M = 0$ and hence $C_2 = 0$

Since
$$EI \frac{d^2 v}{dx^2} = M_x$$

then
$$EI \frac{dv}{dx} = \int \frac{wL^2}{\pi^2} \sin \frac{\pi x}{L} \, dx = -\frac{wL^3}{\pi^3} \cos \frac{\pi x}{L} + C_3$$

and
$$EIv = -\frac{wL^4}{\pi^4} \sin \frac{\pi x}{L} + C_3 x + C_4$$

At $x = 0$, $v = 0$ and hence $C_4 = 0$
At $x = L$, $v = 0$ and hence $C_3 = 0$

Thus
$$EIv = -\frac{wL^4}{\pi^4} \sin \frac{\pi x}{L}$$

At $x = \dfrac{L}{2}$
$$EIv = -\frac{wL^4}{\pi^4}$$

Since
$$\text{total load } W = \frac{2wL}{\pi}$$

$$\text{mid-span deflection} = \frac{WL^3}{2EI\pi^3} \quad \text{(downward)}$$

Problem 2.33

A cantilever beam of span L is to be designed to support a uniformly distributed loading of w per unit length. However, it is found that the deflection caused by this load at the free end is excessive when a beam of uniform rectangular section of breadth b and depth d is used. Investigate by what factor this deflection is reduced if the same amount of material is used as a cantilever of the same breadth b but with the depth varying linearly from zero at the free end to $2d$ at the fixed end.

The standard expression for the deflection at the end of a cantilever beam of uniform section and subjected to a uniform distributed load (Fig. S2.33a) is

$$\Delta_a = \frac{wL^4}{8EI}$$

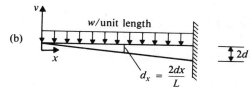

Figure S2.33

For a beam of rectangular section of width b and depth d, then:

$$I = \frac{bd^3}{12}$$

Hence

$$\Delta_a = \frac{3wL^4}{2Ebd^3} \quad \text{(downward)}$$

The deflection at the end of the non-uniform beam (Fig. S2.33b) will be determined using the differential equation of bending, but acknowledging the variations of I along the beam. Thus:

$$d_x = \frac{2dx}{L}$$

Hence

$$I_x = \frac{b}{12}\left(\frac{2dx}{L}\right)^3 = \frac{2bd^3x^3}{3L^3}$$

The bending moment at a point distance x from the left-hand end is

$$M_x = \frac{wx^2}{2}$$

Since

$$\frac{d^2v}{dx^2} = \frac{M_x}{EI_x}$$

substituting for M_x and I_x gives

$$\frac{d^2v}{dx^2} = \frac{3wL^3}{4Ebd^3x}$$

For convenience let
$$k = \frac{3wL^3}{4Ebd^3}$$

Thus
$$\frac{d^2v}{dx^2} = \frac{k}{x}$$

Integrating gives
$$\frac{dv}{dx} = k \log x + C_1$$

At $x = L$, $\dfrac{dv}{dx} = 0$, hence $C_1 = -k \log L$

giving
$$\frac{dv}{dx} = k \log x - k \log L$$

Integrating gives $v = k(x \log x - x) - xk \log L + C_2$

At $x = L$, $v = 0$, hence $C_2 = -kL$
Hence at $x = 0$, $v = C_2 = -kL$

Deflection at the end of the beam $= \dfrac{3wL^4}{4Ebd^3}$ (downward)

A comparison of the two expressions shows that the maximum deflection of the non-uniform beam is one-half of that of the uniform beam.

Problem 2.34

Use the unit load method (in conjunction with tabulated coefficients for product integration) to determine (a) the deflection at the end of the cantilever of the beam in Fig. S2.34a, (b) the rotation at the same position, and (c) the deflection of the cantilever in Fig. S.34b at the position midway between support and load.

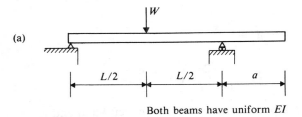

(a)

$$L/2 \qquad L/2 \qquad a$$

Both beams have uniform EI

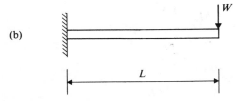

(b)

$$L$$

Figure S2.34

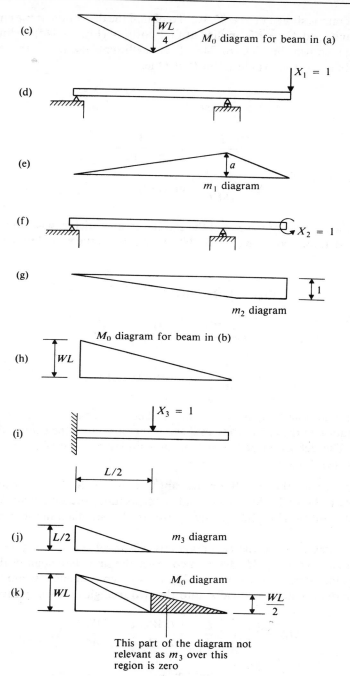

(c) $\dfrac{WL}{4}$ M_0 diagram for beam in (a)

(d) $X_1 = 1$

(e) a m_1 diagram

(f) $X_2 = 1$

(g) 1 m_2 diagram

(h) M_0 diagram for beam in (b) WL

(i) $X_3 = 1$ $L/2$

(j) $L/2$ m_3 diagram

(k) WL M_0 diagram $\dfrac{WL}{2}$

This part of the diagram not relevant as m_3 over this region is zero

Figure S2.34   (*continued*)

(a) The M_0 diagram is shown in Fig. S2.34c. To obtain the deflection at the end of the cantilever, the unit load $X_1 = 1$ is applied at that position (Fig. S2.34d) giving rise to the m_1 diagram shown in Fig. S2.34e. The coefficient used for the product integration is obtained from Table 2.7 in Part One.

$$EI\Delta = \int M_0 m_1 \, dx$$

$$= -\frac{L}{4} \frac{WL}{4} a = -\frac{WL^2 a}{16}$$

$$\Delta = \frac{WL^2 a}{16EI} \quad \text{(upward)}$$

(b) To obtain the rotation at the end of the cantilever, an anticlockwise unit moment $X_2 = 1$ is applied at that position (Fig. S2.34f) giving rise to the m_2 diagram shown in Fig. S2.34g.

$$EI\phi = \int M_0 m_2 \, dx$$

$$= \frac{L}{4} \frac{WL}{4} 1 = \frac{WL^2}{16}$$

$$\phi = \frac{WL^2}{16EI} \quad \text{(anticlockwise)}$$

(The student should note that, as there is no *bending* of the cantilever caused by the load W, the rotation at the end of the cantilever is the same as at the supported end of the cantilever. The deflection at the end of the cantilever is therefore ϕa and this confirms the earlier value.)

(c) The M_0 diagram for the load W on the cantilever in Fig. S2.34b is shown in Fig. S2.34h. To obtain the deflection at the mid-span position, the unit load $X_3 = 1$ is applied at that position (Fig. S2.34i) giving rise to the m_3 diagram shown in Fig. S2.34j.

The product integration of the two diagrams will involve the triangle of the m_3 diagram with that *part* of the M_0 diagram covering the same half-span of the beam, namely, the quadrilateral shown in Fig. S2.34k. For convenience in applying the coefficients, the quadrilateral is divided into the two triangles shown. Hence

$$EI\Delta = \frac{1}{3} \frac{L}{2} WL \frac{L}{2} + \frac{1}{6} \frac{L}{2} \frac{WL}{2} \frac{L}{2} = \frac{5WL^3}{48}$$

$$\Delta = \frac{5WL^3}{48EI} \quad \text{(downward)}$$

Problem 2.35

The column AB in Fig. S2.35a can be regarded as simply supported at A and B for horizontal loads applied at C from the crane. There is a change of section at C—the second moment of area above C is I and below C is $2I$. Calculate the horizontal deflection at the mid-height D caused by the horizontal load P.

The designer has judged that the nature of the foundations is such that there is no significant resistance to rotation of the column at A due to the horizontal load. This is the implication of the statement that the column can be regarded as simply supported at A and B, which of course makes the structure statically determinate.

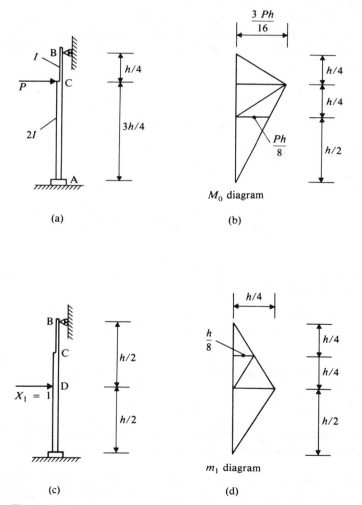

Figure S2.35

The unit load method will be used to determine the deflection at D. To facilitate the calculations when using this method, students should, by this stage, know the values of the more common coefficients for product integration without the necessity for referring to a table. This particular problem is facilitated by knowing that the coefficients for the product integration of:

– triangles *the same way round* is $\frac{1}{3}$
– triangles *the opposite way round* is $\frac{1}{6}$

The M_0 diagram is shown in Fig. 2.35b. To obtain the horizontal movement of the column at its mid-height, a horizontal unit load $X_1 = 1$ is applied at that position (Fig. S2.35c) giving the m_1 diagram shown in Fig. S2.35d.

As the discontinuities (i.e., the peaks) in the M_0 and m_1 diagrams occur at different positions, the product integrations are carried out separately over the three regions: AD, DC, CA. The two quadrilaterals over the region DC are each divided into two triangles as shown.

$$EΔ = \int_{AD} \frac{M_0 m_1}{I_{ad}} \, dx + \int_{DC} \frac{M_0 m_1}{I_{dc}} \, dx + \int_{CB} \frac{M_0 m_1}{I_{cb}} \, dx$$

$$\int_{AD} \frac{M_0 m_1}{I_{ad}} = \frac{1}{2I} \frac{1}{3} \frac{h}{2} \frac{Ph}{8} \frac{h}{4} = \frac{Ph^3}{384I}$$

$$\int_{DC} \frac{M_0 m_1}{I_{dc}} = \frac{1}{2I} \frac{h}{4} \left(\frac{1}{3} \frac{3Ph}{16} \frac{h}{8} + \frac{1}{6} \frac{3Ph}{16} \frac{h}{4} + \frac{1}{6} \frac{Ph}{8} \frac{h}{8} + \frac{1}{3} \frac{Ph}{8} \frac{h}{4} \right) = \frac{11Ph^3}{3072I}$$

$$\int_{CB} \frac{M_0 m_1}{I_{cb}} = \frac{1}{I} \frac{1}{3} \frac{h}{4} \frac{3Ph}{16} \frac{h}{8} = \frac{Ph^3}{512I}$$

Hence

$$EΔ = \frac{Ph^3}{I} \left(\frac{1}{384} + \frac{11}{3072} + \frac{1}{512} \right) = \frac{25Ph^3}{3072I}$$

$$\text{Movement at mid-height} = \frac{25Ph^3}{3072EI} \text{ (to the right)}$$

Problem 2.36

The two adjacent beams in Fig. S2.36a are each simply supported and have the same uniform EI. Determine the discontinuity in line as a result of the rotations at B caused by the loads.

Because the two beams and their loads are identical, the angular discontinuity at B is equal to twice the rotation at the end of one of the beams. This rotation will be determined using the unit load method.

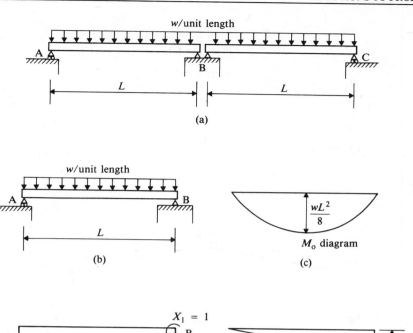

Figure S2.36

The M_0 diagram for the beam in Fig. S2.36b is given in Fig. S2.36c. To obtain the rotation at B, a unit moment $X_1 = 1$ is applied at B (Fig. S2.36d) giving the m_1 diagram in Fig. S2.36e.

Hence

$$EI\phi = \int_{AB} M_0 m_1 \, \mathrm{d}x = \frac{L}{3} \frac{wL^2}{8} 1 = \frac{wL^3}{24}$$

Thus the angular discontinuity from one beam to the next is $\dfrac{wL^3}{12EI}$.

(Similar calculations to this can be used in the analysis of statically indeterminate beam structures as illustrated in Part One.)

Problem 2.37

Obtain the bending moment diagram for the structure in Fig. S2.37a when (a) only the vertical 16 kN load is acting and (b) when only the horizontal 12 kN load is acting. Use the principle of superposition to obtain the bending moment diagram for the combined loading shown in Fig. S2.37a.

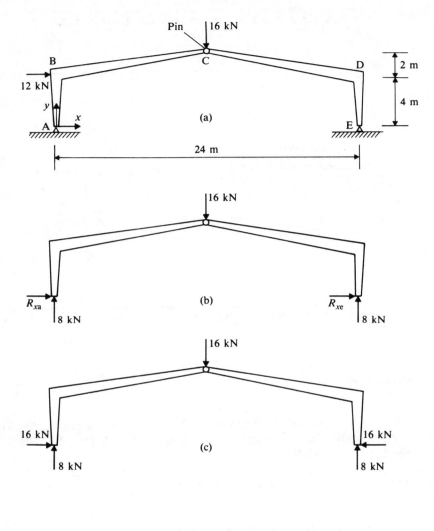

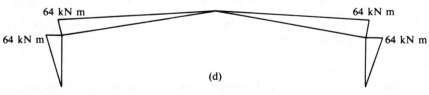

Figure S2.37

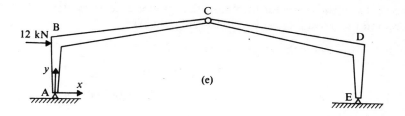

(e)

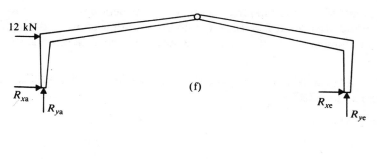

(f)

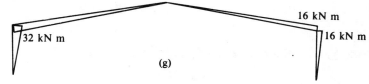

(g)

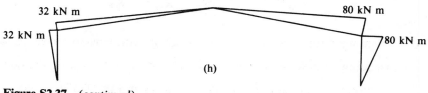

(h)

Figure S2.37   (*continued*)

(a) When only the 16 kN load is acting, symmetry of the frame and the loading allows us to deduce that the vertical reactions at A and E are 8 kN. To complete our knowledge of the external forces on the structure, it is necessary to determine the horizontal reactive forces at A and E (Fig. S2.37b).

Using $\sum M_c = 0$ for the forces to the left of C:

$$(8 \times 12) - (R_{xa} \times 6) = 0$$

$$R_{xa} = 16 \text{ kN}$$

Using $\sum P_x = 0$ for the whole structure:

$$R_{xa} + R_{xe} = 0$$

$$R_{xe} = -16 \text{ kN}$$

A summary of the external forces acting on the structure is given in Fig. S2.37c. Thus, the bending moment at B $= 16 \times 4 = 64$ kN m with tension on the outside. A similar bending moment exists at D and the bending moment diagram is as shown in Fig. S2.37d.

(b) For the case of the horizontal 12 kN load (Fig. S2.37e), let the reactions be as shown in Fig. S2.37f.

Using $\sum M_e = 0$ $\qquad (R_{ya} \times 24) + (12 \times 4) = 0$

$$R_{ya} = -2 \text{ kN}$$

Using $\sum P_y = 0$ $\qquad R_{ya} + R_{ye} = 0$

$$R_{ye} = 2 \text{ kN}$$

Using $\sum M_c = 0$ for forces to the right of C:

$$(2 \times 12) + (R_{xe} \times 6) = 0$$

$$R_{xe} = -4 \text{ kN}$$

Using $\sum P_x = 0$ $\qquad R_{xa} + R_{xe} + 12 = 0$

$$R_{xa} = -8 \text{ kN}$$

Thus:

bending moment at B $= 8 \times 4 = 32$ kN m with tension on the inside

bending moment at D $= 4 \times 4 = 16$ kN m with tension on the outside

The resulting bending moment diagram is shown in Fig. S2.37g.

Using the principle of superposition, the bending moments for the case of both loads applied are as follows:

at B $= 64 - 32 = 32$ kN m (tension on the outside)

at D $= 64 + 16 = 80$ kN m (tension on the outside)

A summary of the bending moment diagram for this combined loading is given in Fig. S2.37h.

Problem 2.38

Explain why the arch can be used for longer spans than the beam. Show that, for a three-pinned parabolic arch of shape defined by the equation

$$y = \frac{4hx(L - x)}{L^2}$$

and subjected to a uniformly distributed load of w/unit horizontal length, the bending moment at every section is zero.

The structural advantage of arches (and portal frames) lies in the development of horizontal reactive forces at the supports which reduce the bending moment in the structure. This effect of the horizontal reactive forces on bending moments is explained with the use of Figs S2.38a, S2.38b and S2.38c. For the simply supported beam AB in Fig. S2.38a subjected to the imposed load W, the bending moment on a section of the beam distance x from the left-hand support (but, say, on the left of the load) is

$$M_x = R_{ya}x$$

For the arch AB in Fig. S2.38b, the supports at A and B are pinned in position. The effect of the load W will now cause horizontal reactive forces at A and B in addition to the vertical reactive forces. The direction of these forces can be deduced by imagining the structural behaviour if the arch had been supported on rollers as in Fig. S2.38c. As a result of the bending moments (with tension on the underside as in a simply supported beam), the supports will tend to move apart. It follows that equal inward-acting forces will develop when the arch is pinned to the supports as in Fig. S2.38b. If these horizontal forces are equal to H, then the bending moment at a section of the arch distance x from the left-hand support (but again to the left of the load) is

$$M_x = R_{ya}x - Hy$$

Hence in this comparison, the reduction in the bending moment as a result of the horizontal forces is Hy. It is such reductions in bending moment which allow the arch to be used for longer spans than the simply supported beam.

The three-pinned arch subjected to a uniformly distributed load is shown in Fig. S2.38d. Symmetry of the arch and the loading allows us to observe that

$$R_{ya} = R_{yb} = \frac{wL}{2}$$

Let the inward-acting horizontal reactions be H (Fig. S2.38e). As the bending moment at the pin must be zero:

$$R_{ya}\frac{L}{2} - Hh - \frac{wL}{2}\frac{L}{4} = 0$$

$$H = \frac{wL^2}{8h}$$

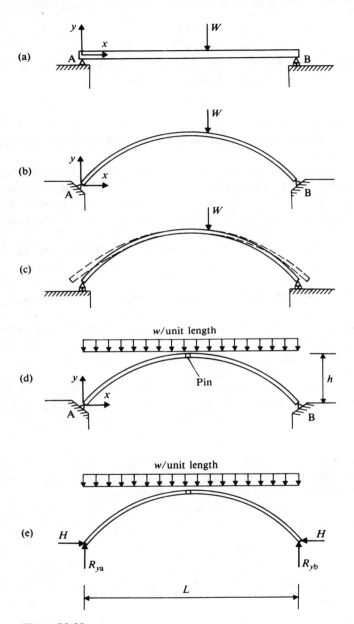

Figure S2.38

Consider a general point on the arch having coordinates x, y. The bending moment at the point is given by

$$M_x = R_{ya}x - Hy - \frac{wx^2}{2}$$

Substituting for R_{ya}, H and y shows that

$$M_x = 0$$

Hence the bending moment at every section of the arch is zero.

Problem 2.39

The horizontal forces at the abutments of a two-pinned parabolic arch of height h and span L caused by a concentrated vertical load W at a horizontal distance x from an abutment is given by

$$H = \frac{5W}{8h}\left(x - \frac{2x^3}{L^2} + \frac{x^4}{L^3}\right)$$

Use this to show that when a uniformly distributed load covers the whole span of such an arch, the bending moment is everywhere zero. The equation determining the shape of the parabolic arch is as in Problem 2.38.

Consider a short length dx of the distributed load (Fig. S2.39a). The load exerted by this short length of load is therefore $w\,dx$. This is, in effect, a concentrated load acting

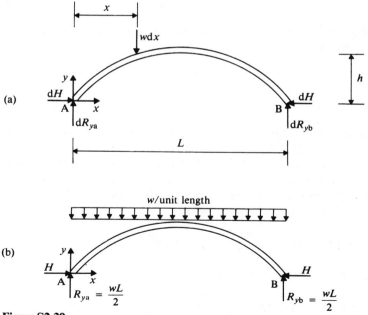

Figure S2.39

on the arch, and the horizontal forces at the supports (say dH) caused by this element of load are therefore given by

$$dH = \frac{5w\, dx}{8h}\left(x - \frac{2x^3}{L^2} + \frac{x^4}{L^3}\right)$$

$$= \frac{5w}{8h}\left(x - \frac{2x^3}{L^2} + \frac{x^4}{L^3}\right) dx$$

The horizontal forces H caused by the whole of the distributed load (Fig. S2.39b) are obtained by the integration over the whole span:

$$H = \int_0^L \frac{5w}{8h}\left(x - \frac{2x^3}{L^2} + \frac{x^4}{L^3}\right) dx$$

$$= \frac{5w}{8h}\left(\frac{x^2}{2} - \frac{x^4}{2L^2} + \frac{x^5}{5L^3}\right)_{x=0}^{x=L}$$

$$= \frac{wL^2}{8h}$$

For a section of the arch at coordinates x, y, the bending moment caused by the distributed load is given by

$$M_x = \frac{wLx}{2} - \frac{wL^2}{8h}y - \frac{wx^2}{2}$$

Substituting the value of y in terms of x as given by the equation for the parabola (see Problem 2.38) gives

$$M_x = \text{zero}$$

Problem 2.40

The statically indeterminate continuous beam ABCDE shown in Fig. S2.40a is supported at B, C, D and E. An analysis shows that the hogging moment at C is 30 kN m and the reaction at E is 20 kN. Draw the bending moment diagram. Calculate the mid-span bending moments and compare these with the values that would have obtained in a statically determinate system of beams if there had been no continuity at C and D.

Bending moment at B $= -\dfrac{10 \times 2^2}{2} = -20\,\text{kN m}$

Bending moment at C $= -30\,\text{kN m}$

Bending moment at D $= (20 \times 5) - \left(10 \times 5 \times \dfrac{5}{2}\right) = -25\,\text{kN m}$

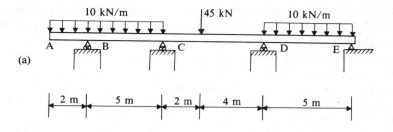

(a)

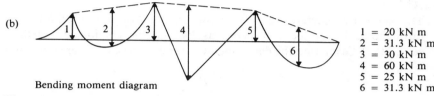

(b)

Bending moment diagram

1 = 20 kN m
2 = 31.3 kN m
3 = 30 kN m
4 = 60 kN m
5 = 25 kN m
6 = 31.3 kN m

Figure S2.40

Maximum 'free' bending moment for span BC $= \dfrac{10 \times 5^2}{8} = 31.3 \text{ kN m}$

Maximum 'free' bending moment for span CD $= \dfrac{45 \times 2 \times 4}{6} = 60 \text{ kN m}$

Maximum 'free' bending moment for span DE $= 31.3 \text{ kN m}$

These values allow the bending moment diagram to be sketched. It is shown in Fig. S2.40b.

Bending moment at mid-span of BC $= 31.3 - 25 = \underline{6.3 \text{ kN m}}$

Bending moment at mid-span of CD $= \left(60 \times \dfrac{3}{4}\right) - \dfrac{(30 + 25)}{2} = \underline{17.5 \text{ kN m}}$

Bending moment at mid-span of DE $= 31.3 - \dfrac{25}{2} = \underline{18.8 \text{ kN m}}$

For the statically determinate system:

Bending moment at mid-span of BC $= 31.3 - \dfrac{20}{2} = \underline{21.3 \text{ kN m}}$

Bending moment at mid-span of CD $= 60 \times \dfrac{3}{4} = \underline{45 \text{ kN m}}$

Bending moment at mid-span of DE $= \underline{31.3 \text{ kN m}}$

These values are all appreciably higher than the corresponding values for the statically indeterminate case, and indeed somewhat higher than the adjacent hogging moments. (Students should not draw too many conclusions at this stage from such a comparison. The matter of which arrangement would be the more desirable is quite complex, although some of the relevant points are discussed in Chapters 4 and 5 of Part One. For example, as far as the economic factor is concerned, if the structure were to be constructed in steel, although bigger steel sections would be required for the statically determinate system, the cost of the simpler joints required at the supports C and D counterbalances this effect. But there are many other interacting factors.)

Problem 2.41

The statically indeterminate beam in Fig. S2.41a has a uniform cross-section throughout. It can be considered as two cantilevers joined together by a pin connection. Given that the deflection at the end of the cantilever shown in Fig. S2.41b is

$$\frac{Wa^2}{2EI}\left(L - \frac{a}{3}\right)$$

use the principle of compatibility of deformations to calculate the vertical force transmitted by the pin.

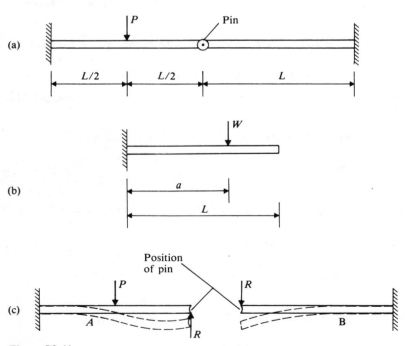

Figure S2.41

Application of the load P to the structure causes the downward deflection of the left-hand cantilever and also, because of the pin-joint between the two, of the right-hand cantilever. In effect the right-hand cantilever is assisting the left-hand cantilever by providing a supporting force (say R) at the position of the pin. The forces acting on the two cantilevers (designated A and B) are therefore as shown in Fig. S2.41c.

The deflected shapes of the cantilevers under the effect of the forces would be as shown by the dotted lines. Since the two cantilevers are connected at their ends, it follows that the deflections at the ends of the two cantilevers must be equal. This enables R to be calculated.

Using the given expression for the deflection at the end of the cantilever in Fig. S2.41b, the deflection at the end of cantilever B due to the force R is obtained by putting $a = L$. Thus:

$$\Delta_b = \frac{RL^3}{3EI}$$

If in the cantilever in Fig. S2.41b, $a = L/2$ then:

$$\text{end deflection} = \frac{5WL^3}{48EI}$$

It follows that the deflection at the end of cantilever A in Fig. S2.41c is

$$\Delta_a = \frac{5PL^3}{48EI} - \frac{RL^3}{3EI}$$

Since

$$\Delta_a = \Delta_b$$

$$\frac{5PL^3}{48EI} - \frac{RL^3}{3EI} = \frac{RL^3}{3EI}$$

$$R = \frac{5P}{32}$$

Problem 2.42

Show that the frame in Fig. S2.42a is statically indeterminate. Obtain the forces in the three members in terms of the displacements of the loaded joint. Use the equilibrium at this joint to obtain the simultaneous equations involving the displacements and so obtain their values. Hence calculate the forces in the members. (If a computer and a suitable planar pin-jointed frame program are available, use these to verify the solution.)

$$\text{Number of members } m = 3$$

$$\text{Number of reactive forces } r = 3 \times 2 = 6$$

$$\text{Number of joints } j = 4$$

As

$$m + r > 2j$$

the structure is statically indeterminate.

The expression for the tensile force in a general member i (Fig. S2.42b) in terms of the x and y components of the displacements at the end m and n of the member is

$$F_i = \left(\frac{AE}{L}\right)_i ((\Delta_{xn} - \Delta_{xm}) \cos \theta_i + (\Delta_{yn} - \Delta_{ym}) \sin \theta_i)$$

For member 1:   Length = 5 m

$$\left(\frac{AE}{L}\right)_1 = \frac{125 \times 200}{5 \times 10^3} = 5 \text{ kN/mm}$$

$$\cos \theta_1 = -\frac{4}{5}$$

$$\sin \theta_1 = \frac{3}{5}$$

For member 2:   Length = 3.57 m

$$\left(\frac{AE}{L}\right)_2 = \frac{250 \times 200}{3.57 \times 10^3} = 14 \text{ kN/mm}$$

$$\cos \theta_2 = 0$$

$$\sin \theta_2 = 1$$

For member 3:   Length = 5 m

$$\left(\frac{AE}{L}\right)_3 = \frac{250 \times 200}{5 \times 10^3} = 10$$

$$\cos \theta_3 = \frac{3}{5}$$

$$\sin \theta_3 = \frac{4}{5}$$

Let the components of the deflection of the loaded joint be Δ_x and Δ_y. Substituting these and the above member properties in the general expression gives

$$F_1 = 5\left((0 - \Delta_x)\left(-\frac{4}{5}\right) + (0 - \Delta_y)\frac{3}{5}\right)$$

$$= 4\Delta_x - 3\Delta_y$$

$$F_2 = 14((0 - \Delta_x)0 + (0 - \Delta_y)1)$$

$$= -14\Delta_y$$

$$F_3 = 10\left((0 - \Delta_x)\frac{3}{5} + (0 - \Delta_y)\frac{4}{5}\right)$$

$$= -6\Delta_x - 8\Delta_y$$

Consider equilibrium at the loaded joint.

E for all three members $= 200 \text{ kN/mm}^2$
Cross-sectional areas: $A_1 = 125 \text{ mm}^2$; $A_2 = A_3 = 250 \text{ mm}^2$

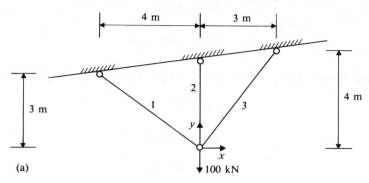

(a)

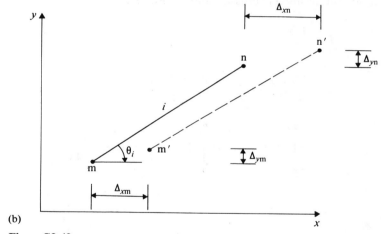

(b)

Figure S2.42

Using $\sum P_x = 0$ $\qquad$ $F_1 \cos \theta_1 + F_3 \cos \theta_3 = 0$

Substituting the previous values in this equation gives

$$17\Delta_x + 6\Delta_y = 0 \tag{2.42a}$$

Using $\sum P_y = 0$ $\qquad$ $F_1 \sin \theta_1 + F_2 \sin \theta_2 + F_3 \sin \theta_3 - 100 = 0$

Substituting the previous values in this equation gives

$$12\Delta_x + 111\Delta_y = -500 \tag{2.42b}$$

Solving the two simultaneous Eqs (2.42a) and (2.42b) gives

$$\Delta_x = \underline{1.65 \text{ mm}}$$

$$\Delta_y = \underline{-4.68 \text{ mm}}$$

Substituting these values in the previous expressions for member forces gives

$$F_1 = \underline{20.7 \text{ kN}}$$

$$F_2 = \underline{65.6 \text{ kN}}$$

$$F_3 = \underline{27.5 \text{ kN}}$$

Problem 3.1

Show that, for the triangular cross-section of Fig. S3.1:

$$I_y = \frac{DB^3}{48}$$

$$h = \frac{2D}{B}\left(\frac{B}{2} - z\right) = D\left(1 - \frac{2}{B}z\right)$$

$$I_y = \int_{-B/2}^{B/2} hz^2 \, dz = 2D \int_0^{B/2} \left(z^2 - \frac{2}{B}z^3\right) dz = \frac{BD^3}{48}$$

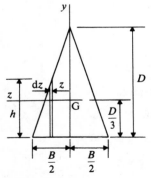

Figure S3.1

Problem 3.2

A universal beam (UB) has a double-symmetrical cross-section of total depth over the flanges at 910 mm and a second moment of area about the major principal axis (i.e., I_z) of 3751×10^6 mm^4. The beam has a mass of 223 kg/m and is simply supported over a span of 9 m.

Calculate the maximum tensile and compressive bending stresses in the beam as a 400 kN wheel load moves slowly across the span.

From Eq. (3.8) of Part One:

$$\sigma_x = -\frac{My}{I}$$

The maximum tensile and compressive stresses will occur when M has its maximum value, i.e., when $M = M_{max}$.

The cross-section of the beam is symmetrical with a total depth of 910 mm. The centroid is therefore located at the midpoint of the web, i.e., at 455 mm from the top and bottom of the cross-section. The maximum tensile stress will occur when $M = M_{max}$ and $y = -455$ mm. The maximum compressive stress will occur when $M = M_{max}$ and $y = +455$ mm. The bending moment on the beam at any section is the sum of the bending moment due to the uniformly distributed self-weight of the beam, w/unit length $= (233 \times 9.81 \times 10^{-3})$ kN/m, and the imposed wheel load, $W = 400$ kN.

w causes maximum bending moment at the centre of the span.
W will also cause maximum bending moment at the centre of the span when W is located at the centre.

Therefore
$$M_{max} = \frac{wl^2}{8} + \frac{Wl}{4} \text{ where } l = \text{span}$$

i.e.,
$$M_{max} = \left(\frac{223 \times 9.81 \times 10^{-3}}{8} \times 9^2\right) + \left(\frac{400 \times 9}{4}\right) = 922.2 \text{ kN m}$$

Therefore, the maximum bending tensile stress, at $y = -455$ mm is

$$\sigma_{x\,max} = -\frac{922.2 \times 10^6}{3751 \times 10^6} \times (-455) = \underline{109 \text{ N/mm}^2}$$

and the maximum bending compressive stress, at $y = 445$ mm is

$$\sigma_{x\,min} = \underline{-109 \text{ N/mm}^2}$$

Problem 3.3

Find the shear stress distribution on a triangular cross-section such as that shown in Fig. S3.1 if $D = 90$ mm, $B = 60$ mm, and a vertical shear force of 150 kN acts along Gy. See Fig. S3.3.

The relevant equation is Eq. (3.27) of Part One, namely, $\tau = \dfrac{S}{It}\,\bar{A}\bar{y}$, which gives the value of the shear stress τ at y_1 from Gz where the width of the section is t and $\bar{A}\bar{y}$ is the moment about Gz of the area of the cross-section lying at $y > y_1$. S is the shear force at the section applied along Gy.

For the cross-section, illustrated in Fig. S3.3, the second moment of area about Gz (see Eq. (3.17) of Part One) is

$$I = \frac{BD^3}{36} = \frac{60 \times 90^3}{36} = \underline{121.5 \times 10^4 \text{ mm}}$$

Let t be the width of the cross-section at distance y_1 from Gz. Then, area abc lying at $y > y_1 = \bar{A} = (60 - y_1)\dfrac{t}{2}$

The moment of area abc about Gz

$$= \bar{A}\bar{y} = (60 - y_1)\frac{t}{2}\left\{y_1 + \frac{1}{3}(60 - y_1)\right\}$$

i.e.,

$$\bar{A}\bar{y} = \frac{t}{2}\left(1200 + 20y_1 - \frac{2}{3}y_1^2\right)$$

therefore

$$\tau = \frac{1}{2}\frac{S}{I}\left(1200 + 20y_1 - \frac{2}{3}y_1^2\right)$$

Putting

$$S = (150 \times 10^3)\,\text{N} \quad \text{and} \quad I = 121.5 \times 10^4 \text{ mm}^4$$

$$\tau = 0.062\left(1200 + 20y_1 - \frac{2}{3}y_1^2\right)$$

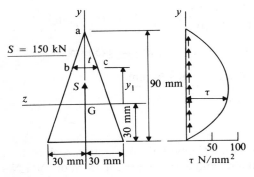

Figure S3.3

When $y_1 = 60$ mm, $\tau = 0$ and when $y_1 = -30$ mm, $\tau = 0$

y_1, mm	-20	-10	0	10	20	30	40	50
τ, N/mm²	33	58	74	83	83	74	58	33

The distribution of τ is shown on Fig. S3.3.

Problem 3.4

Two thin-walled tubes, of the same length but of different mean diameters, are assembled concentrically as shown on Fig. S3.4. The ends of both tubes are fixed to rigid circular plates, one at each end. The outside tube is steel, of mean diameter $2d$, wall thickness t, and shear modulus G_s. The inside tube is copper, of mean diameter d, wall thickness t, and shear modulus G_c. If a torque T is applied to this assembly, find, in terms of G_s and G_c,

(i) the ratio of the shear stress in the steel tube to that in the copper tube,
(ii) the ratio of the torque transmitted by the steel tube to that transmitted by the copper tube.

The tubes are described as thin-walled and only the mean diameter is given for each. Therefore, we can assume that the shear stress is constant across the wall of the steel tube at a value of τ_s and constant across the wall of the copper tube at a value of τ_c.

Because the ends of the tubes are fixed together through the rigid discs, then, when T is applied, both tubes must experience the same twist θ.

The polar second moments of area for the copper and steel tubes, from Eq. (3.40) of Part One, are respectively:

$$J_c = 2\pi\left(\frac{d}{2}\right)^3 t = \frac{\pi d^3 t}{4} \quad \text{and} \quad J_s = 2\pi d^3 t$$

(i) From Eq. (3.37) of Part One: $\dfrac{\tau}{r} = \dfrac{G\theta}{L}$

i.e., $\dfrac{\theta}{L} = \dfrac{\tau}{r}$

Because θ is the same for both tubes over the same length, say L, then

$$\frac{\tau_c}{d/2}\frac{1}{G_c} = \frac{\tau_s}{d}\frac{1}{G_s}$$

therefore $\dfrac{\tau_s}{\tau_c} = \dfrac{2G_s}{G_c}$

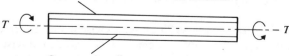

Steel tube, diameter $2d$, wall thickness t

T $-T$

Copper tube, diameter d, wall thickness t

Figure S3.4

(ii) From Eq. (3.37) of Part One:

$$\frac{T}{J} = \frac{G\theta}{L}$$

If T_c is the torque transmitted by the copper tube and T_s is the torque transmitted by the steel tube, then because θ/L is the same for both tubes:

$$\frac{T_c}{J_c}\frac{1}{G_c} = \frac{T_s}{J_s}\frac{1}{G_s}$$

i.e.,

$$\frac{4T_c}{\pi d^3 t}\frac{1}{G_c} = \frac{T_s}{2\pi d^3 t}\frac{1}{G_s}$$

therefore

$$\frac{T_s}{T_c} = \frac{8G_s}{G_c}$$

Problem 3.5

At a point in a steel plate, the state of stress is defined by F where

$$F = \begin{bmatrix} 150 & 75 & 0 \\ 75 & 50 & 0 \\ 0 & 0 & 0 \end{bmatrix} \text{ N/mm}^2 \text{ with respect to the axes } xyz$$

Use Mohr's circle of stress to find the locations of the principal planes, the directions of the principal axes and the values of the principal stresses. Sketch the element bounded by the principal planes and write the principal stress array.

The stress array F is repeated in Fig. S3.5a and illustrated in Fig. S3.5b.

To construct the stress circle, first establish the axes σ, $\tau_{\bar{x}\bar{y}}$ with $\tau_{\bar{x}\bar{y}}$ positive downward as shown in Fig. S3.5c. Plot the point R at coordinates (150, 75) to represent the normal and shear stresses (σ_x, τ_{xy}) on face bc of the element. Similarly, plot the point S to represent the stresses on face ab. Construct the circle on RS as diameter.

In Fig. S3.5c, the distance OC = 0.5(150 + 50) = 100 N/mm²

Radius of the circle $r = \text{CR} = \sqrt{\left(\frac{150 - 50}{2}\right)^2 + 75^2} = 90.1 \text{ N/mm}^2$

The maximum principal stress $\sigma_1 = \text{OP}_1 = (100 + 90.1) = \underline{190.1 \text{ N/mm}^2}$

The minimum principal stress $\sigma_2 = \text{OP}_2 = (100 - 90.1) = \underline{9.9 \text{ N/mm}^2}$

The direction of σ_1 (at P_1) on the circle is 2α anticlockwise from the direction of σ_x (at R). Therefore, the direction of σ_1 on the element is α anticlockwise from the direction of σ_x on the element. The direction of σ_2 (at P_2) on the circle is 180° different from that of σ_1 (at P_1) and therefore 90° different on the element.

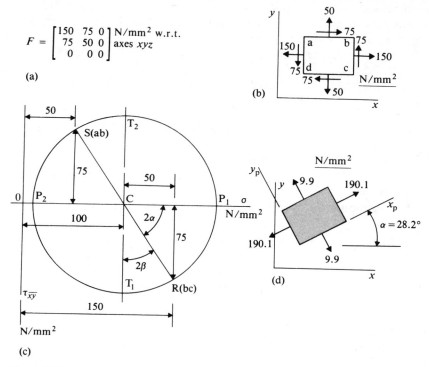

$$F = \begin{bmatrix} 150 & 75 & 0 \\ 75 & 50 & 0 \\ 0 & 0 & 0 \end{bmatrix} \begin{array}{l} \text{N/mm}^2 \text{ w.r.t.} \\ \text{axes } xyz \end{array}$$

(a)

(b)

(c)

(d)

Figure S3.5

From the geometry of the circle:

$$2\alpha = \tan^{-1} \frac{75}{50} \qquad \therefore 2\alpha = 56.3° \qquad \therefore \underline{\alpha = 28.2°}$$

The principal planes coincide with the directions of the principal stresses.
The required element, bounded by the principal planes, is shown in Fig. S3.5d.
Note that the shear stress is zero on a principal plane. The principal stress array F_p is

$$F_p = \begin{bmatrix} 190.1 & 0 & 0 \\ 0 & 9.9 & 0 \\ 0 & 0 & 0 \end{bmatrix} \begin{array}{l} \text{N/mm}^2 \text{ with respect to axes at } 28.2° \\ \text{anticlockwise from } xy \end{array}$$

Problem 3.6

(Do Problem 3.5 before Problem 3.6.) For the state of stress defined in Problem 3.5, use Mohr's circle of stress to find the magnitudes and directions of the maximum and minimum shear stresses and the direct stresses acting on the maximum and minimum shear planes. Write the maximum and minimum shear stress arrays F_{T1} and F_{T2}, and sketch the element bounded by the maximum and minimum shear planes, showing all the stresses acting on the element.

The state of stress F is the same as in Problem 3.5, so the Mohr's stress circle is that in Fig. S3.5c.

The maximum shear stress is represented by the point T_1 in Fig. S3.5c and acts along a plane at $\beta = \frac{1}{2}(90 - 2\alpha) = 16.9°$, clockwise from the plane bc in Fig. S3.5b. The normal stress, $\sigma_{\bar{x}}$, on this plane is equal to $OC = 100$ N/mm^2 and $\tau_{\bar{x}\bar{y}} = r = 90.1$ N/mm^2.

The point T_2 on the circle, separated from T_1 by 180°, represents the stresses $\sigma_{\bar{y}}$ and $\tau_{\bar{x}\bar{y}}$, acting at 90° to $\sigma_{\bar{x}}$ and $\tau_{\bar{x}\bar{y}}$ on the element, which is drawn in Fig. S3.6a. Clearly, $\sigma_{\bar{y}} = 100$ N/mm$^2 = \sigma_{\bar{x}}$.

The maximum shear stress array F_{T1} is therefore given by

$$F_{T1} = \begin{bmatrix} 100 & 90.1 & 0 \\ 90.1 & 100 & 0 \\ 0 & 0 & 0 \end{bmatrix} \quad \begin{array}{l} \text{N/mm}^2 \text{ with respect to axes at } 16.9° \\ \text{clockwise from } xy \end{array}$$

As a further exercise, find the minimum shear stress array

$$F_{T2} = \begin{bmatrix} 100 & -90.1 & 0 \\ -90.1 & 100 & 0 \\ 0 & 0 & 0 \end{bmatrix} \quad \begin{array}{l} \text{N/mm}^2 \text{ with respect to axes at } 73.1° \\ \text{anticlockwise from } xy \end{array}$$

It is illustrated on Fig. 3.6b. Note that F_{T2} (Fig. S3.6b) defines precisely the same element carrying the same stresses as F_{T1} (Fig. S3.6a). This demonstrates the fact that it is irrelevant whether one finds F_{T1} or F_{T2} provided that their respective orientations to the xy axes are specified.

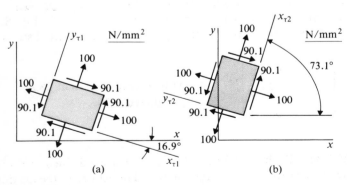

Figure S3.6

Problem 3.7

Each of the strain arrays below represents the state of strain at a point on a plate lying in the xy plane. The strains are with respect to the xy axes. In each case, find the directions of the principal strains, the principal strain array and the maximum shear strain in the plane of the plate.

In each case, find the maximum shear strain array and sketch the deformed shape of the element suffering maximum shear strain.

(i) $S_1 = \begin{bmatrix} 400 & 200 & 0 \\ 200 & -200 & 0 \\ 0 & 0 & \varepsilon_z \end{bmatrix} \times 10^{-6}$ (ii) $S_2 = \begin{bmatrix} 0 & -200 & 0 \\ -200 & 400 & 0 \\ 0 & 0 & \varepsilon_z \end{bmatrix} \times 10^{-6}$

(iii) $S_3 = \begin{bmatrix} 200 & 0 & 0 \\ 0 & 200 & 0 \\ 0 & 0 & \varepsilon_z \end{bmatrix} \times 10^{-6}$ (iv) $S_4 = \begin{bmatrix} -200 & 200 & 0 \\ 200 & -600 & 0 \\ 0 & 0 & \varepsilon_z \end{bmatrix} \times 10^{-6}$

The following solutions for S_1, S_2, S_3 and S_4 should be read in conjunction with Fig. S3.7, parts (i), (ii), (iii) and (iv), respectively.

(i) The strain array S_1 is represented pictorially by Fig. S3.7, part (i), (b). This is helpful when drawing the strain circle shown on Fig. S3.7, part (i), (c). The point R represents the strain combination $\varepsilon_x = 400 \times 10^{-6}$, $\tau_{\bar{x}\bar{y}}/2 = 200 \times 10^{-6}$. The x in brackets is placed after R to remind us that we have plotted the strain ε_x. Similarly,

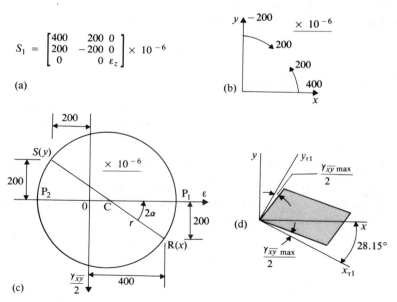

Figure S3.7(i)

$S(y)$ represents the strain combination at $90°$ anticlockwise from the direction of x, namely, $\varepsilon = \varepsilon_y = -200 \times 10^{-6}$ and $\tau_{\bar{x}\bar{y}}/2 = \tau_{xy}/2 = -200 \times 10^{-6}$. Draw the circle on RS as diameter. From the circle:

$$OC = \frac{1}{2}(400 - 200) \times 10^{-6} = 100 \times 10^{-6}$$

$$r = 10^{-6}\sqrt{(300^2 + 200^2)} = 360.6 \times 10^{-6}$$

therefore $\qquad \varepsilon_1 = OP_1 = (100 + 360.6) \times 10^{-6} = \underline{460.6 \times 10^{-6}}$

$$\varepsilon_2 = OP_2 = (100 - 360.6) \times 10^{-6} = \underline{-260.6 \times 10^{-6}}$$

On the circle, move 2α anticlockwise from the direction of x, represented by R, to reach the direction of the maximum principal strain, represented by P_1.

$$2\alpha = \tan^{-1}\left(\frac{200}{300}\right) = 33.69° \quad \therefore \alpha = \underline{16.85°}$$

The principal strain array in the xy plane is therefore

$$S_p = \begin{bmatrix} 460.6 & 0 & 0 \\ 0 & -260.6 & 0 \\ 0 & 0 & \varepsilon_z \end{bmatrix} \times 10^{-6}$$

with respect to axes $x_p y_p$ located at $16.85°$ anticlockwise from xy

The maximum shear strain $\gamma_{\bar{x}\bar{y}\,max}$ is in the axes $x_{T_1} y_{T_1}$ where the direction of x_{T_1} is represented on the circle by T_1, that is, at $(90 - 2\alpha)°$ clockwise from R on the circle, and therefore at $(45 - \alpha)° = 28.15°$ clockwise from the direction of x on the element.

Also $\qquad \dfrac{\tau_{\bar{x}\bar{y}\,max}}{2} = r = 360.6 \times 10^6$

The maximum shear strain array is therefore

$$S_{T1} = \begin{bmatrix} 100 & 360.6 & 0 \\ 360.6 & 100 & 0 \\ 0 & 0 & \varepsilon_z \end{bmatrix} \times 10^{-6}$$

with respect to axes $x_{T_1} y_{T_1}$ located at $28.15°$ clockwise from xy

The corresponding deformed element is sketched in Fig. S3.7, part (i), (d).

(ii) See Fig. S3.7, part (ii), where the strain circle is shown in Fig. S3.7, part (ii), (c). Note that, in this case, $\tau_{\bar{x}\bar{y}}$ when $\theta = 0$ is equal to τ_{xy}, which is negative. Therefore the point R is plotted above the ε-axis.

$$OC = 200 \times 10^{-6}, \; r = 10^{-6}\sqrt{(200^2 + 200^2)} = 282.84 \times 10^{-6}$$

$$\therefore \varepsilon_1 = OP_1 = 482.84 \times 10^{-6}$$

$$\varepsilon_1 = OP_2 = -82.84 \times 10^{-6}$$

$$2\alpha = \tan^{-1} 1.0, \qquad \therefore 2\alpha = 45°, \qquad \therefore \alpha = 22.5°$$

$$S_2 = \begin{bmatrix} 0 & -200 & 0 \\ -200 & 400 & 0 \\ 0 & 0 & \varepsilon_z \end{bmatrix} \times 10^{-6}$$

(a)

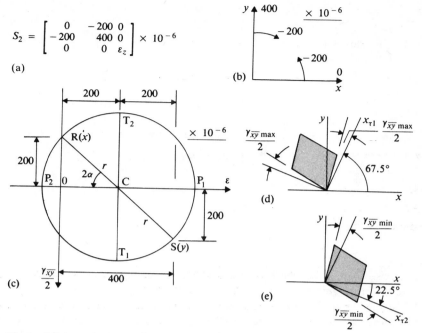

(b)

(c)

(d)

(e)

Figure S3.7 (ii)

The principal strain array in the xy plane is therefore

$$S_p = \begin{bmatrix} 482.84 & 0 & 0 \\ 0 & -82.84 & 0 \\ 0 & 0 & \varepsilon_z \end{bmatrix} \times 10^{-6}$$

with respect to axes $x_p y_p$ located at 67.5° clockwise from xy

$$\frac{\gamma_{\bar{x}\bar{y}\,max}}{2} = r = 282.84 \times 10^{-6}$$

The maximum shear strain is in the axes $x_{T_1} y_{T_1}$ located at 67.5° anticlockwise from xy, x_{T_1} being represented by the point T_1 on the circle, when $\varepsilon_{\bar{x}} = \varepsilon_{\bar{y}} = 200 \times 10^{-6}$. Thus:

$$S_{T1} = \begin{bmatrix} 200 & 282.8 & 0 \\ 282.8 & 200 & 0 \\ 0 & 0 & \varepsilon_z \end{bmatrix} \times 10^{-6}$$

with respect to axes $x_{T_1} y_{T_1}$ located at 67.5° anticlockwise from xy

This strain array is demonstrated by the deformed element in Fig. S3.7, part (ii), (d).

It is worth noting that if, instead of moving anticlockwise from R to T_1 on the circle, we moved clockwise through 45° to the point T_2, then x_{T_2} would be located at 22.5°

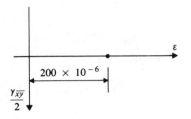

$$S_3 = \begin{bmatrix} 200 & 0 & 0 \\ 0 & 200 & 0 \\ 0 & 0 & \varepsilon_z \end{bmatrix} \times 10^{-6}$$

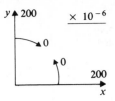

Circle reduces to
a point.
No shear strain
in any direction

Figure S3.7 (iii)

clockwise from x on the element and the strain indicated would be $\gamma_{\bar{x}\bar{y}\,min}/2$, that is, equal in magnitude to $\gamma_{\bar{x}\bar{y}\,max}/2$ but negative.

The corresponding element is sketched in Fig. S3.7, part (ii), e and, as can be seen, the deformed elements of Figs S3.7, part ii, (d) and S3.7, part (ii), (e) fit together precisely because, of course, they can be regarded as parts of a single deformed element at the point.

(iii) Figure S3.7, part (iii). In this case, the points R and S coincide and the strain circle reduces to a point. Therefore, the strain array S_3 applies in all directions in the plane xy and there is no shear strain in any direction. The strain array S_3 describes the case of, for example, a plate subjected to equal tensile stresses in any two directions at right angles to each other. It is an interesting special case because it means that if a square was drawn on the plate at any orientation to the directions of the applied stresses, the square would simply get bigger without any shear distortion. The effect of applying equal compressive stresses perpendicular to each other would have a similar but opposite effect.

(iv) The strain circle corresponding to S_4 is drawn on Fig. S3.7, part (iv). From the circle:

$$OC = \frac{1}{2}(-600 - 200) \times 10^{-6} \qquad = -400 \times 10^{-6}$$

$$r = 10^{-6}\sqrt{(200^2 + 200^2)} \qquad = 282.8 \times 10^{-6}$$

$$\varepsilon_1 = OP_1 = (-400 + 282.8) \times 10^{-6} = -117.2 \times 10^{-6}$$

$$\varepsilon_2 = OP_2 = (-400 - 282.8) \times 10^{-6} = -682.8 \times 10^{-6}$$

$$2\alpha = \tan^{-1} 1.0, \qquad \therefore \qquad \alpha = 22.5°$$

$$S_4 = \begin{bmatrix} -200 & 200 & 0 \\ 200 & -600 & 0 \\ 0 & 0 & \varepsilon_z \end{bmatrix} \times 10^{-6}$$

(a)

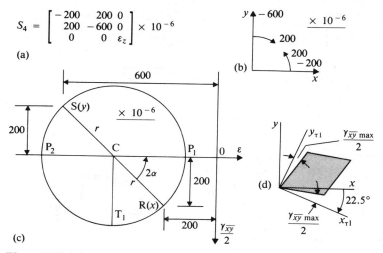

(b)

(c)

(d)

Figure S3.7 (iv)

The principal strain array is therefore

$$S_p = \begin{bmatrix} -117.2 & 0 & 0 \\ 0 & -682.8 & 0 \\ 0 & 0 & \varepsilon_z \end{bmatrix} \times 10^{-6}$$

with respect to axes $x_p y_p$ located at $22.5°$ anticlockwise from xy

The maximum shear strain array is

$$S_{T1} = \begin{bmatrix} -400 & 282.8 & 0 \\ 282.8 & -400 & 0 \\ 0 & 0 & \varepsilon_z \end{bmatrix} \times 10^{-6}$$

with respect to axes $x_{T_1} y_{T_1}$ located at $22.5°$ clockwise from xy

The deformed element experiencing $\gamma_{\bar{x}\bar{y}\,max}$ is sketched in Fig. S3.7, part (iv), (d).

Problem 3.8

A rectangle is drawn on a flat sheet of thin rubber. The sheet is then subjected to strains $\varepsilon_x = 3 \times 10^{-3}$, $\varepsilon_y = -2 \times 10^{-3}$, $\gamma_{xy} = 2 \times 10^{-3}$. If the angles of the rectangle do not change when these strains are applied, at what orientation to the xy axes was the rectangle drawn?

The strain array with respect to the axes xy is

$$S = \begin{bmatrix} 3.0 & 1.0 & 0 \\ 1.0 & -2.0 & 0 \\ 0 & 0 & \varepsilon_z \end{bmatrix} \times 10^{-3}$$

The corresponding Mohr's circle of strain is drawn on Fig. S3.8.

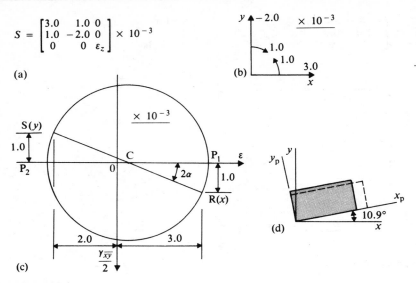

$$S = \begin{bmatrix} 3.0 & 1.0 & 0 \\ 1.0 & -2.0 & 0 \\ 0 & 0 & \varepsilon_z \end{bmatrix} \times 10^{-3}$$

(a)

(b)

(c)

(d)

Figure S3.8

If the angles of the rectangle do not change, then the sides of the rectangle must be parallel to the principal axes, that is, their directions are represented by the points P_1 and P_2 on the circle.

From the circle:

$$OC = \frac{1}{2}(3.0 - 2.0) \times 10^{-3} \qquad = 0.5 \times 10^{-3}$$

$$2\alpha = \tan^{-1}\frac{1.0}{2.5} = 21.8°, \quad \therefore \alpha = 10.9°$$

Therefore, the rectangle was drawn in $x_p y_p$, that is, with one of its sides oriented at 10.9° anticlockwise from the x-axis as shown in Fig. S3.8d.

Problem 3.9

Show that, for a state of *plane stress*:

$$\varepsilon_z = -\frac{v}{(1-v)}(\varepsilon_x + \varepsilon_y)$$

For plane stress in the xy plane, $\sigma_z = 0$.

From Eq. (3.121) of Part One:

$$\varepsilon_z = -\frac{v}{E}(\sigma_x + \sigma_y)$$

From Eq. (3.121a) of Part One:

$$\sigma_x = \frac{E}{(1 - v^2)} (\varepsilon_x + v\varepsilon_y)$$

$$\sigma_y = \frac{E}{(1 - v^2)} (v\varepsilon_x + \varepsilon_y)$$

Substituting these expressions for σ_x and σ_y into the equation for ε_z gives

$$\varepsilon_z = -\frac{v}{E} \frac{E}{(1 - v^2)} (\varepsilon_x + v\varepsilon_y + v\varepsilon_x + \varepsilon_y)$$

i.e.,

$$\varepsilon_z = -\frac{v}{(1 - v^2)} (1 + v)(\varepsilon_x + \varepsilon_y) = -\frac{v}{(1 - v)} (\varepsilon_x + \varepsilon_y)$$

Problem 3.10

Show that, for a state of *plane strain*:

(i) $\sigma_x = \dfrac{E}{(1 + v)(1 - 2v)} \{(1 - v)\varepsilon_x + v\varepsilon_y\}$

(ii) $\sigma_y = \dfrac{E}{(1 + v)(1 - 2v)} \{v\varepsilon_x + (1 - v)\varepsilon_y\}$

(iii) $\sigma_z = \dfrac{vE}{(1 + v)(1 - 2v)} (\varepsilon_x + \varepsilon_y)$

Expressions (i), (ii) and (iii) above are obtained from the manipulation of the expressions for ε_x and ε_y and σ_z in Eqs (3.123) and (3.124) of Part One.

From Eq. (3.124):

$$\sigma_z = v(\sigma_x + \sigma_y)$$

Substituting for σ_z into the expressions for ε_x and ε_y in Eq. (3.123) gives

$$E\varepsilon_x = \sigma_x(1 - v^2) - v\sigma_y(1 + v) \tag{3.10a}$$

$$E\varepsilon_y = -v\sigma_x(1 + v) + \sigma_y(1 - v^2) \tag{3.10b}$$

Multiplying Eq. (3.10a) by $(1 - v)/v$ gives

$$E\varepsilon_x \frac{(1 - v)}{v} = \sigma_x(1 - v^2) \frac{(1 - v)}{v} - \sigma_y(1 - v^2)$$

which, when added to Eq. (3.10b) gives

$$E\varepsilon_x \frac{(1 - v)}{v} + E\varepsilon_y = \sigma_x(1 - v^2) \frac{(1 - v)}{v} - v\sigma_x(1 + v) \tag{3.10c}$$

Multiplying Eq. (3.10c) by v gives

$$E\{(1 - v)\varepsilon_x + v\varepsilon_y\} = \sigma_x(1 + v)(1 - 2v)$$

$$\sigma_x = \frac{E}{(1 + v)(1 - 2v)} \{(1 - v)\varepsilon_x + v\varepsilon_y\}$$

which is identical with (i) above.

Similarly, the expression for σ_y, as in (ii) above, can be derived.

The expression for σ_z as in (iii) above is then obtained by substituting (i) and (ii) into $\sigma_z = v(\sigma_x + \sigma_y)$.

Problem 3.11

The state of stress at a point in a material with $E = 100 \text{ kN/mm}^2$ and $v = 0.25$ is given by F (below). Draw the stress circle corresponding to the stress array F. From the stress circle, deduce the directions of the maximum and minimum principal stresses, find the values of the principal stresses and, using Hooke's Law, calculate the principal strains.

Check the answers against those given in Example 3.14 of Part One (which dealt with the problem by first finding the strain state S corresponding to F and then drawing the strain circle).

$$F = \begin{bmatrix} -60 & -40 & 0 \\ -40 & 50 & 0 \\ 0 & 0 & 0 \end{bmatrix} \text{N/mm}^2 \text{ with respect to the axes } xyz$$

The corresponding Mohr's circle of stress is shown on Fig. S3.11. It has radius r.

From the geometry of the circle:

$$\text{OC} = \frac{1}{2}(\sigma_x + \sigma_y) = \frac{1}{2}(-60 + 50) \qquad\qquad = -5 \text{ N/mm}^2$$

$$r = \sqrt{(55^2 + 40^2)} \qquad\qquad = 68.01 \text{ N/mm}^2$$

$$\tan 2\alpha = -40/55, \qquad \therefore 2\alpha = 143.97°, \qquad \therefore \alpha = 71.99°$$

$$\sigma_1 = \text{OP}_1 = (-5 + 68.01) \text{ N/mm}^2 \qquad\qquad = 63 \text{ N/mm}^2$$

$$\sigma_2 = \text{OP}_2 = (-5 - 68.01) \text{ N/mm}^2 \qquad\qquad = -73 \text{ N/mm}^2$$

Thus, on the element, σ_1 occurs at 71.99° clockwise from the direction of σ_x, and σ_2 occurs at $(90 - 71.99)° = 18.01°$ anticlockwise from the direction of σ_x.

The principal strains occur in the same directions as the principal stresses. Thus, from Eq. (3.121) of Part One:

$$\varepsilon_1 = \frac{1}{100 \times 10^3}(\sigma_1 - v\sigma_2) = \frac{1}{100 \times 10^3}(63 + 0.25 \times 73) = 0.8126 \times 10^{-3}$$

$$\varepsilon_2 = \frac{1}{100 \times 10^3}(\sigma_2 - v\sigma_1) = \frac{1}{100 \times 10^3}(-73 - 0.25 \times 63) = -0.8876 \times 10^{-3}$$

These values of ε_1 and ε_2 agree with those calculated in Example 3.14 of Part One.

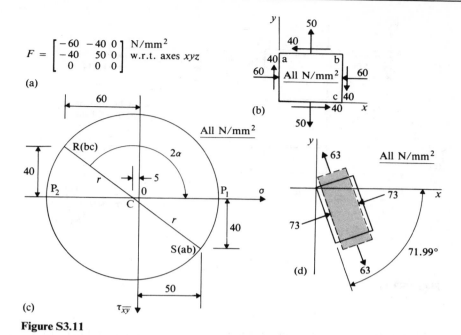

$$F = \begin{bmatrix} -60 & -40 & 0 \\ -40 & 50 & 0 \\ 0 & 0 & 0 \end{bmatrix} \begin{array}{l} \text{N/mm}^2 \\ \text{w.r.t. axes } xyz \end{array}$$

(a)

Figure S3.11

The element experiencing principal stresses and principal strains is shown in Fig. S3.11d. It does not experience any shear stress or shear strain because it lies in the principal axes.

Problem 3.12

The beam used in Example 3.15 of Part One is subjected to a concentrated load of 250 kN at the middle of the 10 m span. If it can be assumed that the shear force at any cross-section of the beam is distributed as a uniform shear stress in the web, find the maximum tensile stress in the web of the beam. For the beam, $I = 674 \times 10^6$ mm^4 and the web is 430 mm deep and 10 mm thick (see Fig. 3.62e of Part One). The bending moment and shear force diagrams for the loaded beam are shown in Figs S3.12a and S3.12b.

The maximum tensile stress in the web due to bending alone will occur at the point where the web meets the flange at the bottom of the beam and at the middle of the span where $M = 625$ kN m (Fig. S3.12b). The magnitude of the shear is 125 kN. It does not matter whether we combine the positive or the negative shear with the maximum bending moment to find the maximum tensile stress. (This point will become apparent when we draw the Mohr's circle of stress.) Suppose, therefore, we use $S = 125$ kN, i.e., just to the right of C, the middle point of the beam.

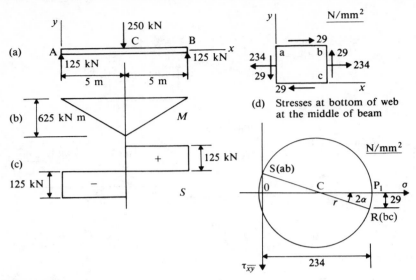

Figure S3.12

Maximum tensile stress due to bending is therefore given by Eq. (3.133) of Part One as

$$\sigma = \frac{625 \times 10^6}{674 \times 10^6} \times 215 = 234 \text{ N/mm}^2$$

The shear stress throughout the depth of the web (assuming a uniform distribution as allowed by the question) is τ where

$$\tau = \frac{125 \times 10^3}{430 \times 10} = 29.1 \text{ N/mm}^2$$

The stress array at the bottom of the web just to the right of the middle of the beam is therefore given by F, where

$$F = \begin{bmatrix} 234 & 29 & 0 \\ 29 & 0 & 0 \\ 0 & 0 & 0 \end{bmatrix} \text{ N/mm}^2 \text{ with respect to the axes } xyz$$

The stressed element is illustrated in Fig. S3.12d and the corresponding stress circle in Fig. S3.12e.

$$\text{On the circle, OC} = \frac{234}{2} = 117 \text{ N/mm}^2$$

$$r = \sqrt{(117^2 + 29^2)} = 120.5 \text{ N/mm}^2$$

∴ maximum tensile stress is given by

$$\sigma_1 = OP_1 = OC + r = 237.5 \text{ N/mm}^2$$

which will act in a direction α anticlockwise from the x direction, where

$$\alpha = \frac{1}{2} \tan^{-1} \frac{29}{117} = 7°$$

This example illustrates an important point, namely, that the influence of the shear stress on the maximum bending stress is usually small in the web of a beam. In other words, the maximum bending stress derived from Eq. (3.133) of Part One would usually be used for design purposes.

Problem 3.13

(i) Show that, if a column has a solid, circular cross-section of diameter D, the point of application of a compressive load P must not be greater than $D/8$ from the centre of the cross-section if there is to be no tensile stress at any point on the cross-section. (ii) If the column has a hollow circular cross-section of outside diameter D and inside diameter d, show that the radius of the core is equal to $(D^2 + d^2)/8D$.

(i) Figure S3.13 illustrates the solid cross-section with the load P applied in the x direction (i.e., *into* the diagram) at a distance e_z from the centre G. The minimum compressive stress will obviously occur at L. We want to find the limiting value of e such that the stress at L does not become tensile, that is, when it is just changing from compressive to tensile, that is, when it is zero.

The governing equation is Eq. (3.145) of Part One, that is, Eq. (3.142) of Part One when P is known to be negative (compressive). Thus:

$$\sigma_x = -\frac{P}{A} - \frac{Pe_y}{I_z} y - \frac{Pe_z}{I_y} z$$

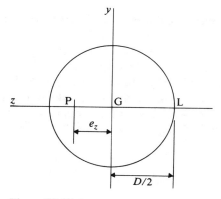

Figure S3.13

In this case, $A = \pi D^2/4$:

$$I_z = I_y = \frac{\pi R^4}{4} = \frac{\pi}{64} D^4$$

Therefore the limiting value of e_z is given by

$$0 = -\frac{P}{\pi D^2/4} - \frac{P e_z}{\pi D^4/64}\left(-\frac{D}{2}\right)$$

$$\therefore e_z = D/8$$

Clearly, symmetry means that the core is a circle of radius equal to $D/8$.

(ii) The argument to find the core of a hollow circular cross-section, of outside diameter D and inside diameter d, is precisely similar to that used above for the solid cross-section. The only things that change are the values of I, which now equals $\frac{\pi}{64}(D^4 - d^4)$, and the value of A, which now equals $\frac{\pi}{4}(D^2 - d^2)$.

Therefore, again using Eq. (3.145) of Part One, the limiting value of e_z to define the radius of the core, is given by

$$0 = -\frac{P}{\frac{\pi}{4}(D^2 - d^2)} - \frac{P e_z}{\frac{\pi}{64}(D^4 - d^4)}\left(-\frac{D}{2}\right)$$

$$\therefore \quad 0 = -\frac{1}{(D^2 - d^2)} + \frac{8D e_z}{(D^2 + d^2)(D^2 - d^2)}$$

$$\therefore e_z = (D^2 + d^2)/8D$$

Problem 3.14

Figure 3.73 of Part One shows a box-section column with a bracket welded to it. The bracket carries a load of 50 kN at a point on the Gz axis and 150 mm from the centroid G of the column cross-section. A second load $\bar{P}$ acts through G. Find the maximum allowable value of $\bar{P}$ if the compressive stress must not exceed 150 N/mm^2 (i.e. $\sigma_x \not< -150$ N/mm^2) at any point on the column cross-section.

The loading on the column, reproduced in Fig. S3.14a, is equivalent to that shown in Fig. S3.14b, namely, an axial load of $-(\bar{P} + 50)$ kN acting through the centroid of the column cross-section plus a moment M_y equal to (50×0.15) kN m about the principal axis Gy of the column.

The σ_x-distribution is given by Eq. (3.144) of Part One, in which

$$P = -(\bar{P} + 50)\ \text{kN} = -(\bar{P} + 50) \times 10^3\ \text{N}$$

$$A = (125 \times 100) - (100 \times 75) = 5000\ \text{mm}^2$$

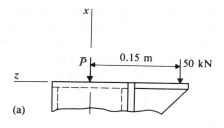

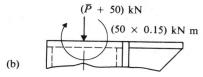

(b)

Figure S3.14

$$M_z = 0$$

$$M_y = (50 \times 0.15) \text{ kN m} = (50 \times 0.15) \times 10^6 \text{ N mm}$$

$$I_y = \frac{1}{12}(100 \times 125^3 - 75 \times 100^3) = 1002.6 \times 10^4 \text{ mm}^4$$

From Eq. (3.144):

$$\sigma_x = \frac{-(\bar{P} + 50) \times 10^3}{5000} + \frac{50 \times 0.15 \times 10^6}{1002.6 \times 10^4} \times z$$

i.e.,

$$\sigma_x = -\frac{\bar{P}}{5} - 10 + 0.748 \times z$$

The maximum compressive stress will occur in the column wall at $z = -\dfrac{125}{2}$ mm. Therefore, at $z = -62.5$ mm, $\sigma_x \not< -150$ N/mm². The limiting value of $\bar{P}$ is therefore given by

$$-\frac{\bar{P}}{5} - 10 - 0.748 \times 62.5 \not< -150, \text{ that is, } \underline{\bar{P} \not> 469 \text{ kN}}$$

Problem 3.15

The beam ABCD in Fig. 3.74a of Part One carries a vertical load of 20 kN which can move to any point between A and D. The beam is simply-supported at B and C and has a hollow box cross-section with relative dimensions as shown in Fig. 3.74b. Deduce the actual dimensions of the cross-section if the bending stress in the beam must not exceed 250 N/mm² in magnitude.

The maximum bending stress will occur at the section of the beam experiencing the maximum bending moment. The first task, therefore, is to find the location of the load which causes the maximum bending moment at any point along the beam.

When the load is at A, the bending moment diagram is shown in Fig. S3.15a. This gives the maximum negative bending moment at B of -40 kN m because if the load moves between A and B, the bending moment at B will always be of smaller magnitude. If the load moves between B and C, the maximum positive bending moment of 20 kN m will occur under the load when it is at mid-span, as shown on Fig. S3.15b.

The design bending moment is therefore -40 kN m.

The cross-section is symmetrical and its centroid G is at the centre of the section as shown in Fig. S3.15c. It will bend about the principal axis Gz.

Second moment of area about Gz is given by I, where

$$I = \frac{1}{12}[10k(20k)^3 - 8k(17k)^3] = 3391.3k^4 \text{ mm}^4$$

From Eq. (3.8) of Part One:

$$\sigma_x = -\frac{M}{I}y$$

The maximum positive (tensile) value of σ_x will occur when $M = -40$ kN m and $y = 10k$. The maximum negative (compressive) value of σ_x will occur when $M = -40$ kN m and $y = -10k$. They will have the same magnitude, therefore let us

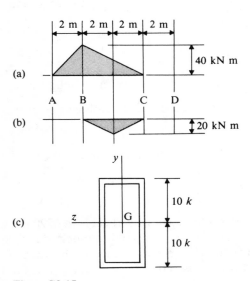

Figure S3.15

use the maximum tensile value of σ_x to determine k. Thus, at the limit:

$$\sigma_{x\,max} = 250 \text{ N/mm}^2$$

$$\therefore 250 = \frac{-40 \times 10^6}{3391.3k^4} \times (-10k)$$

i.e.,

$$k^3 = \frac{40 \times 10^6 \times 10}{3391.3 \times 250} \text{ mm}^3$$

$$\therefore k = 7.78 \text{ mm}$$

Problem 3.16

(i) Show that, for a thin-walled circular tube of mean diameter d and wall thickness t, the second moment of area of the cross-section about the diameter is approximately $\pi d^3 t/8$.

(ii) Figure 3.75 of Part One shows the elevation of a mast made of two thin-walled tubes A and B. Tube A has a mean diameter of 100 mm and a wall thickness of 3 mm. Tube B has a mean diameter of 150 mm. The mast has a total height of 3.5 m and, applied through the centre-line at the top, carries a load of 6 kN inclined at 60° to the horizontal. If the maximum *vertical* compressive stress is not to exceed 200 N/mm² in either tube, find the maximum length of tube A and the minimum wall thickness of tube B.

(i) Figure S3.16, part (i) shows a thin-walled tube cross-section of mean diameter d and mean radius r. Because $J = I_z + I_y = \text{constant}$ and $I_z = I_y = I$, then $J = 2I$. The simplest way to find I is to find J first and then $I = J/2$. J is the polar second moment of area, that is, the second moment of area about Gx.

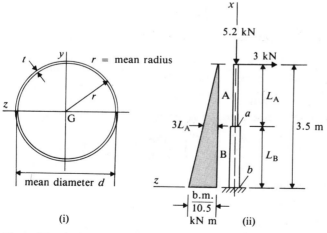

Figure S3.16

$$\text{Length of tube wall (on cross-section)} \simeq 2\pi r$$

$$\text{Area of tube wall} = A \simeq 2\pi r t$$

$$\text{First moment of } A \text{ about } Gx = Ar$$

$$\text{Second moment of } A \text{ about } Gx = J = Ar^2$$

$$\therefore \ J \simeq 2\pi r^3 t \text{ and } I = J/2 \simeq \pi r^3 t = \frac{\pi d^3 t}{8}$$

(ii) The load at 6 kN acting on the mast in Fig. 3.75 of Part One can be resolved into $6 \cos 60° = 3$ kN acting horizontally, and $6 \sin 60° = 5.2$ kN acting vertically, as shown on Fig. S3.16, part (ii).

The vertical load of 5.2 kN, acting through the x-direction centroidal axis of the mast, will cause uniform compressive stresses in tubes A and B. (The uniform stress in tube A will be different from that in tube B, of course.)

The horizontal load of 3 kN will cause the mast to bend, with compressive stress down the right-hand side of the mast and tensile stress down the left-hand side. The bending moment diagram is shown in Fig. S3.16, part (ii).

The maximum length of tube A
The maximum compressive stress in tube A will occur in the wall at point a where the axial compressive load and the maximum bending moment on tube A both cause compressive stress.

For tube A, $I_A = \dfrac{\pi}{8} d_A^3 t_A = \dfrac{\pi}{8} \times 100^3 \times 3 = 1178 \times 10^3 \text{ mm}^4$

Cross-sectional area of tube A $= A_A \simeq \pi d_A t_A = \pi \times 100 \times 3 = 942 \text{ mm}^2$

At the bottom of tube A, $M_y = -(3L_A \times 10^6)$ N mm, where L_A is in metres

At the point a, $e_z = -100/2 = -50$ mm

From Eq. (3.144) of Part One: the x-direction stress at a is given by

$$\sigma_{xa} = -\frac{5.2 \times 10^3}{942} - \frac{3L_A \times 10^6}{1178 \times 10^3} \times 50 = -200 \text{ N/mm}^2$$

when L_A has its maximum value.
Therefore, the maximum value of L_A is given by

$$L_A = \left(200 - \frac{5.2 \times 10^3}{942}\right)\frac{1178 \times 10^3}{3 \times 10^6 \times 50} = 1.53 \text{ m}$$

The minimum wall thickness of tube B
Let this be t_B (mm).

For tube B, $I_B = \dfrac{\pi}{8} d_B^3 t_B = \dfrac{\pi}{8} \times 150^3 t_B = 1325 \times 10^3 t_B \text{ mm}^4$

Cross-sectional area of tube B $\simeq A_B = \pi d_B t_B = \pi \times 150 \times t_B = 471 t_B$ mm²

At the bottom of tube B, $M_y = -(3 \times 3.5) \times 10^6 = -10.5 \times 10^6$ N mm

At the point b, $e_z = -150/2 = -75$ mm

From Eq. (3.144) of Part One: the value of σ_x at the point b is given by

$$\sigma_{xb} = -\frac{5.2 \times 10^3}{471 t_B} - \frac{3 \times 3.5 \times 10^6}{1325 \times 10^3 t_B} \times 75 = -200 \text{ N/mm}^2$$

when the wall thickness of tube B has its minimum allowable value.

$$\therefore t_B = \frac{1}{200}\left(\frac{5.2 \times 10^3}{471} + \frac{3 \times 3.5 \times 10^6 \times 75}{1325 \times 10^3}\right) = \underline{3.0 \text{ mm}}$$

Problem 3.17

A shaft consists of two concentric thin-walled circular tubes fixed to each other at their ends through stiff circular discs. The outer tube is copper and has a mean diameter of 100 mm and a wall thickness of 3 mm. The inner tube is of steel. Find the mean diameter and the wall thickness of the steel tube such that, when a torque is applied to the assembly, the consequent shear stresses in the steel and the copper are in the ratio 2:1 and the torque is shared equally between the two tubes. For steel, $G = 83 \times 10^3$ N/mm². For copper, $G = 26 \times 10^3$ N/mm².

The solution to this problem is found by the use of Eq. (3.37) of Part One and the fact that both tubes must twist by the same amount because they are fixed together at their ends. That is, (θ/L)—see Eq. (3.37)—is the same for the copper tube and the steel tube. (Notice that the value of L is not given in the question. That is because, as both tubes are of equal length, L is irrelevant.)

Let d and t be the mean diameter and wall thickness respectively of the steel tube.

For the steel tube, $J_s \simeq \dfrac{\pi}{8} d^3 t$ (see Problem 3.16)

For the copper tube, $J_c \simeq \dfrac{\pi}{8} \times 100^3 \times 3 = 1178 \times 10^3$ mm⁴

Equation (3.37) is $\dfrac{T}{J} = \dfrac{\tau_r}{r} = G\left(\dfrac{\theta}{L}\right)$ and this, of course, applies to both tubes.

$$\therefore \left(\frac{\theta}{L}\right) = \frac{T}{JG} = \frac{\tau_r}{rG}$$

Let T_c = torque taken by copper tube.
Let T_s = torque taken by steel tube.

Then
$$\left(\frac{\theta}{L}\right) = \frac{T_c}{J_c G_c} = \frac{T_s}{J_s G_s} \tag{3.17a}$$

and
$$\left(\frac{\theta}{L}\right) = \frac{\tau_c}{\left(\frac{100}{2}\right)G_c} = \frac{\tau_s}{\left(\frac{d}{2}\right)G_s}$$
(3.17b)

Also, we are told that
$$T_s = T_c$$
(3.17c)

and
$$\tau_s = 2\tau_c$$
(3.17d)

Feeding Eq. (3.17c) and Eq. (3.17d) into Eq. (3.17a) and Eq. (3.17b), we get

$$\frac{T_c}{(1178 \times 10^3) \times (26 \times 10^3)} = \frac{T_c}{\left(\frac{\pi d^3 t}{8}\right) \times (83 \times 10^3)}$$
(3.17e)

and
$$\frac{\tau_c}{50 \times 26 \times 10^3} = \frac{2\tau_c}{\left(\frac{d}{2}\right) \times 83 \times 10^3}$$
(3.17f)

From Eq. (3.17b), $d = (2 \times 50 \times 26 \times 2)/83 = \underline{62.7 \text{ mm}}$

From Eq. (3.17a), $t = \dfrac{1178 \times 10^3 \times 26 \times 8}{\pi \times 62.7^3 \times 83} = \underline{3.8 \text{ mm}}$

Problem 3.18

At a point in the web of a plate girder, a given set of loads produces a vertical shear stress of 90 N/mm^2 together with a horizontal compressive stress of 120 N/mm^2. Assuming that an increase in the loads causes the stresses to increase in the same proportion, by what factor can the loads be increased before the maximum shear stress at the point in the web reaches 135 N/mm^2.

If we stipulate that the web lies in the xy plane, then the stress array F at the specified point is

$$F = \begin{bmatrix} 0 & 90 & 0 \\ 90 & -120 & 0 \\ 0 & 0 & 0 \end{bmatrix} \text{ N/mm}^2 \text{ with respect to the } xyz \text{ axes}$$

The corresponding Mohr's circle of stress is drawn in Fig. S3.18.

The radius of the circle, $r = \sqrt{(90^2 + 60^2)} = 108.12 \text{ N/mm}^2$.

$$\therefore \tau_{\bar{x}\bar{y}\,\text{max}} = CT_1 = r = 108.12 \text{ N/mm}^2$$

To reach 135 N/mm^2, this shear stress must be multiplied by a factor of $135/108.12 = \underline{1.25}$.

This is the required factor by which the applied loads can be increased because, as is demonstrated by the dashed circle in Fig. S3.18, if τ_{xy} and σ_y are multiplied by 1.25, to

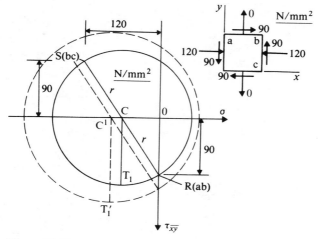

Figure S3.18

become 112.5 N/mm^2 and 150 N/mm^2 respectively, the new maximum shear stress $C'T' = \sqrt{(112.5^2 + 75^2)} = 135$ N/mm^2.

Problem 3.19

A uniform beam is of symmetrical I-section with flanges 125 mm × 15 mm and a web 300 mm × 10 mm. It is supported as a cantilever in a horizontal position and carries a vertical load of 200 kN at its tip, acting in the plane of the web.

Estimate the magnitudes and directions of the principal stresses and the maximum shear stress in the web at a section distance 0.5 m from the applied load:

(a) at a point on the central longitudinal axis,
(b) at a point near the junction of the web and the top flange.

Using the parallel axis theorem (Eq. (3.11) of Part One) to find the second moment of area I about Gz (see Fig. S3.19):

$$I = 2(125 \times 15 \times 157.5^2) + 2\left(\frac{125 \times 15^3}{12}\right) + \frac{10 \times 300^3}{12} = 115.6 \times 10^6 \text{ mm}^4$$

At the section distance 0.5 m from the free end of the cantilever, see Fig. S3.19b:

$$\text{Shear force} = S = (200 \times 10^3) \text{ N}$$

$$\text{Bending moment} = M = -100 \times 10^6 \text{ N mm}$$

The shear stress τ at any distance y_1 from Gz is given by Eq. (3.27) of Part One as

$$\tau = \frac{S}{It}\,A\bar{y}$$

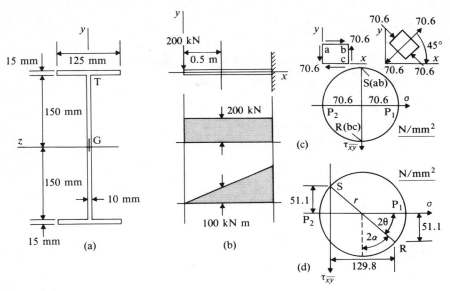

Figure S3.19

The bending stress at any distance y from Gz is given by Eq. (3.8) of Part One as

$$\sigma_x = -\frac{M}{I} y$$

(a) *At the mid-height of the web* (Gz axis)

$$\tau = \frac{200 \times 10^3}{115.6 \times 10^6 \times 10}((125 \times 15 \times 157.5) + (150 \times 10 \times 75)) = 70.6 \text{ N/mm}^2$$

$$\sigma_x = 0$$

i.e., $F = \begin{bmatrix} 70.6 & 0 \\ 0 & 0 \end{bmatrix}$ N/mm² with respect to the axes xy

The Mohr's circle of stress is drawn on Fig. S3.19c from which it is apparent that the principal stresses are 70.6 N/mm² and −70.6 N/mm² acting respectively at 45° anticlockwise from the direction of x, and 45° clockwise from the direction of x. The maximum shear stress is clearly 70.6 N/mm².

(b) *At a point near the junction of the web and the top flange* (point T on Fig. S3.19a)

$$\tau = \frac{200 \times 10^3}{115.6 \times 10^6 \times 10}(125 \times 15 \times 157.5) = 51.5 \text{ N/mm}^2$$

$$\sigma_x = -\frac{(-100 \times 10^6)}{115.6 \times 10^6} \times 150 = 129.8 \text{ N/mm}^2$$

i.e., $F = \begin{bmatrix} 129.8 & 51.1 \\ 51.1 & 0 \end{bmatrix}$ N/mm² with respect to the axes xy

The corresponding Mohr's stress cricle is drawn on Fig. S3.19d.

The maximum shear stress $= \tau_{max} = \sqrt{\left[\left(\dfrac{129.8}{2}\right)^2 + 51.1^2\right]} = \underline{82.6 \text{ N/mm}^2}$

acting along planes at α clockwise from x and y where

$$\alpha = \frac{1}{2} \tan^{-1}\left(\frac{129.8}{2 \times 51.1}\right) = \underline{25.9°}$$

The principal stresses are given by σ_1 and σ_2 where

$$\sigma_1 = \frac{129.8}{2} + 82.6 = \underline{147.5 \text{ N/mm}^2}$$

$$\sigma_2 = \frac{129.8}{2} - 82.6 = \underline{-17.7 \text{ N/mm}^2}$$

The major principal axis is located at θ anticlockwise from the direction of x where

$$\theta = \frac{1}{2} \tan^{-1}\left(\frac{51.1 \times 2}{129.8}\right) = \underline{19.1°}$$

Problem 3.20

A rectangular plate (Fig. 3.76 of Part One) of dimensions $3l \times l$ lies in the xy plane and is subjected to uniform stress in its own plane. As a result, the area of the plate increases by 0.1 per cent and the length of its perimeter also increases by 0.1 per cent. Also, the angle bad increases by 1.5×10^{-3} radians.

Deduce the principal strain array for the plate and sketch an element subjected to principal strains in the xy plane. (Ignore products of strain as negligibly small.)

The plate is shown in Fig. 3.20a and is labelled abcd.

Let ε_x and ε_y be the direct strains experienced by the plate and let γ_{xy} be the shear strain.

Increase in area $= \Delta A = 3l(1 + \varepsilon_x) \times l(1 + \varepsilon_y) - 3l^2$

$$= 3l^2(\varepsilon_x + \varepsilon_y + \varepsilon_x\varepsilon_y)$$

Ignoring $(\varepsilon_x\varepsilon_y)$, then $\Delta A = 3l^2(\varepsilon_x + \varepsilon_y)$

$$\therefore \frac{\Delta A}{A} = \frac{3l^2(\varepsilon_x + \varepsilon_y)}{3l^2} = \varepsilon_x + \varepsilon_y = 0.001 \tag{3.19a}$$

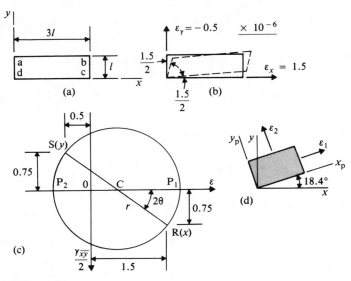

Figure S3.20

Increase in perimeter $= \Delta L = 6l(1 + \varepsilon_x) + 2l(1 + \varepsilon_y) - 8l$

$$= (6\varepsilon_x + 2\varepsilon_y)l$$

$$\therefore \frac{\Delta L}{L} = \frac{(6\varepsilon_x + 2\varepsilon_y)l}{8l} = 0.001$$

$$\therefore 3\varepsilon_x + \varepsilon_y = 0.004 \tag{3.19b}$$

From Eqs (3.19a) and (3.19b), $\varepsilon_x = 1.5 \times 10^{-3}$ and $\varepsilon_y = -0.5 \times 10^{-3}$

Figure S3.20b shows the deformed shape of the plate. We are told that the angle bad increases by 1.5×10^{-3} radians. It follows that the angle cda (lying in the xy axes) reduces by 1.5×10^{-3} radians.

Therefore, by definition of positive shear strain, $\gamma_{xy} = 1.5 \times 10^{-3}$ radians.

The corresponding Mohr's circle of strain is drawn on Fig. S3.20c.

From the circle:

$$OC = \frac{1}{2}(1.5 - 0.5) \times 10^{-3} = 0.5 \times 10^{-3}$$

$$r = \sqrt{(1.0^2 + 0.75^2)} \times 10^{-3} = 0.75 \times 10^{-3}$$

$$\therefore \varepsilon_1 = (0.5 + 1.25) \times 10^{-3} = \underline{1.75 \times 10^{-3}}$$

$$\varepsilon_2 = (0.5 - 1.25) \times 10^{-3} = \underline{-0.75 \times 10^{-3}}$$

$$2\alpha = \tan^{-1}\left(\frac{0.75}{1.0}\right) = 36.9°, \quad \therefore \ \underline{\alpha = 18.4°}$$

The principal strain array S_p is therefore,

$$S_p = \begin{bmatrix} 1.75 & 0 & 0 \\ 0 & -0.75 & 0 \\ 0 & 0 & \varepsilon_z \end{bmatrix} \times 10^{-3} \quad \begin{array}{l} \text{with respect to axes } x_p y_p \\ \text{located at } 18.4° \\ \text{anticlockwise from } xy \end{array}$$

The element is shown in Fig. S3.20d.

Problem 3.21

The plate shown in Fig. 3.77 of Part One is 3 mm thick and, when unloaded, just fits between constraints which allow movement in the y direction but not in the x direction. A compressive load P is applied in the y direction, causing a strain of -200×10^{-5} in the direction m at 45° to the axes.

Draw the Mohr's circles of stress and strain, find the value of P, and write the principal stress array. Sketch an element of the plate subjected to the maximum shear strain and show its orientation to the x-axis.

For the plate, $E = 210 \times 10^3 \ \text{MN/m}^2$ and $v = 0.3$

There is no shear on the plate in the x direction. Therefore, the x and y axes are principal axes. The plate is shown in Fig. S3.21a. There will be a strain in the y direction, but the strain in the x direction is zero, because the constraints at the sides

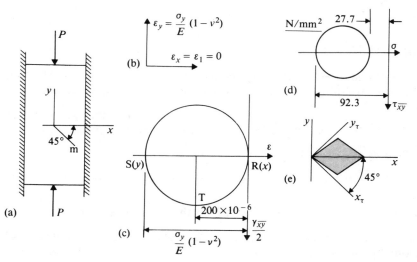

Figure S3.21

will not allow the plate to expand in the x direction (as it would, due to the Poisson's ratio effect, if the constraints were not present).

There will be a stress in the y direction due to the presence of P. There will also be a stress in the x direction because, in order to prevent the plate from getting wider, the constraints must apply forces.

Let the x and y-direction stresses be σ_x and σ_y (which are the principal stresses) and the corresponding strains be ε_x and ε_y (principal strains). Then from Eq. (3.121) of Part One:

$$\varepsilon_x = \frac{1}{E} (\sigma_x - v\sigma_y) = 0 \tag{3.21a}$$

$$\varepsilon_y = \frac{1}{E} (\sigma_y - v\sigma_x) \tag{3.21b}$$

From Eq. (3.21a), $\sigma_x = v\sigma_y$. Therefore, from Eq. (3.21b):

$$\varepsilon_y = \frac{1}{E} (\sigma_y - v^2\sigma_y) = \frac{\sigma_y}{E} (1 - v^2)$$

Remembering that ε_x and ε_y are principal strains, we can now draw the Mohr's circle of strain in terms of σ_y. It is clear that ε_y will be compressive (because P is shown as a compressive load). The state of strain is depicted on Fig. S3.20b and the corresponding circle of strain is shown on Fig. S3.20c.

The given information is that the strain at $45°$ clockwise from the x direction is -200×10^{-6}. This, therefore, is represented at the point T on the strain circle. Thus:

$$\frac{\sigma_y}{E} (1 - v^2) = -2 \times 200 \times 10^{-6}$$

$$\sigma_y = \frac{-2 \times 200 \times 10^{-6}}{(1 - 0.3^2)} \times 210 \times 10^3 = -92.3 \text{ N/mm}^2$$

and $\qquad \sigma_x = 0.3 \times (-92.3) = -27.7 \text{ N/mm}^2$

$\sigma_x = \sigma_1$ and $\sigma_y = \sigma_2$, where σ_1 and σ_2 are the principal stresses. The Mohr's circle of stress is, therefore, as drawn in Fig. S3.21d, the principal stress array being F_p, where

$$F_p = \begin{bmatrix} -92.3 & 0 \\ 0 & -27.7 \end{bmatrix} \begin{array}{l} \text{N/mm}^2 \text{ with respect to the axes } xy \text{ which are} \\ \text{also the principal axes} \end{array}$$

The load P is given by

$$P = (\text{cross-sectional area of plate}) \times \sigma_x = (100 \times 3 \times 92.3) \text{ N}$$

i.e., $\qquad \underline{P = 27.7 \text{ kN in the direction shown}}$

The element experiencing the maximum shear strain is orientated at $45°$ clockwise from the x-axis and experiences direct strains of -200×10^{-6} and a shear strain of 400×10^{-6}. The element is sketched in Fig. S3.21e.

Problem 3.22

A tension specimen of aluminium alloy has a gauge length of 200 mm and a diameter of 12 mm. A load of 37.5 kN develops a tensile stress in the bar equal to the proportional limit of the material. At this load, the gauge length has increased to 200.914 mm and the diameter has decreased to 11.983 mm.

Determine the following properties of the material: the modulus of elasticity, the proportional limit, Poisson's ratio, the shear modulus and the bulk modulus.

At the proportional limit (i.e., the stress up to which the tensile strain increases linearly with the application of tensile stress in a tensile test):

$$\sigma_p = \frac{37.5 \times 1000}{\pi \times 12^2 \times \frac{1}{4}} = \underline{331.6 \text{ N/mm}^2}$$

The strain at σ_p is equal to

$$\varepsilon_p = \frac{0.914}{200} = 4.57 \times 10^{-3}$$

$$\therefore E = \frac{\sigma_p}{\varepsilon_p} = \frac{331.6 \times 10^3}{4.57} = \underline{72\,560 \text{ N/mm}^2}$$

Recalling the definition of Poisson's ratio v (Section 3.6 of Part One), then

$$v = \frac{-(11.983 - 12)/12}{4.57 \times 10^{-3}} = \underline{0.31}$$

The shear modulus G is expressed in terms of E and v in Eq. (3.131) of Part One. Therefore

$$G = \frac{72\,560}{2(1 + 0.31)} = \underline{27\,695 \text{ N/mm}^2}$$

The bulk modulus K is expressed in terms of E and v by Eq. (3.132) of Part One. Therefore

$$K = \frac{72\,560}{3(1 - 2 \times 0.31)} = \underline{63\,649 \text{ N/mm}^2}$$

Problem 3.23

A column of cruciform cross-section (Fig. 3.78 of Part One) is made from two I-sections of size 467 mm × 193 mm. The column is fabricated by cutting one of the I-sections down the centre of the web and welding the resulting T-sections A and B to either side of the other I-section as shown.

Obtain the equation of the line QQ defining the limit of the zone within which a compressive load of 1300 kN, applied parallel to the x-axis, must act if the compressive stress at the point e is not to exceed 250 N/mm², i.e., $\sigma_x \not< -250$ N/mm². (The increase in depth due to cutting and welding of the reconstituted section AB, and the area of the welds, may be neglected.)

The properties of each I-section about its own principal axes are $I_{max} = 457 \times 10^6$ mm⁴, $I_{min} = 22 \times 10^6$ mm⁴, and the cross-sectional area of each is $A = 12.52 \times 10^3$ mm².

Indicate the entire area within which the compressive load must lie if the specified stress is not to be exceeded at any point on the cross-section.

The column cross-section is drawn again on Fig. S3.23a. Use Eq. (3.135) of Part One, remembering that, in setting up this equation, P was assumed to be a compressive load. Therefore, we put $P = 1300 \, \text{kN}$ when using Eq. (3.145).

For the cross-section:

$$\text{Total area} = 2 \times 12.52 \times 10^3 = 25.04 \times 10^3 \, \text{mm}^2$$

and
$$I_z = I_y = (457 + 22) \times 10^6 = 479 \times 10^6 \, \text{mm}^4$$

The coordinates of the point e are $z = 233.5 \, \text{mm}$, $y = 96.5 \, \text{mm}$.

If the load P is applied at eccentricities e_z and e_y (with respect to the x-axis of the column), then the limiting condition (σ_x at the point e not less than $-250 \, \text{N/mm}^2$) for

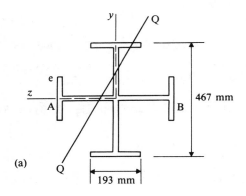

(a)

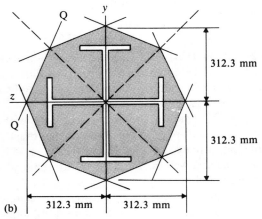

(b)

Figure S3.23

the location of P is given by Eq. (3.145) as

$$-250 = -\frac{1300 \times 10^3}{25.04 \times 10^3} - \frac{1300 \times 10^3}{479 \times 10^6} \times 96.5e_y - \frac{1300 \times 10^3}{479 \times 10^6} \times 233.5e_z$$

i.e.,

$$\frac{e_z}{312.3} + \frac{e_y}{756.3} = 1$$

When $e_y = 0$, $e_z = 312.3$ mm from the equation above.

When $e_z = 0$, $e_y = 756.3$ mm.

The correct position of line QQ is drawn on Fig. S3.23b. It crosses the z-axis at 312.3 mm and, if extended, would cross the y-axis at 756.3 mm. Because there are seven other corners on the cross-section similar to e, then symmetry means that there must be seven other boundaries corresponding to QQ, each one protecting one of the corners from a stress less than -250 N/mm². The load must, therefore, be confined within the envelope formed by these eight lines.

Problem 4.1

The maximum compressive stress in the concrete compression zone at a particular section of a reinforced concrete beam where the bending moment is M can be obtained from the standard beam equation

$$f_c = \frac{Md_n}{I_t}$$

where the depth of the compression zone d_n is given by

$$d_n = d(\sqrt{(m^2r^2 + 2mr)} - mr)$$

and the second moment of area of the transformed section I_t is given by

$$I_t = \frac{bd_n^3}{3} + mrbd(d - d_n)^2$$

Show that the combination of these three equations result in the following expression:

$$f_c = \frac{2M}{bd_n(d - d_n/3)}$$

Rearranging the equation for d_n:

$$\sqrt{(m^2r^2 + 2mr)} = \frac{d_n}{d} + mr$$

Squaring both sides of this equation and then rearranging again gives

$$mr = \frac{d_n^2}{2d(d - d_n)}$$

Substituting this expression for mr into the equation for I_t gives

$$I_t = \frac{bd_n^3}{3} + bd(d - d_n)^2 \times \frac{d_n^2}{2d(d - d_n)}$$

$$= \frac{bd_n^2}{2}(d - d_n/3)$$

Substituting this expression in the beam equation gives

$$f_c = \frac{2M}{bd_n(d - d_n/3)}$$

Problem 4.2

The steelwork beam–column connection illustrated in Fig. S4.2 is assumed to transmit shear force only. For the case where the thickness of the flange of the column is 12 mm, the thickness of the end-plate of the beam is 10 mm, and the bolts are 19 mm diameter, calculate the maximum design shear force if the maximum permissible shear stress in the bolts is 160 N/mm² and the maximum permissible bearing stress on the bolts is 430 N/mm².

Maximum shear force per bolt based on shear stress $= \dfrac{\pi}{4} \times 19^2 \times 160 \times 10^{-3}$

$$= 45.4 \, \text{kN}$$

The maximum shear force based on the bearing stresses will be determined by the end-plate, this being thinner than the flange of the column. Hence:

Maximum shear force per bolt based on bearing stress $= 19 \times 10 \times 430 \times 10^{-3}$

$$= 82 \, \text{kN}$$

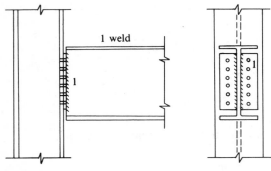

Figure S4.2

In order to satisfy the limiting values in both shear and bearing:

Maximum shear force per bolt   = 45.4 kN

Total number of bolts in the connection = 12

Maximum permissible shear force $= 12 \times 45.4 = \underline{545\ kN}$

Problem 4.3

The crane gantry girder shown in Fig. S4.3 is made up from a Universal Beam and a channel, these having the sectional properties indicated. Calculate the elastic moduli of the girder about its horizontal axis. A horizontal force applied by the crane is considered to be resisted by the composite top flange. Calculate the relevant elastic modulus of this top flange.

In order to obtain the second moment of area of the compound girder, it is first necessary to find the position of the centroid of the section. For this, we will take first moments of area about the bottom edge. If $\bar{y}$ is the height of this centroid above the bottom edge:

$$10^2(85.3 + 33.2)\bar{y} = 10^2\left(85.3 \times \frac{457}{2}\right) + 10^2(33.2 \times 444.6)$$

$$\bar{y} = 289\ mm$$

Using the parallel axis theorem, the second moment of area of the compound girder about the horizontal axis through the centroid is

$$I_{zz} = (28\ 522 \times 10^4) + (85.3 \times 10^2) \times (289 - 457/2)^2 + (159 \times 10^4)$$
$$+\ 33.2 \times 10^2 \times (444.6 - 289)^2$$

$$= 398 \times 10^6\ mm^4$$

	I-beam		Channel	
Depth		457 mm	Depth	229 mm
Thickness of flanges		15 mm	Thickness of web	7.6 mm
Thickness of web		9 mm	Centroid p	20 mm
Area		85.3 cm^2	Area	33.2 cm^2
I_{zz}		28 522 cm^4	I_{zz}	2610 cm^4
I_{yy}		829 cm^4	I_{yy}	159 cm^4

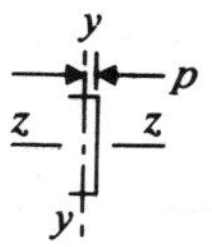

Figure S4.3

$$\text{Section modulus of the top flange} \quad = \frac{398 \times 10^6}{(464.6 - 289)} = \underline{2.27 \times 10^6 \text{ mm}^3}$$

$$\text{Section modulus of the bottom flange} = \frac{398 \times 10^6}{289} = \underline{1.38 \times 10^6 \text{ mm}^3}$$

The second moment of area of the web of the Universal beam about the y–y axis

$$= \frac{427 \times 9^3}{12} = 0.026 \times 10^6 \text{ mm}^4$$

Hence the second moment of area of the two flanges of the universal beam about the y–y axis is

$$= 8.29 \times 10^6 - 0.026 \times 10^6 = 8.264 \times 10^6 \text{ mm}^4$$

Thus the second moment of area of one of the flanges $= 4.132 \times 10^6 \text{ mm}^4$

We know that the second moment of area of the channel about the y–y axis of the compound girder (the z–z axis of the channel section) is $26.1 \times 10^6 \text{ mm}^4$.

Hence the total second moment of area of the top flange of the compound girder is

$$(26.1 + 4.13) \times 10^6 = 30.23 \times 10^6 \text{ mm}^4$$

(Note the high contribution of the channel section to this second moment of area.)

$$\text{Section modulus of the top flange} = \frac{30.23 \times 10^6}{229/2} = \underline{0.264 \times 10^6 \text{ mm}^3}$$

Check these answers by referring to the table of properties of gantry girders given in a Handbook on Structural Steelwork issued by steelwork organizations.

Problem 4.4

For beam and slab structures in reinforced concrete, a beam resisting sagging bending moments is considered to be of the form shown in Fig. S4.4. Assuming linear elastic

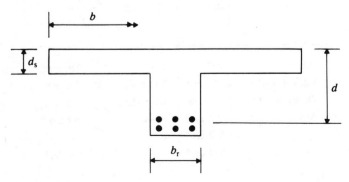

Figure S4.4

conditions, obtain the qudratic equation for the depth of the compression zone when the neutral axis falls below the slab.

For linear elastic conditions we can use the transformed section method whereby the area of reinforcement A_s is transformed to an area of concrete mA_s, where m is the modular ratio E_s/E_c.

If d_n is the depth of compression zone below the top edge of the section, then equating the moment of the tensile area about the neutral axis to that of the compression area, we obtain

$$mA_s(d - d_n) = b_r d_n \frac{d_n}{2} + (b - b_r)d_s(d_n - d_s/2)$$

This equation can be rearranged to give the following quadratic in d_n:

$$b_r d_n^2 + 2(d_s(b - b_r) + mA_s)d_n - (b - b_r)d_s^2 - 2mA_s d = 0$$

Index

Aggregates, 325, 326
Angle (*see* Steelwork, sections)
Angle of twist, 12, 13, 219–223
Angular displacement (*see* Angular rotation)
Angular rotation, 13, 17, 27, 28, 38, 40, 41, 48, 49,
    57, 58, 100, 101, 146, 155–159, 164–168,
    174, 183–185, 311, 521–523, 527, 530
Applied moment (*see* Moment)
Arches, 19, 67, 177–180
Area:
    centroid of, 199, 354
    first moment of, 199, 202, 340, 441, 579, 581
    polar second moment of, 205–206
    second moment of, 152–173, 182–185, 200–210,
        340–343, 518–531, 577–580
Axes (*see* Coordinate axes)
Axial forces, 6–17, 68, 345, 453–455
Axial stress (*see* Stress, direct)

Balloon roof, 16
Beam (*see* Bending members; Continuous beam;
    Simply supported beam)
Bending members, 7–13, 20–24, 33, 37–41,
    362–373, 397–401
    (*see also* Bending moment)
Bending moment, 9–13, 42–44, 121–124, 193–196
    in conjunction with axial force, 291–299,
        345–348, 354–356
    curvature due to, 195–200
    diagrams, 124–130, 166–171, 175–177, 181, 184,
        197, 289, 494–511, 517–523, 527–535
    diagrams using superposition, 143–145
    neutral axis, 151, 196–200, 207, 293–296,
        338–340
    relationship with curvature, 152
    relationship with shear force, 131
    sign convention for, 42–44, 123, 126
    strains due to, 196–199
    stresses due to, 21, 195–200
Bolted connections, 315–317
Bracing (*see* Steelwork)
Bracket, 139–142, 497–499

Brickwork (*see* Shear walls)
Bridge, 19, 35–39, 59, 120, 323, 343, 352, 392–405
Buckling (*see* Steelwork)
Bulk modulus of elasticity, 282–284
    relations with other constants, 284–285

Cables:
    design considerations of structures supported
        by, 119
    shape of, 117–119, 121
    theory of, 116–117
    use of, 35–37, 115
Cantilever, 23, 128–130, 132–134, 142, 155, 160,
    163–168, 194, 211, 256, 288, 333, 358,
    371–374, 405, 460–463, 471–474, 521–528,
    538–540
Castellated beam, 310, 312–313
Cement, 325
Centroid (*see* Area, centroid of)
Centroidal axes, 69
Channel (*see* Steelwork, sections)
Code of Practice, 414
Column, 26
    with eccentric load, 296–300
Column matrix, 77–78
Compatibility, 187
Complex frame, 81–82
Components of force, 44–46
Composite action, 364–366, 377
Compound frame, 80
Compression members:
    arrangements of, 18
    (*see also* Arches; Column)
    bending stiffness of, 17, 33
    cross-sectional shape of, 26
Compressive force, 6–8, 13, 17–18
Components of force, 44–46
Computer analysis, 106
Concrete:
    aggregates for, 325
    cement, 325
    compaction of, 401

Concrete (*contd.*)
  corrosion of reinforcement in, 356
  creep of, 330
  curing of, 408
  deformation of, 329–331
  deterioration of, 359
  encasement of steelwork, 320
  formwork for, 327, 401–403
  micro-cracking in, 328
  mixing of, 324
  placing of, 334, 401
  precast, 352–356, 403–407
  pumping of, 334, 402
  reinforced (*see* Reinforced concrete)
  setting of, 326
  strength of, 327–329
Concurrent forces, 53, 111
Connections (*see* Steelwork)
Constitutive relations (*see* Stress–strain
    relationships)
Continuous beam, 38–40, 59, 65, 181–185, 191,
    323, 333, 399, 404, 508–510, 538–540
Contractor, 3
Contraflexure, 144
Coordinate axes, 40–42, 309
Core:
  of a cross-section, 298–300
  structure, 384
Corrosion, 317, 356
Counterbracing, 379
Couple, 53, 143
Cross-section:
  centroid of area of, 199, 354
  core of, 298–300
  principal axes of a beam, 202–209
  moment of area of (*see* Area)
  radii of gyration of, 208
  section moduli of, 207–210, 354–356
Curvature, 132, 196
  (*see also* Radius of curvature)
Cylinder:
  thick-walled, 223–225
  thin-walled, 223–224

Deflection of beams, 21, 38, 40, 399
  differential equation method of, 154–159, 516,
    519–520, 523–526
  strain energy method of, 159–161, 516–519
  unit load method of, 161–171, 516–517,
    520–523, 526–531
Deflection of pin-jointed frames, 27–28, 48–49, 379
  graphical method of, 84–87, 471–474, 477–480
  strain energy, method of, 88–91

stiffness method (use of matrices), 98–104,
    469–471, 477–479, 480–482
  unit load method of, 91–98, 469–477, 488–489
Deformations, 12–13
  (*see also* Concrete, deformation of; Reinforcing
    bars, deformations on; Deflection of beams;
    Deflection of pin-jointed frames)
Design:
  phase, 3
  philosophy, 412–414
Determinacy, 79–82, 111, 120, 173–175, 180–188
Diagonal bracing (*see* Steelwork)
Diagonal cracking:
  brickwork, 382
  concrete, 335–336
Direct force (*see* Force, tensile; Force,
    compressive)
Direct stress (*see* Stress, direct)
Displacements (*see* Deflection of beams;
    Deflection of pin-jointed frames; Rotation
    of beams)
Ductility, 304–309
Durability, 304

Eccentric load, 292–299
Elastic, 305, 307
  analysis of prestressed concrete, 351–356
  analysis of reinforced concrete, 336–343
  (*see also* Stress analysis; Strain analysis)
Elasticity (*see* Modulus of elasticity)
Equilibrium, 7–10, 20–21, 29–31
  in bending members, 7–12, 20–21, 123–131, 178,
    338–339
  calculation of reactions using, 7–9, 58–63, 70,
    117, 124, 128, 141, 449–451, 453–455, 463,
    490, 495, 497, 499–500, 503–504, 508, 510,
    518, 534
  of concurrent forces, 51–53, 73–79, 84–85, 108,
    111, 114, 437–441, 451
  of direct forces, 8–11, 15–18, 29–31, 51–53,
    70–79, 84–85, 114, 437–441, 451, 457–465,
    482–489
  equations of, 46, 70, 108, 111, 121, 180
  polygon of forces, 50–52, 493
  unstable, 17, 63–65
Equivalent force system, 40, 53–55
Extension of members, 12–13, 28, 100
External force, 6–7, 123
  (*see also* Equilibrium, calculations of reactions
    using)

Factor of safety, 414–415
Failure, 336, 408

Fire protection, 318–321
First moment of area (*see* Area, first moment of)
Fixed support, 56–57, 174
Flexibility (*see* Force, approach)
Floors, 362–373
    (*see also* Reinforced concrete slabs)
Force, 6
    approach (to analysis of indeterminate
        structures), 180–186
    bending moment, 7, 122–123 (*see also* Bending
        moment)
    components, 44–46
    compressive, 6–8, 13, 17–18
    equivalent systems, 53–55
    external (*see* External force)
    internal, 9, 193, 227
    polygon, 50–52, 493
    shear (*see* Shear force)
    sign convention for, 42–44
    tensile, 6–11, 14–15, 28
    twisting moment, 7, 22–24
Formwork (for concrete), 326–327, 401–403, 409
Frames (*see* Pin-jointed frames; Rigid-jointed
    frames; Space frames)
Free-body diagrams, 8, 10–11, 13, 15–16, 18, 21,
    24–25, 27, 30, 40, 43–44, 52, 56, 58, 60,
    64–65, 72–75, 118, 122, 125, 127, 129, 131,
    133, 140, 147, 151, 179, 184, 194, 197, 199,
    211, 215, 289, 322, 437, 444, 447, 449, 452,
    454–455, 458, 461, 464, 466, 490–491, 493,
    498, 540

Graphical Method:
    determination of forces by the, 50–52
    determination of deflections by the, 84–88,
        471–474, 477–480
Gravity loading, 6, 54, 69, 120

Hinge, 27–28, 57–58
Hogging bending, 40–41, 132, 137–139, 143–147,
    155, 160, 164, 168, 170, 174, 176, 181–185,
    372, 397
Hollow shaft:
    thick-walled, 223–225
    thin-walled, 223–224
Homogeneous materials, 273
Hooke's Law, 198, 277, 279–281
Horizontal shear, 211–214
Hydrostatic stress, 284

Indeterminacy (*see* Statically indeterminate
    structures)

Instability, 408
    (*see also* Steelwork, buckling of)
Intensity:
    of deformation, 12–14 (*see also* Strain)
    of force, 10–12 (*see also* Stress)
    load, 441
Internal force, 9, 193, 227–228
Inverse (of matrix), 78, 103
In-situ concrete, 352–353 (*see also* Concrete,
    formwork for)
Isotropic materials, 273

John Hancock Center, 385–386
Joints:
    deflection of, 28, 82–105, 469–482, 488–489
    equilibrium of, 15–16, 27, 29–31, 52, 73–79,
        112–114, 437, 449–451, 456–469, 482–489
    method of, 73–79, 458–463, 488–489
    pinned, 27–31, 40–41, 68–69 (*see also* Pin-
        jointed frames)
    rigid, 40–41, 173–177 (*see also* Rigid-jointed
        frames)
    for space frames, 110

K truss, 31–32
Kernel (*see* Core of a cross-section)

Lattice girder, 387–388 (*see also* Pin-jointed
    frames)
Linear:
    deformation, conditions for, 48
    elastic material, 305
    elastic analysis (*see* Elastic)
Load, 16, 54 (*see also* External force)
Local buckling, 321–323
Long-line prestressing bed, 351

Materials, 304
    isotropic, 273
    (*see also* Concrete; Steel)
Matrix, 46–47, 75–79, 98–104, 277, 281, 465, 469,
    471
Mechanics:
    of solids, definition of, 5
    of structures, definition of, 5
Mechanism, 28–29
Micro-cracking (in concrete), 328
Modular ratio, 341
Modulus of elasticity, 14, 284–285, 306, 329–330,
    341
    (*see also* Shear modulus; Bulk modulus)
Modulus of rigidity (*see* Shear modulus)

Mohr's circle:
  of strain, 264–271
  of stress, 243–253
  of three-dimensional stress, 253–255
Moment, 45, 139–140, 454–455
  (*see also* Bending moment; Twisting moment)
Moment of area (*see* Area)
Moment of inertia (*see* Area, second moment of)
Multi-storey building, 33, 362–387

Negative face, 42–43, 123, 230
Neutral axis, 151, 196–200, 207, 293–296
Normal stress (*see* Stress, direct)

Parallel axes theorem, 200–203
  (*see also* Area, second moment of)
Parallelogram of forces, 50–51
Perpendicular axis theorem, 205–206
Pinned joint, 27–31, 40–41, 68–69
Pin-jointed frames, 27–31, 68–69
  deflections of, 82–105
  diagonal bracing in, 376–381, 389–392
  internal forces in, 70–79
  reactions for, 59–61, 70
Pinned support, 57–58
Planar statics, 50–55
Plane of bending, 195
Plane cross-sections, 196–199
Plane strain (*see* Strain)
Plane stress (*see* Stress)
Plastic deformation, 306–308
Poisson's ratio, 273–284
  relations with other constants, 284–285
Polar second moment of area, 205–206
Polygon of forces, 50–51, 493
Portal frame, 40–41, 61–63, 189, 387–388, 398–399
Positive face, 42–43, 123, 229–230
Post-tensioning (*see* Prestressed concrete
    techniques)
Power (transmitted in torsion), 224–226
Pratt truss, 31–32, 390
Precast concrete, 403–408, 411
Prestressed concrete, 343
  advantages of, 343–344
  elastic analysis of, 351–356
  principles of, 344–348
  techniques, 348–351
Pretensioning (*see* Prestressed concrete
    techniques)
Principal:
  axes,
    of a beam, 202–209
    in stress analysis, 237–241

planes, 237–241, 267
strains, 267–272
stresses, 237–245, 248–253
Principle of superposition, 48–49, 143–145
Prismatic member, 196
Product integrals, 167
Profiled steel sheeting, 409–410
Properties of materials, 304
Pure bending, 196–200
Pure shear, 253
Purlins, 390–391

Radisson Hotel, method of construction, 366–388
Radius of curvature, 151–152, 154–155, 195–200
Radius of gyration, 208
Reactions, 7–9
  (*see also* Equilibrium, calculation of reactions
      using)
Rectangular beam:
  bending stresses in, 195–200
  shear stresses in, 210–218
Reinforced concrete, 331
  analysis of, 336–343
  bond between steel and concrete in, 332–335
  end walls, 382
  flexural cracking of, 331–333
  problem of shear in, 335–336
  reinforcing bars for, 308, 409
  shrinkage effects in, 356–359
  slabs, 362–363, 372–373
  temperature effects in, 357–359
Reinforcing bars, 409
  deformation on surface of, 334
  stress–strain characteristics of, 308
  testing of, 305
Relations between elastic constants, 284–285
Resident engineer, 4
Righting moment, 446
Rigidity modulus (*see* Shear modulus)
Rigid-joint, 173 (*see also* Joints)
Rigid-jointed frames, 173–177, 374–376
Roller support, 58–59, 176
Roof bracing, 389–392
Rotation:
  of axes (*see* Transformation of axes)
  beam (angular displacement), 164–169, 183–185
  relative (in pin-jointed frames). 27–28

Safety factor, 414–415
Sagging bending, 132
Second moment of area (*see* Area, second moment
    of)
Section modulus, 207–210, 354–356

Sections, method of, 70–73
Self-weight, 354–355
Serviceability, 413–415
Shaft (*see* Torsion)
Shear cracking, 335–336, 382
Shear connectors, 365–366
Shear force, 9–10, 121–124
    diagrams, 124–130, 215, 289, 494, 498, 501, 505, 507, 509, 511, 513
    relationship with bending moment, 131
    relationship with load, 131
    sign convention for, 42–43
Shear modulus, 14, 221–223
Shear strain (*see* Strain, shear)
Shear stress, 12, 195, 210–213, 228–230
    complementary, 212–213, 230–231
    formula for beams, 214
    horizontal, 211–212 (*see also* Shear stress, complementary)
    in I-section beams, 216–219
    on inclined planes, 232–235
    maximum and minimum values of, 241–243
    in rectangular beams, 215–216
    sign convention for, 229–230
    due to torsion, 219–226
Shear walls, 23, 25, 382–383
Shortening of members, 12–14
Shrinkage (in reinforced concrete), 356–359
Sign convention, 40–44, 123, 229–230
Simply supported beam, 20–21, 37–38, 54, 121–128, 133–151, 157–159, 161–163, 165–166, 168–171, 176–177, 189, 197, 215, 289, 331–335, 344–347, 353–356, 397, 494–508, 510–520, 523–524, 526–531
Slabs, 22–23, 362–363, 372–373
Slope, 154–159, 194–196
Space frames, 31, 33, 107
    analysis of, 112–114
    conditions for determinacy of, 111
    corollaries for concurrent forces in, 111
    joints for, 110
    one-way spanning, 393
    types of support for, 108
Stability, 17
Statics, 50
Statically indeterminate structures, 180
    advantages and disadvantages of, 188–190
    analysis of, 180–188
    examples of, 65, 81, 181, 186, 191
    fundamentals of, 188
Steel, 304
    corrosion of, 317–318
    properties of, 304–309
    testing of, 304–307
Steelwork, 304
    brittle fracture of, 323–324
    connections, 68–69, 107, 110, 312, 314–317
    fire protection, 318–321
    frames, 30, 33, 319–320, 364, 366–367, 377–383, 386–392, 410
        bracing of, 376–381, 393–396
        roof bracing in, 389–391
    local buckling of, 321–322
    sections, 309–310, 313–314
    shear connectors for, 365–366
Step functions, 134–136, 158–159
Stiffness, 14–15, 189
Strain, 12–14
    analysis of, 255–271
    array, 259–260, 264, 266–272
    distribution in a beam due to a bending moment, 196–199
    energy, 88–91, 159–161
    hardening, 306–307
    Mohr's circle of, 264–271
    principal, 267–272
    at rupture, 306, 329
    shear, 13–14, 220–221, 258–264
    sign convention for, 256–257
    in terms of displacements, 257–259
Stress, 10–12, 21
    analysis of plane, 232–253
    array, 230, 232, 235–237, 241, 244, 247–254
    axial, 251–252, 273
    definition of, 226–229
    direct, 11–12, 195, 228
    distribution in a beam due to a bending moment, 21, 195–200
    distribution in a beam due to a shear force, 215–219
    distribution in a shaft due to torsion, 219–226
    general state of uniform, 229–232
    on inclined planes (*see* Stress, analysis of plane)
    Mohr's circle of, 243–253
    normal (*see* Stress, direct)
    principal, 237–245, 248–253
    shear (*see* Shear stress)
    three-dimensional, 229–232, 253–255, 280–284
Stress–strain relationships, 14, 272–284, 306–309, 329–331
Stress resultants, 6, 195
Strength, 413–415
Structural engineering project, 3–4
Strut, 27
Superposition, 48–49, 143–145
Supports, 55–59

Temperature effects, 318, 357–359
Tensile force, 6–11, 14–16, 28
Tensile stress, 10–12 (*see also* Stress)
Tensile test, 304–307
Thick-walled cylinders, 223–225
Thin-walled cylinders, 223–224
Three-dimensional structures (*see* Space frames)
Torque (*see* Torsion)
Torsion, 12–13, 22–24, 36–37, 219–220
    of hollow tube, 223–225
    power transmitted in, 224, 226
    of solid circular shaft, 219–223, 226
    strains due to, 221
    stresses due to, 219–226
    of thin-walled tubes, 223–224
Transformation of axes:
    for strains, 259–260, 262–271
    for stresses, 232–240, 243–255
Triangle of forces, 50–51, 493
Trusses, 27
    arrangements of members in, 27–33
    calculations of reactions for, 59–61, 70
    deflections of, 82–105
    Howe, 31–32
    internal forces in, 70–79
    K, 31–32
    Pratt, 31–32
    for single-storey buildings, 388–393
    Warren, 32, 388, 394–396
Tube:
    system of resisting wind, 384–387
    torsion of (*see* Torsion of hollow tube)

Twist (*see* Angle of twist)
Twisting moment (*see* Torsion)

Uniaxial stress, 273–274, 304–305
Uniformly distributed load, 54
Unit load method (*see* Deflection of beams;
        Deflection of pin-pointed frames; Rotation,
        beam)
Universal beam, 309
Universal column, 313
Unstable structures, 64–65, 408

Virtual force (*see* Virtual work)
Virtual work, 91–95
    application to bending structures, 161–171
    application to pin-pointed frames, 91–98
Volumetric modulus of elasticity, 282–284
Volumetric strain, 282–284

Wall, shear, 23, 25, 382–383
Warren truss, 31–32, 388, 394–396
Waterloo Bridge, 399–400
Weld, 312–313
Welded connections, 314–317
Williott diagram, 85–87, 478
Williott–Mohr diagram, 88
Work (*see* Strain energy; Virtual work)
World Trade Center, 385–387

Yield stress, 306–309
Young's Modulus, 14
    (*see also* Modulus of elasticity)